M

**BARRON'S**

Mechanics
Table saw

# ELECTRO

## FOURTH EDITION

**Rex Miller**
Professor Emeritus, State University College
Buffalo, New York

**Mark R. Miller**
Chair and Associate Professor,
Texas A & M University-Kingsville
Kingsville, Texas

# THE
# EASY
# WAY

**BARRON'S**

JAN 2 7 2005

*All inquiries should be addressed to:*
Barron's Educational Series, Inc.
250 Wireless Boulevard
Hauppauge, New York 11788
*http://www.barronseduc.com*

Library of Congress Catalog Card No. 2002016321
International Standard Book No. 0-7641-1981-8

Library of Congress Cataloging-in-Publication Data
Miller, Rex
    Electronics the easy way / Rex Miller, Mark R. Miller. — 4th ed.
        p.    cm.
    Includes index.
    ISBN 0-7641-1981-8
    1. Electronics.    I. Miller, Mark R.    II. Title.
    TK7816 .M53    2002
    621.381—dc21

                                        2002016321

PRINTED IN THE UNITED STATES OF AMERICA

10  9  8  7  6  5

# CONTENTS

# PREFACE

Electronics is everywhere, and it touches everyone. Everyone knows electrical engineers need to know electronics, but there are also millions of employees of electrical utilities, telephone companies, the broadcast industry, and TV, radio, computer, and appliance repair industries, as well as electronic hobbyists who use their knowledge of electronics every day.

Where have these people learned the electronics they know? Many have taken courses in high school and/or college, but many others are self-taught, drawn to electronics by their interest in the field.

*Electronics the Easy* Way is as close as you can come to getting electronics made simple enough to understand on your own.

This book has been prepared both for those who have no background whatsoever in electronics and for those who already know something about it and would like to put their bits and pieces together in a logical, sequential fashion. It can serve as a supplement to any textbook you may already have, inasmuch as it will expand on some ideas and present a new approach for others.

One of the most interesting points about electronics is that it is a growing field. Advances are made every year. Although the foundation of electronics never changes, new areas are always opening up. Just a few years ago, people were proud of their transistor radios that were so much more compact than those manufactured with vacuum tubes. Now transistors are old hat, and integrated circuits have made electrical circuitry even smaller.

In deference to this new knowledge, this book covers microwaves, fiber optics, lasers, computers, video recorders, cable TV, and digital electronics. Traditional topics have not been neglected either. Plenty of space is devoted to audio amplifiers, radio frequency amplifiers, power amplifiers, and stereo amplifiers. Other chapters look at circuits, inductance, capacitance, alternating current, resonance, changing alternating current to direct current, semiconductors, oscillations, superheterodyne receivers, television, and radar, as well as motors and power. Mathematics have been kept to a minimum.

Looking toward the future in electronics can be exciting, too, so the last chapters include robotics, space manufacturing, cellular phones, pagers, facsimile (fax) machines, and printers.

This book is designed to give you enough information to whet your appetite so you will dig deeper into the area and become as fascinated with it as we have, so it does not necessarily treat electronics comprehensively.

Electronics is not something you get the first time through, so you will have to concentrate on what you read and then re-read it a few times until it soaks in. Remember, the time spent will be well worth the effort since almost any job today involves electricity or electronics in some degree.

Enjoy yourself. There is no more fascinating a subject with such dynamic growth and great promise for the future.

**Rex Miller**
**Mark Miller**

# Chapter 1

# CAREERS

This chapter is about opportunity: The opportunity to learn about one of the fastest growing industries today — electronics. Electronics touches all aspects of modern life.

There is no sure way to master electronics. It takes time and effort on the part of the learner. Since many jobs require a knowledge of electronics, it is never something that you cannot use. The knowledge acquired can serve you in many ways. This book has been developed with you as the focus. It provides certain basic information you will need in almost any field.

The information covered here will aid you in deciding whether you want to go further in studying electricity and electronics. To help you decide, this chapter previews a number of job opportunities. These opportunities are open to people who have a solid background in electricity and electronics.

One or more of the positions and careers described in this chapter may interest you. If so, the opportunities you select should guide you in further study of electricity and electronics. If you choose a career in this field, be aware that further study and work will be needed. You will need to know more than you have learned in this book. You will also need more than you will learn in a single, beginning course.

In each area there is a review of the requirements and the nature of the work. There are discussions about the opportunities you might find. In addition, there are names and addresses of sources of further information.

## Electrical Engineering

Electrical engineers are concerned with jobs that relate to the generation of electricity, its transmission, and its use. They are also specialists in the many aspects of electronics with specialty areas in computers, circuit design, equipment design, manufacture of printed circuits, and manufacture of semiconductor materials and devices. Electrical engineers can do any number of jobs. but most will specialize in electronics and some aspect of it. Others will specialize in the generation of electricity and work with utility companies. Some will work with the distribution of electricity, while still others spend their time in research laboratories developing new devices and processes. The electrical engineer has often been described as a practical mathematician. Therefore, the electrical engineer must have a good math background and is required to take many math courses in college.

The major areas of work in this branch of engineering include electronics and electrical equipment manufacturing. Other opportunities lie in communications, power, illumination, and transportation. Electrical engineers usually specialize in one of these areas.

Actually, these areas are so broad that many engineers specialize in subdivisions of one area. They may specialize in computers, or they may want to work with electrical motors and generators.

Electrical engineers have many types of responsibilities, which may include research, design, and development. Other opportunities lie in manufacturing or administration. In addition, some engineers are technical sales representatives. Many are self-employed and operate their own businesses.

Engineers are located everywhere there is electricity and/or electronics, and that is everywhere. Opportunities are open in many types of organizations. Electrical and electronics equipment manufacturers employ many engineers. So do the electric light and power, aircraft and missile, telephone, and broadcasting industries. Many electrical engineers work for government agencies, colleges, universities, or consulting firms. Some are self-employed as consulting engineers.

Most jobs are located in industrial centers. These areas tend to be heavily populated. However, there are also opportunities in rural areas, with electric light and power companies, telephone companies, and broadcast stations. Satellite TV and cable TV systems now employ a number of electrical engineers for aid in designing and maintaining equipment. The design of lasers for communications and fiber optics systems for telephone companies are the latest in the list of opportunities. These jobs are usually located in large cities.

At present there are not enough engineers to answer the call for designing new products and researching old ones. Rapid growth is expected in this profession. Increasing demand for qualified engineers is expected in the computer field, fiber optics, and satellite communications. Also growing is the area of communications and military electronics. Newer types of power generation equipment — solar, geothermal, and nuclear — will add to the new demands.

Compensation in the field of electrical engineering, with a bachelor's degree, is usually the highest of all the engineering professions. In most cases a bachelor's degree is the minimum requirement. However, an associate degree from an accredited two-year college with a specialty in electronics is also a good starting point. These graduates are in high demand for their skills and acquired knowledge. In most cases, the employer approaches students who are still in college. Both salary and fringe benefits are discussed at that time. For more information contact:

Institute of Electrical and Electronic Engineers
1828 L Street, NW, Suite 1202
Washington, DC 20036-5104
(http://www.ieee.org/)

# Newer Employment Opportunities

Electronics does not stand still as a field of employment. It continues to offer more exciting job opportunities than any other field. In the past 50 years the job opportunities in electronics have grown very rapidly. This industry provides a job market that offers easy placement and rapid advancement for those who are willing and able to keep up with continual change. For many years electronics was limited to the AM radio. After World War II, the field changed quickly and expanded to include the new fields of radar, television, and the computer. However, the industry was still limited to a hard-wired vacuum tube technology. Technicians were required to expand their knowledge base of circuits and systems to take advantage of the expanding job market. The increased demands of new circuits required new skills for using the new and more elaborate forms of electronic test equipment.

# The Integrated Circuit

The integrated circuit has changed the whole field of electronics. This has developed since 1960 and has continued to change every year. It has reached an amazing level of development. The cost of circuits has shrunk, and circuit reliability has developed to a very high level. Today applications are expanding. Analog instruments of a decade ago are available today in all or partially digital form and at a much lesser expense.

Along with this rapid development has come some new job titles. A few of them are listed here. These are some of the titles that have developed for *only printed circuit boards*. There are many others associated with semiconductor production that may be found in the appendix. For a complete description of the job titles take a look at the Appendix.

Supervisor, printed circuit board testing
Supervisor, printed circuit board assembly
Group Leader, printed circuit board assembly
Group Leader, printed circuit board quality control
Wave Soldering Machine Operator, printed circuit boards
Functional Tester, printed circuit boards
Electronic Circuit Tester, printed circuit boards
Inspector, printed circuit boards
Reworker, printed circuit boards
Production Repairer, printed circuit board assembly
Touch-up Screener, printed circuit board assembly
Solder-Leveler, printed circuit boards
Lamination Assembler, printed circuit boards
Preassembler, printed circuit boards

# Electronics Engineer

The electronics engineer conducts research and development activities concerned with the design, manufacture, and testing of electronic components, products, and problems, and in the development of applications of products to commercial, industrial, medical, military, and scientific uses. The engineer designs electrical circuits, electronic components, and integrated systems, using ferro-electric materials, dielectric phosphors, photo-conductive materials, and thermo-electric materials. This type of engineering may be further broken down into job titles that may be described as Design Engineer, Products Engineer, Test Engineer, Electronics Research Engineer, Electronics Products and Systems Sales Engineer.

## Systems Engineer, Electronic Data Processing

This type of engineer analyzes data-processing requirements to determine an electronic data processing system that will provide system capabilities required for projects or workloads. This type of job may also call for planning the layout of new system installation or the modification of an existing system. The utilization of knowledge acquired about both electronics and data-processing principles and equipment is applied.

The engineer confers with data-processing and project managerial personnel to obtain data on limitations and capabilities of an existing system and the capabilities required for data-processing projects and the workload proposed. The engineer analyzes data to determine, recommend, and plan layout for the type of computer and peripheral equipment, or modifications to existing equipment and system, that will provide the capability for the proposed project or workload, its efficient operation, and the effective use of allotted space. This job may call for power supply requirements and configuration. Recommendations for the purchase of equipment to control dust, temperature, and humidity in areas of system installation may also be required. This type of engineer may specialize in one area of system application or in one type or make of equipment. Consulting for equipment manufacturers can also be part of the job description of a systems engineer.

# Electric Utilities

There are approximately 3700 electrical utilities in the United States. Some of these are privately owned. Others are owned by federal, state, or local governments. Each utility has the ability to generate, distribute, and maintain its networks. There are power-distribution networks that make it possible for utilities to buy power from one another.

## How the Future Looks in the Utilities

There are currently over 500,000 people working for utility companies. Many thousands of new employees enter this field every year. Some of these jobs are available because people leave the industry; other jobs open as a result of continuing growth. Because of changes taking place, excellent job potential should continue.

## Where the Jobs Are Located

About 10 percent of the utilities' jobs are in generating plants. About 40 percent are related to the transmission and distribution of power. About 20 percent of the workers are involved in repair and maintenance jobs. Some 15 percent are in customer service. An estimated 9 percent of the employees hold engineering and technical positions. It is this 9 percent that we will focus on here.

**Power Plant Jobs.** Power plant jobs are directly related to the generation of power. Responsibilities include control panel operation, checking equipment, and keeping records. Specific job titles include boiler operators, turbine operators, auxiliary equipment operators, control room operators, and watch engineers. In this field promotions usually result from experience. Beginning employees start with cleanup and other simple jobs. They advance as they become familiar with the equipment and with responsibilities of other jobs. It also helps to understand basic electricity and some electronics.

**Transmission and Distribution Jobs.** Most of these jobs involve physical as well as mental ability. The list includes load dispatcher, substation operator, line installer, line maintenance worker, troubleshooter, and cable splicer.

**Consumer Service Occupations.** Persons who work in consumer service jobs perform a number of technical, managerial, and office jobs. The greatest number are clerical in nature. Some technicians install, test, and repair meters. In some areas, utility employees also repair electric appliances and machinery. Administrative and office jobs include the preparation of work orders. These work orders can involve installation, maintenance, or termination of service. Other office workers are involved in billing, collection, and record keeping. For more information contact:

International Brotherhood of Electrical Workers
1825 L Street, NW, Suite 1202
Washington, DC 20036-5104
(http://www.ibew.org)

American Public Power Association
2301 M Street, NW
Washington, DC 20037
(http://www.appanet.org/cgi-bin/wwwais)

# The Telephone Industry

This is one of the largest industries in the country. It was previously dominated by one major company. This has now been broken up, and any number of opportunities are available to those who wish to open their own phone business on any of a number of levels.

Verizon is one of the larger companies in this industry. There are over 1000 other smaller companies. With competition heating up over the breakup of AT&T, many doors are being opened for entry-level and more sophisticated jobs in the business.

## How the Future Looks in the Telephone Industry

With the advent of inexpensive automobile phones and satellite communications, the telephone industry will be a booming industry for some time. Complete telephone companies with their own share of satellite time and their own microwave links are being formed. These firms will need specialists in a number of electrical and electronics areas. This industry has expanded rapidly in the past, and growth is expected to continue.

## Where the Jobs Are Located

In the telephone industry it is estimated that one-third of all employees are telephone operators. Telephone operators have opportunities to advance into other positions in traffic areas. Today, positions as telephone operators are open to men and women.

Some 25 percent of telephone utility workers are in installation and service positions. They install, repair, and maintain telephones. They also set up wires and cables, switching equipment, and message accounting systems. Positions in this area include central office installers, line construction crews, and installation/repair specialists. For more information contact:

American Public Power Association
2301 M Street, NW
Washington, DC 20037
(http://www.appanet.org/egi-bin/wwwais)

Communication Workers of America
501 Third Street, NW
Washington, DC 20001-2797
(http://www.cwa-union.org)

The Fiber Optic Association
Box 230851
Boston, MA 02123-0851
(http://world.std.com/~foa/)

Independent Electrical Contractors, Inc.
2010-A Eisenhower Avenue
Alexandria, VA 22314
(http://www.ieci.org)

The National Association of Radio and Telecommunication Engineers
Post Office Box 678
Medway, MA 02053
(http://www.narte.org/)

Optical Society of America
2010 Massachusetts Avenue, NW
Washington, DC 20036
(http://www.osa.org/)

United States Telephone Association
1401 H Street, NW, Suite 600
Washington, DC 20005-2164
(http://www.usta.org)

# TV, Radio, and Appliance Repair

Thousands of people are employed at shops that service TV sets, radios, computers, stereo equipment, and a wide range of home appliances. This is a popular area for the self-employed person who likes to be his or her own boss. It is easy and relatively inexpensive to open a business in TV repair and appliance repair. The demand for quality repair work is always high.

## A Place for the Self-Taught

Many enterprising individuals in this field are self-taught. After a basic education in electricity and electronics, many have worked through manuals on their own. They have learned by buying kits and assembling their own radios, TV sets, stereo components, and even small computers. Others have specialized in the home installation of satellite TV receivers. It is customary for repairers to own their own hand tools. Employers usually furnish needed instruments and meters. However, many technicians own these tools as well.

Formal training for TV and other electrical repair work is readily available. There are many public and private vocational and trade schools as well as many correspondence schools. Many young persons also learn basic electronic repair skills in military schools. These skills can be valuable for individuals returning to civilian life.

It takes, generally speaking, two to three years of on-the-job training or vocational school training to qualify as a TV repair technician. In some states or cities, written tests are required. In some areas, individuals or businesses must be licensed to offer these services.

## The Future of the TV, Radio, and Appliance Repair Field

The outlook is for continued growth in these fields. As more TV sets and consumer electronics are sold, the need will increase accordingly. As TV sets and appliances become increasingly more complex, service requirements also increase in complexity.

The future prospects for television and electronics repair may not be too good. The complexity of the equipment is increasing. That means more expensive and elaborate testing procedures are called for. It also means that more knowledge will be required.

Another factor affecting the future of a technician is the fact that people can buy cheap new equipment that is imported, which is often cheaper than repairing the old equipment. This sets upper limits on repair bills and repair feasibility. It also presses the technician, often making him/her work cheaply to make a repair cost-effective. In effect the technician is limited as to what he/she can charge, even though the repair may be very difficult.

Another consideration is parts. Parts for the many imported varieties are often difficult to obtain or prohibitive in cost. A person considering becoming a technician should be aware of some of the pressures presently in the field and slated to continue or increase in the future.

For more information contact:

EIA
2500 Wilson Blvd.
Arlington, VA 22201-3834
(http://www.eia.org)

Edison Electric Institute
701 Pennsylvania Avenue, NW
Washington, DC 20004-2616
(http://www.eei.org)

# The Broadcast Industry

There are about 7650 commercial radio and 730 commercial TV stations in the United States. There are also more than 1000 noncommercial radio stations and 260 noncommercial TV stations. These broadcast facilities were all licensed to transmit signals through the air. In addition, the cable TV industry is growing rapidly. There are more than 3000 cable TV systems serving over 10,000 communities. Satellite distribution of broadcast materials is opening new areas of opportunity.

## Where the Jobs Are Located

There are four national networks for radio with over 900 employees. Four TV networks and cable companies employ about 20,000 people. Many thousands of additional people are employed at individual stations. A qualified, licensed technician must be present or on call whenever a radio or TV station is transmitting.

Broadcasters hire many technicians to install, operate, and maintain broadcast transmitters. At least one person on duty at every station must have a license. To qualify for the license the person must take a test given by the Federal Communications Commission (FCC). This is rapidly changing as the government is getting out of the licensing business. Thus, some special education or training is usually required to qualify as a broadcast technician. This usually involves a high school education that is followed by study in a vocational or technical school.

## How the Future Looks in the Broadcast Industry

Broadcasting is limited in its opportunities. This is considered a glamor industry. So, there are usually many more applicants than there are jobs available. In addition, employment in large stations usually requires membership in at least one union. To break into broadcasting, many individuals start at small stations in remote areas. These positions tend to be low-paying. As experience is gained, qualified people often move to larger stations or networks.

The easiest way to learn more about broadcasting is to visit a radio or TV station in your area. You will usually find someone willing to discuss your interests and potential local opportunities. You may also obtain information from:

Federal Communications Commission
Policy Analysis Branch
1919 M Street, NW
Washington, DC 20554
(http://www.fcc.org)

National Association of Broadcasters
1771 N Street, NW
Washington, DC 20036-2891
(http://www.nab.org)

Society of Broadcast Engineers
8445 Keystone Crossing, Suite 140
Indianapolis, IN 46240-2454
(http://www.sbe.org)

# Summary

Electrical engineering offers many opportunities to those qualified for the positions opening in the newer fields of electronics. At the present there are not enough engineers to do the work needed to be done.

Electrical engineers have many types of responsibilities. They include research, design, and development of new products. Other opportunities lie in manufacturing and administration. In addition, many engineers are technical sales representatives and consulting engineers who are self-employed.

Engineers work for governmental agencies, colleges, universities, private industry, and consulting firms.

Electrical utilities offer positions for engineers and technicians. New opportunities are being created every day in the generation and distribution of power. Immediate expansion will not be as great as previously projected, but the future still looks bright.

The telephone industry is undergoing rapid change, and many new systems are being installed. These will offer opportunities for entry-level jobs and some at the higher level. Electronics specialists will be in great demand as the telephone industry expands its portable phone and satellite communications operations.

The fields of TV, radio, and electric appliance repair seem to be in great need of quality technicians. Work is easily arranged to suit the individual. It is a good place to find self-employed individuals who can handle the business as well as the technical aspects of the field. As equipment becomes more complicated and the consumer buys more digital electronics, the need for specialists will increase rapidly.

The broadcast industry is a field where job opportunities are limited since it is a glamor industry which attracts large numbers of applicants. You usually must start in an apprenticeship in a smaller station or community before you are offered a job in a larger market. Technicians and engineers in this field will find a better atmosphere as far as opportunities are concerned. The technician will have to keep up with the latest digitized equipment and become more self-educated in these newer fields of electronics.

# Review Questions

1. What does an electrical engineer do?
2. Where does the electrical engineer work?
3. What opportunities are there for engineers?
4. What are the opportunities in the telephone industry?
5. What are the opportunities in the TV repair field?
6. How does the broadcast industry look for the future?
7. What education do you need to be an engineer?
8. What do you need to know to become a technician?
9. Where can you obtain information on electrical engineering?
10. What does the job market look like in the electrical utilities field?

# Chapter 2

# INTRODUCTION TO ELECTRONICS

The Greeks are believed to have discovered electricity in the process of conducting some of their experiments. While working with a piece of amber (the fossilized resin from an ancient species of tree), which is translucent and golden in color, they found that if the amber was rubbed briskly, it would exhibit an attraction for tiny bits of light-weight material. You have probably seen this happen when you run a comb through your hair on a dry day and pick up bits of paper with it. The Greeks believed that the forces of amber were at work in this phenomenon. Similarly, the Romans found that lignite (a form of coal) could be rubbed and could produce, by friction, the same reaction. However, it is the Greeks who have been given credit for the first experiments that later led in 1600 to the work of William Gilbert (an Englishman) with friction and static electricity. Gilbert wrote a book on the substances with which he had experimented. He showed that amber was not the *only* such material that produced an attraction for the bits of paper.

Gilbert is given credit for coining the word *electrics*. The Latin word *electrum* is derived from the Greek *elektron*, which in turn means "amber." Gilbert was influenced by the Latin being studied at the time. In all probability this is how the name worked its way into print and history.

Gilbert has been called the father of electricity since he was the first to classify objects that would produce an electrostatic field when rubbed. He called these substances *electrics*.

*Electronics* is the application of electrical principles. Electrical principles are derived from the uses and generation of electric energy. Therefore, electricity is necessary for the proper operation of electronic devices and electronic circuits. In order to understand electronics it is first necessary to know how electricity is generated, distributed, and put to work in circuits.

## Basic Atomic Structure

*Electricity* is defined as the flow of electrons along a conductor. A conductor is an object that allows electrons to pass easily. That means electrons must be organized and pushed toward a goal. This is done in a number of ways. But first, we must know what an electron is before we can start working with it.

Elements are the most basic materials in the universe. There are 106 elements including some that have been made in the laboratory. Elements such as iron, copper, gold, lead, and silver have been found in nature. Eleven others have been made in the laboratory. Every known substance — solid, liquid, or gas — is composed of elements.

An electron is the smallest part of an atom. An atom is the smallest particle of an element that retains all the properties of that element. Each element has its own kind of atom. That is, hydrogen atoms are alike, and they are different from the atoms of all other elements. However, all atoms have certain things in common. They all have an inner part, the nucleus. The nucleus is composed of very small particles called *protons* and *neutrons*. An atom also has an outer part, consisting of other small particles. These very small particles are called *electrons*. The electrons orbit around the nucleus (see Figure 2-1).

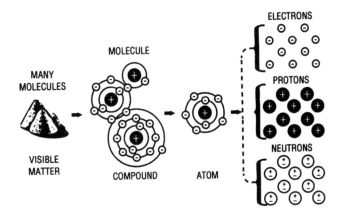

Figure 2-1. Molecular structure. The negative (–) particles are electrons.

# Neutrons

Neutrons have no electric charge, but protons are positively charged. Because of these charges, protons and electrons are particles of energy. That is, these charges form an electric field of force within the atom. These charges are always pulling and pushing one another; this action produces energy in the form of movement.

The atoms of each element have a definite number of electrons, and they have the same number of protons. A hydrogen atom has one electron and one proton (see Figure 2-2). The aluminum atom has thirteen of each (see Figure 2-3A). The opposite charges — negative electrons and positive protons — attract each other and tend to hold electrons in orbit. As long as this arrangement is not changed, an atom is electrically balanced.

A closer look at the atom shows that the electron orbits the nucleus in shells. A *shell* is made by an electron orbiting the nucleus. An electron rotating around the nucleus makes

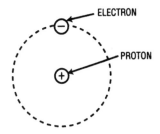

Figure 2-2. The hydrogen atom has one electron and one proton.

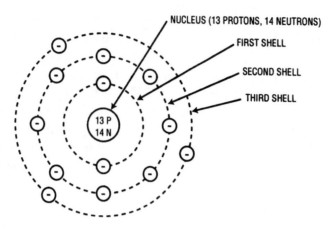

**Figure 2-3A.** The aluminum atom has 13 electrons.

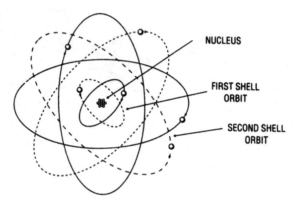

**Figure 2-3B.** Electron orbits and shells. Note how the rotation produces the appearance of a shell.

a ring around it, and the ring may be elliptical in shape. At the same time the *plane* of the electron shifts a few degrees, so that another ring is produced, and then another and another, so that finally there is an electron tracing the path of a complete sphere or shell. Some electrons have a greater distance away from the nucleus and they also shift their planes or orbits of rotation. They are arranged in a similar fashion that forms a second shell (see Figure 2-3B). The maximum number of shells for any known element is *seven* (see Figure 2-3C). Shells are labeled alphabetically from *K* through *Q*, starting with the innermost shell. Each shell has a definite maximum limit as to the number of electron orbits it can have. For instance, the *K shell* has a maximum number of electrons of two. The second shell, *L*, has eight, the third shell, *M*, 18, and so on. Copper has 29 electrons, so it has 2, 8, 18, and 1. The 1 electron sitting out so far away from the holding force of the nucleus is easily pushed along in various directions and can be moved into an adjacent atom's orbit. Then that atom's electron is moved to the next, and so on. This means there is a movement of electrons. The flow of electrons is defined as electricity.

Keep in mind that the electrons sitting in the last shell are called *valence electrons* and the electrical properties of a material are dependent on the number of such electrons. A material with eight valence electrons produces an inert material. Atoms with fewer than four valence electrons tend to give up or move electrons, and the fewer the valence elec-

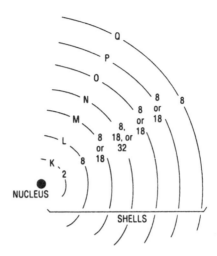

**Figure 2-3C.** Identification of atomic shells and the number of electrons in each shell.

trons, the greater this tendency. On the other hand, atoms with more than four electrons in their orbits or last shell have a tendency to acquire one or more additional electrons. In elements with atomic valences of four, adjacent atoms form into a *crystal structure* sharing their electrons in *covalent bonds*. Such bonds fill the valence shell, and the material is electrically inert. This creates the *semiconductor properties* that are the basis of solid state (transistor) electronics.

# Free Electrons

Electrons in copper drift along in a random fashion when at room temperature.

Heat is only one of the types of energy that can cause electrons to be forced from their orbits. A magnetic field can also be used to cause electrons to move in a given direction. Light energy and pressure on a crystal are also used to generate electricity by forcing electrons to flow along a given path. When electrons leave their orbits, they move from atom to atom at random, drifting in no particular direction. Electrons that move in such a way are referred to as *free electrons*. However, a force can be used to cause them to move in a given direction. That is how electricity (the flow of electrons along a conductor) is generated. A conductor is any material that has many free electrons by virtue of its physical makeup.

# Electric Energy

So far you have read about electrons being very small. Just how small are they? Well, electrons are incredibly small. The diameter of an electron is about 0.000 000 000 000 22 (or $2.2 \times 10^{-13}$) in. You may wonder how anything so small can be a source of energy. Much of the answer lies in the fact that electrons move at nearly the speed of light, or 186,000 miles per second (mi/s). In metric terms that is 300 million meters per second (m/s). As you can see from their size, billions of them can move at once through a wire. The combination of speed and concentration together produces great energy.

As you study more about electricity, electron theory, and electronics, you will discover that the electron is thought to be a quantum mechanical pattern, rather than a little solid ball. However, for purposes of ease in understanding and definition, we will refer to it here as *a particle of energy*.

# Current Flow

When a flow of electrons along a conductor occurs, this is commonly referred to as *current flow*. Thus, you can see that the movement of electrons is related to current electricity.

# Conductors

A material through which electricity passes easily is called a *conductor* because it has free electrons. In other words, a conductor offers very little resistance or opposition to the flow of electrons.

All metals are conductors of electricity to some extent. Some are much better than others. Silver, copper, and aluminum let electricity pass easily. Silver is a better conductor than copper. However, copper is used more frequently because it is cheaper. Aluminum is used as a conductor where light weight is important.

Why are some materials good conductors? One of the most important reasons is the presence of *free electrons*. If a material has many electrons which are free to move away from their atoms, that material will be a good conductor of electricity.

Although free electrons usually move in a haphazard way, their movement can be controlled. The electrons can be made to move in the same direction, and this flow is called *electric current*.

Conductors may be in the form of bars, tubes, or sheets. The most familiar conductors are wire. Many sizes of wire are available. Some are only the thickness of a hair. Other wire may be as thick as your arm. To prevent conductors from touching at the wrong place they are usually coated with a plastic or cloth material. This covering on the conductor is called an *insulator*.

# Insulators

An insulator is a material with very few, if any, free electrons. No known material is a perfect insulator. However, there are materials that are such poor conductors that they are classified as insulators. Glass, dry wood, rubber, mica, and certain plastics are insulating materials.

# Semiconductors

So far you have looked at insulators and conductors. In between the two extremes are semiconductors. Semiconductors in the form of transistors, diodes, and integrated circuits or chips are used every day in electronic devices. Now is the time to place them in their proper category.

Materials used in the manufacture of transistors and diodes have a conductivity halfway between that of a good conductor and a good insulator. Therefore, the name *semi-*

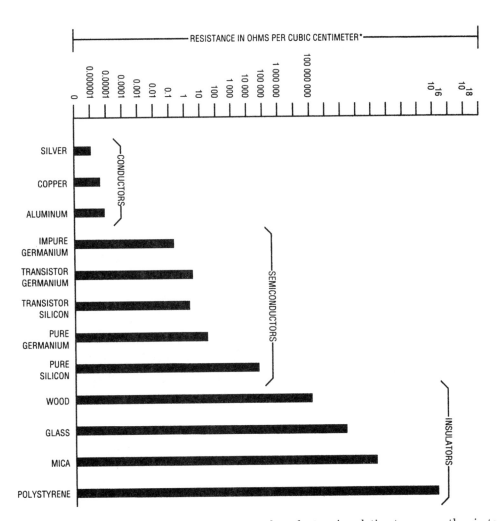

**Figure 2-4.** Location of insulators, semiconductors, and conductors in relation to one another in terms of inherent resistance.
*A cubic centimeter (cm$^3$) is one centimeter (0.3937 inches) long, one centimeter wide, and one centimeter high.

conductor is given them. Germanium and silicon are the two most commonly known semiconductor materials. Through the introduction of small amounts of other elements (called impurities) these nearly pure (99.999 999%) elements become *limited* conductors. The opposite of conductors is resistors. *Resistors* are devices used to give a measured amount of opposition or resistance to the flow of electrons. This opposition to current flow is measured in ohms ($\Omega$) and indicates the amount of resistance a piece of material offers to the flow of electrons. Take a look at Figure 2-4 to see how these semiconductor materials are placed between good conductors and poor conductors.

# Voltage and Current

In order to be able to measure the movement of electrons along a conductor, it is necessary to have units of measurement. This is somewhat difficult since you cannot see, taste, smell, or hear electricity. Feeling it is possible, but dangerous. This means that some way must be devised to be able to detect its presence, its direction of flow, and its magnitude. Before you can measure amount, you must have some type of unit to measure electricity in. (Electricity is defined as the flow of electrons along a conductor.)

# Volts

We measure the difference of potential between two plates in a battery in terms of volts (V). It is actually *electric pressure* exerted on electrons in a circuit. A circuit is a pathway for the movement of electrons. An external force exerted on electrons to make them flow through a conductor is known as *electromotive force*, or emf. It is measured in volts. Electric pressure, potential difference, and emf mean the same thing. The words *voltage drop* and *potential drop* can be interchanged.

# Current

For electrons to move in a particular direction, it is necessary for a potential difference to exist between two points of the emf source. If 6,250,000,000,000,000,000 (or $6.25 \times 10^{18}$) electrons pass a given point in one second, there is said to be one *ampere* (A) of current flowing. The same number of electrons stored on an object (a static charge) and not moving is called a *coulomb* (C).

As mentioned above, current is measured in amperes. However, in electronics it is sometimes necessary to use smaller units of measurement.

The *milliampere* is abbreviated as mA. It is one-thousandth (0.001) of an ampere. The *microampere* is abbreviated as μA. It is one-millionth (0.000 001) of an ampere. Note the Greek letter *mu* (μ) is used for micro.

You may want to become familiar with the Greek alphabet. Table 2-1 defines the terms used in electricity and electronics and their corresponding Greek letters.

Current flow is assumed to be from negative (–) to positive (+) in our explanations here. Electron flow is negative (–) to positive (+), and we assume that current flow and electron flow are one and the same. It makes explanations simpler as we progress into electronics. The *conventional* current flow is the opposite, or positive (+) to negative (–).

An ammeter is used to measure current flow in a circuit. A milliammeter is used to measure smaller amounts, while the microammeter is used to measure very small amounts of current. (See Table 2-2 for metric prefixes.)

A voltmeter is used to measure voltage. In some instances it is possible to obtain a meter which will measure both voltage and current plus resistance. This is called a *multimeter*, or *volt-ohm-milliammeter* (VOM).

# Power

*Power* is defined as the *rate* at which work is done. It is expressed in metric measurement terms of watts (W) for power and joules (J) for energy or work. A *watt* is the power that gives rise to the production of energy at the rate of one joule per second (W = J/s). A *joule*

**TABLE 2-1**

**THE GREEK ALPHABET USED IN ELECTRICITY AND ELECTRONICS**

| Name | Capital | Small | Used to Designate |
|------|---------|-------|-------------------|
| Alpha | A | $\alpha$ | Angles, area, coefficients, and attenuation constant. |
| Beta | B | $\beta$ | Angles and coefficients. |
| Gamma | $\Gamma$ | $\gamma$ | Electrical conductivity and propagation constant. |
| Delta | $\Delta$ | $\delta$ | Angles, increment, decrement, and determinants. |
| Epsilon | E | $\epsilon$ | Dielectric constant, permittivity, and base of natural logarithms. |
| Zeta | Z | $\zeta$ | Coordinates. |
| Eta | H | $\eta$ | Efficiency, hysteresis, and coordinates. |
| Theta | $\Theta$ | $\upsilon\ \theta$ | Angles and angular phase displacement. |
| Iota | I | $\iota$ | Coupling coefficient. |
| Kappa | K | $\kappa$ | |
| Lambda | $\Lambda$ | $\lambda$ | Wavelength. |
| Mu | M | $\mu$ | Permeability, amplification factor, and prefix *micro*. |
| Nu | N | $\nu$ | |
| Xi | $\Xi$ | $\xi$ | |
| Omicron | O | $o$ | |
| Pi | $\Pi$ | $\pi$ | Pi = 3.1416. |
| Rho | P | $\rho$ | Resistivity and volume charge density. |
| Sigma | $\Sigma$ | $\sigma\ \varsigma$ | Summation. |
| Tau | T | $\tau$ | Time constant and time-phase displacement. |
| Upsilon | Y | $\upsilon$ | |
| Phi | $\Phi$ | $\phi\ \varphi$ | Magnetic flux and angles. |
| Chi | X | $\chi$ | Angles. |
| Psi | $\Psi$ | $\psi$ | Dielectric flux. |
| Omega | $\Omega$ | $\omega$ | Resistance in ohms and angular velocity. |

is the work done when the point of application of force of one newton is displaced a distance of one meter in the direction of the force ($J = N \cdot m$).

It has long been the practice in this country to measure work in terms of horsepower (hp). Electric motors are still rated in horsepower and probably will be for some time inasmuch as the United States did not adopt the metric standards for everything.

Power can be electric or mechanical. When a mechanical force is used to lift a weight, *work* is done. The rate at which the weight is moved is *power*. *Horsepower* is defined in terms of moving a certain weight over a certain distance in one minute (e.g., 33,000 lb lifted 1 ft in 1 min equals 1 hp). Energy is consumed in moving a weight or when work is done. The findings in this field have been equated with the same amount of work done by electric energy. It takes 746 W of electric power to equal 1 hp.

The horsepower rating of electric motors is arrived at by taking the voltage and multiplying it by the current drawn under full load. This power is measured in watts. In other words, 1 V times 1 A equals 1 W. When put into a formula it reads:

$$\text{Power} = \text{volts} \times \text{amperes} \quad \text{or} \quad P = E \times I$$

where $E$ = voltage, or emf, and $I$ = current, or intensity of electron flow.

## TABLE 2-2

### METRIC PREFIXES AND POWERS OF TEN

**Metric Prefixes**

| Multiple | Prefix | Abbrev. | Multiple | Prefix | Abbrev. |
|----------|--------|---------|----------|--------|---------|
| $10^{12}$ | tera | T | $10^{-1}$ | deci | d |
| $10^{9}$ | giga | G | $10^{-2}$ | centi | c |
| $10^{6}$ | mega | M | $10^{-3}$ | milli | m |
| $10^{4}$ | myria | My | $10^{-6}$ | micro | $\mu$ |
| $10^{3}$ | kilo | k | $10^{-9}$ | nano | n |
| $10$ | deka | D | $10^{-12}$ | pico | p |

$$1 = 10^0$$
$$10 = 10^1$$
$$100 = 10^2$$
$$1\,000 = 10^3$$
$$10\,000 = 10^4$$
$$100\,000 = 10^5$$
$$1\,000\,000 = 10^6$$

Likewise, powers of ten can be used to simplify decimal expressions. The submultiples of 10 from 0.1 to 0.000001, with their equivalents in powers of ten, are:

$$0.1 = 10^{-1}$$
$$0.01 = 10^{-2}$$
$$0.001 = 10^{-3}$$
$$0.000\,1 = 10^{-4}$$
$$0.000\,01 = 10^{-5}$$
$$0.000\,001 = 10^{-6}$$

**Scientific Notation (Powers of Ten)**

Large numbers can be simplified by using scientific notation (powers of ten). For example, the multiples of 10 from 1 to 1,000,000, with their equivalents in powers of ten, are:

**Kilowatt.** The kilowatt is commonly used to express the amount of electric energy used or available. The term *kilo* (k) means one thousand (1000). A kilowatt (kW) is one thousand watts.

When the kilowatt is used in terms of power dissipated or consumed by a home for a month, it is expressed in kilowatthours. The unit kilowatthour is abbreviated as kWh. It is the equivalent of one thousand watts used for a period of one hour. Electric bills are figured or computed on an hourly basis and then read in the kWh unit. The entire month's time is equated to one hour's time.

Milliwatt is a term you will encounter when working with electronics. The *milliwatt* (mW) means one-thousandth (0.001) of a watt. The milliwatt is used in terms of some very small amplifiers and other electronic devices. For instance, a speaker used on a portable transistor radio will be rated as 100 milliwatts, or 0.1 W. Transistor circuits are designed in milliwatts, but power line electric power is measured in kilowatts. Keep in mind that *kilo* means 1000 and *milli* means 0.001.

# Resistance

Any time there is movement there is resistance. This resistance is useful in electric and electronic circuits. Resistance makes it possible to generate heat, control electron flow, and supply the correct voltage to a device.

Resistance in a conductor depends on four factors: material, length, cross-sectional area, and temperature.

**Material.** Some materials offer more resistance than others. It depends upon the number of free electrons present in the material.

**Length.** The longer the wire or conductor, the more resistance it has. Resistance is said to vary *directly* with the length of the wire.

**Cross-Sectional Area.** Resistance varies inversely with the size of the conductor in cross section. In other words, the larger the wire, the smaller the resistance per foot of length.

**Temperature.** For most materials, the higher the temperature, the higher the resistance. However, there are some exceptions to this in devices known as *thermistors*. Thermistors change resistance with temperature. They *decrease* in resistance with an increase in temperature. Thermistors are used in certain types of meters to measure temperature.

Resistance is measured by a unit called the *ohm*. The Greek letter omega ($\Omega$) is used as the symbol for electrical resistance.

# Resistors

*Resistors* are devices that provide measured amounts of resistance. They are valuable when it comes to making sure that the proper amount of voltage is present in a circuit. They are useful when generating heat.

Resistors are classified as either *wirewound* or *carbon-composition*. The symbol for a resistor of either type is —⋀⋀⋀—

Wirewound resistors are used to provide sufficient opposition to current flow to dissipate power of 5 W or more. A watt is a unit of electric power. A watt is equal to one volt times one ampere.

Wirewound resistors are made of wire that has controlled resistance per unit length.

Resistance causes a voltage drop across a resistor when current flows through it. The voltage is dropped or dissipated as heat and must be eliminated into the air.

Some variable resistors can be varied but can also be adjusted for a particular setting. Resistors are available in various sizes, shapes, and wattage ratings.

Carbon-composition resistors are usually found in electronics devices. They are of low wattage. They are made in $^1/_8$-W, $^1/_4$-W, $^1/_2$-W, 1-W, and 2-W sizes. The physical size determines the wattage rating or their ability to dissipate heat (see Figure 2-5).

Carbon-composition resistors are usually marked according to their ohmic value with a *color code*. The colors are placed on the resistors in rings (see Figure 2-6).

Table 2-3 shows the values for reading the color code of carbon-composition resistors.

Take a close look at a carbon-composition resistor. The bands should be to your left. Read from left to right. The band closest to one end is placed to the left so you can read it from left to right. The first band gives the first number according to the color code. In this case (Figure 2-6) it is red, or 2. The second band gives the next number, which is violet, or 7. The third band represents the multiplier or divisor.

If the third band is a color in the 0 to 9 range in the color code, it states the number of zeros to be added to the first two numbers. Orange is 3; so the resistor in Figure 2-6 has a value of 27,000 $\Omega$ of resistance.

The 27,000 $\Omega$ is usually written as 27 k$\Omega$. The k stands for thousand; it takes the place of three zeros. In some cases, resistors are referred to as 27 M$\Omega$ (which means 27,000,000, or 27 million $\Omega$), because the M stands for *mega*, and that is the unit for million.

If there is no fourth band, the resistor has a tolerance rating of ±20% (± means plus or minus). If the fourth band is silver, the resistor has a tolerance of ±10%. If the fourth band is gold, the resistor has a tolerance of ±5%.

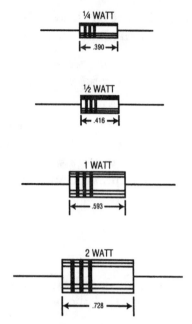

**Figure 2-5.** Wattage ratings of carbon-composition resistors. All measurements shown here are in inches.

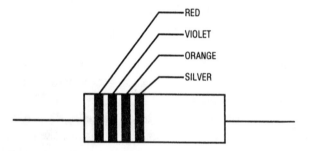

**Figure 2-6.** A 27,000-ohm (Ω) resistor.

### TABLE 2-3

### RESISTOR COLOR CODE

| | |
|---|---|
| 0 Black | 5 Green |
| 1 Brown | 6 Blue |
| 2 Red | 7 Violet |
| 3 Orange | 8 Gray |
| 4 Yellow | 9 White |

Silver and gold may also be used for the *third* band. In this case, according to the color code, the first two numbers (obtained from the first two color bands) must be divided by 10 or 100. Silver means divide the first two numbers by 100. Gold means divide the first two numbers by 10. For example, if the bands of the resistor are red, yellow, and gold, then the value is 24 divided by 10, or 2.4 Ω. If the third band is silver and the two colors are yellow and orange, then the 43 is divided by 100 to produce the answer of 0.43 Ω. Keep in mind, though, that the fourth band will still be either gold or silver to indicate the tolerance.

Resistors marked with the color code are available in hundreds of size and wattage rating combinations. Wattage rating refers to the wattage or power consumed by the resistor.

# Ohm's Law

A German physicist by the name of Georg Ohm discovered the relationship between voltage, current, and resistance in 1827. He found that in any circuit where the only opposition to the flow of electrons is resistance, there is a relationship between the values of voltage, current, and resistance. The strength or intensity of the current is directly proportional to the voltage and inversely proportional to the resistance.

It is easier to work with Ohm's law when it is expressed in a formula. In the formula, $E$ represents emf, or voltage; $I$ is the current, or the intensity of electron flow; $R$ stands for resistance. The formula is $E = I \times R$. It is used to find the emf (voltage) when the current and the resistance are known.

To find the current, when the voltage and resistance are known, use

$$I = \frac{E}{R}$$

To find the resistance, when the voltage and current are known, use

$$R = \frac{E}{I}$$

# Using Ohm's Law

Ohm's law is very useful in electrical and electronics work. You will need it often to determine the missing value. In order to make it easy to remember the formula take a look at Figure 2-7. Here the formulas are arrived at by placing your finger on the unknown and the other two will have their relationship displayed.

The best way to become accustomed to solving problems is to start with something simple, such as:

1. If the voltage is given as 100 V and the resistance is 25 Ω, it is a simple problem and a practical application of Ohm's law to find the current in the circuit. Use

$$I = \frac{E}{R}$$

Substituting the values in the formula,

$$I = \frac{100}{25}$$

means 100 is divided by 25 to produce 4 A for the current.

2. If the current is given as 2 A (you may read it on an ammeter in the circuit), and the voltage (read from the voltmeter) is 100 V, it is easy to find the resistance. Use

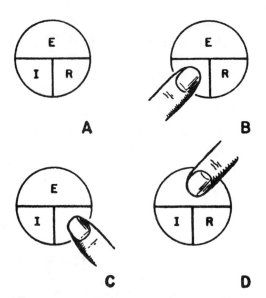

**Figure 2-7.** Ohm's law. Place your finger on the unknown value and the remaining two letters will give the formula to use for finding the unknown value.

$$R = \frac{E}{I}$$

Substituting the values in the formula,

$$R = \frac{100}{2}$$

means 100 divided by 2 equals 50 Ω for the circuit.

3. If the current is known to be 10 A, and the resistance is found to be 50 Ω (measured before the circuit is energized), it is then possible to determine how much voltage is needed to cause the circuit to function properly. Use

$$E = I \times R$$

Substituting the values in the formula,

$$E = 10 \times 50$$

means 10 times 50 produces 500 or that it would take 500 V to push 10 A through 50 Ω of resistance.

# Circuits

There are a number of different types of circuits. Circuits are the pathways along which electrons move to produce various effects.

The *complete* circuit is necessary for the controlled flow or movement of electrons along a conductor (see Figure 2-8). A complete circuit is made up of a source of electricity, a conductor, and a consuming device. This is the simplest of circuits. The flow of electrons through the consuming device produces heat, light, or work.

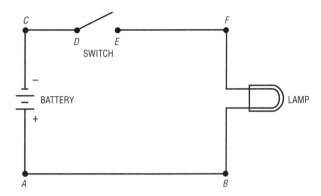

**Figure 2-8.** A simple circuit with a switch.

In order to form a complete circuit, these rules must be followed:

1. Connect one side of the power source to one side of the consuming device: *A* to *B*. (See Figure 2-8.)
2. Connect the other side of the power source to one side of the control device, usually a switch: *C* to *D*. (See Figure 2-8.)
3. Connect the other side of the switch to the consuming device it is supposed to control: *E* to *F*. (See Figure 2-8.) When the switch is closed the circuit is complete.

However, when the switch is open, or not closed, there is no path for electrons to flow, and there is an *open circuit* condition where no current flows.

This method is used to make a complete path for electrons to flow from that side of the battery with an excess of electrons to the other side which has a deficiency of electrons. The battery has a negative (–) charge where there is an excess of electrons and a positive (+) charge where there is a deficiency of electrons. Yes, you read it right: the – means excess and + means deficiency. This is due to the fact that we are using the current flow and electron flow as both the same and from – to + in the circuit.

A single path for electrons to flow is called a *closed*, or *complete*, *circuit*. However, in some instances the circuit may have more than one consuming device. In this situation we have what is called a *series circuit* if the two or more resistors or consuming devices are placed one after the other as shown in Figure 2-9.

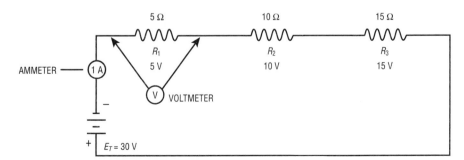

**Figure 2-9.** A series circuit with three resistors.

## Using the Meter to Measure

The voltmeter is placed across the resistor whose voltage drop is being measured. See Figure 2-9. If the meter is placed across the 10-$\Omega$ resistor $R_2$, the voltage indicated would be 10 V, and across resistor $R_3$ the meter would read 15 V. Note, however, that the ammeter is placed in series with the resistors and power source. The ammeter can become part of the circuit. The voltmeter is placed across the resistor and is not a permanent part of the circuit. If the voltmeter is placed across the power source (the battery) it would read the total voltage $E_T$ of 30 V, or the sum of the individual voltage drops around the circuit loop from the – terminal of the battery to the + terminal of the battery. See the following discussion of the series circuit.

## Series Circuit

Figure 2-9 shows a series circuit. The three resistors are connected in series, or one after the other, to complete the path from one terminal of the battery to the other. The current flows through each of them before returning to the positive terminal of the battery.

There is a law concerning the voltages in a series circuit. *Kirchhoff's voltage law* states that the sum of all voltages across resistors or loads is equal to the applied voltage. Voltage drop is considered across the resistor. Figure 2-9 shows the current flow through three resistors. The voltage drop across $R_1$ is 5 V. Across $R_2$ the voltage drop is 10 V. And, across $R_3$ the voltage drop is 15 V. The sum of the individual voltage drops is equal to the total or applied voltage of 30 V. $E_T$ means total voltage. It may also be written as $E_A$ for applied voltage or $E_S$ for source voltage.

To find the total resistance in a series circuit, just add the individual resistances or $R_T = R_1 + R_2 + R_3$. In this instance (Figure 2-9) the total resistance is 5 + 10 + 15, or 30 $\Omega$.

## Parallel Circuits

In a parallel circuit each resistance is connected directly across the voltage source or line. There are as many separate paths for current flow as there are branches (see Figure 2-10).

The voltage across all branches of a parallel circuit is the same. This is because all branches are connected across the voltage source. Current in a parallel circuit depends on the resistance of the branch. Ohm's law can be used to determine the current in each branch. You can find the total current for a parallel circuit by simply adding the individual currents. When written as a formula it reads

$$I_T = I_1 + I_2 + I_3 + \cdots$$

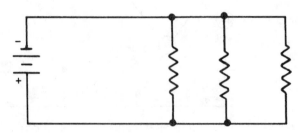

**Figure 2-10.** A parallel circuit.

The total resistance of a parallel circuit cannot be found by adding the resistor values. Two formulas are used for finding the total resistance ($R_T$). If there are *only two* resistors in parallel, a simple formula can be used:

$$R_T = \frac{R_1 \times R_2}{R_1 + R_2}$$

If there are more than two resistors in parallel, you can use the following formula. This formula may also be used with two resistors in parallel. In fact it can be used for *any* number of resistors.

$$\frac{1}{R_T} = \frac{1}{R_1} + \frac{1}{R_2} + \frac{1}{R_3} + \frac{1}{R_4} + \cdots$$

One thing should be kept in mind in parallel resistances: The total resistance is always less than the smallest resistance.*

As branches are added to a parallel circuit, the voltage across each branch is the same. However, the current divides according to the resistance in the branch. The total current is equal to the sum of the individual currents. Inasmuch as current and resistance are inversely related, that means if the currents are added then the total or equivalent resistance of the parallel circuit decreases with the increase in current. In order to account for this decrease even though more resistance is added to the circuit, the mathematical answer lies in the reciprocal ($1/R$) formula. The reciprocal of the sum of the reciprocals of the individual resistors in the circuit produces the desired mathematical result and Ohm's law is satisfied when applied to the total circuit values and when used for individual values within the branch circuits.

# Using Ohm's Law to Solve Circuit Problems

Examples of problem solving for series circuits using Ohm's law will show how the circuits use the various physical laws of nature.

Keep in mind that in a series circuit three rules aid in the solution of these circuit problems. They deal with the three factors found in any circuit — voltage, current, and resistance.

## Rules of Series Circuits

1. Current in all resistors is the *same* as the total current.

$$I_T = I_{R_1} = I_{R_2} = I_{R_3}$$

2. Voltage *divides* according to the resistance of each individual resistor. Or, the sum of the individual voltage drops across the resistors equals the applied voltage.

$$E_A = E_{R_1} + E_{R_2} + E_{R_3} + \cdots$$

3. Total resistance if found by *adding* the individual resistances:

$$R_T = R_1 + R_2 + R_3 + \cdots$$

---

*This is not true if one of the resistances is *negative*. The condition occurs only in active circuits, so for *most* applications the statement is true enough to be used for quick checks of your math.

## Example 1

What is the total resistance of four resistors connected in series if the resistors have resistances of 10, 20, 30, and 40 $\Omega$?

1. Determine what is to be found. In this case it is the total resistance.
2. Note that the formula for finding total resistance in a series circuit is

$$R_T = R_1 + R_2 + R_3 + R_4$$

3. Substitute the resistances for the letters:

$$R_T = 10 + 20 + 30 + 40$$

4. Sum up the resistances to produce 100 $\Omega$.


Keep in mind that voltage drop refers to the voltage *across* an individual resistor.


## Example 2

What is the total resistance of the circuit shown in Figure 2-11?

1. Determine what is to be found. In this case it is the total resistance.
2. Note that the formula for finding the total resistance in a series circuit is

$$R_T = R_1 + R_2 + R_3$$

3. Substitute the resistances for the letters in the formula:

$$R_T = 20 + 30 + 50$$

4. Sum up the resistances to produce 100 $\Omega$.


Using Ohm's law, it is possible to determine the voltage drop across each resistor if the total current is given.

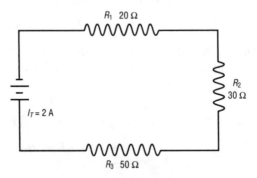

Figure 2-11.

### Example 3

What is the voltage drop across each resistor in Figure 2-11 if the total current is 2 amperes?

1. Determine what you know at this point. Check the circuit to see what the resistor values are: They are 20, 30, and 50 $\Omega$.
2. Check what the current is. Current is given at 2 amperes total.
3. Check the rules for the series circuit for current. The total current is equal to the current through each resistor in series, which means that the current through each resistor is 2 amperes.
4. Therefore, you know the current through each resistor, and you know the resistance. The formula (Ohm's law) for finding the voltage drop across each resistor is simply $E = I \times R$.

So,

$$E_{R_1} = I \times R_1 \quad \text{or} \quad E_{R_1} = 2 \times 20 \text{ or } 40 \text{ V}$$
$$E_{R_2} = I \times R_2 \quad \text{or} \quad E_{R_2} = 2 \times 30 \text{ or } 60 \text{ V}$$
$$E_{R_3} = I \times R_3 \quad \text{or} \quad E_{R_3} = 2 \times 50 \text{ or } 100 \text{ V}$$

## Series-Parallel Circuits

The series-parallel circuit is a combination of the series and the parallel arrangement. Figure 2-12 shows an example of the series-parallel circuit. It takes a minimum of three resistances to make a series-parallel circuit. This type has to be reduced to a series equivalent before it can be solved in terms of resistance. The parallel portions are reduced to the total for that part of the circuit, and then the equivalent resistance is added to the series part to obtain the total resistance.

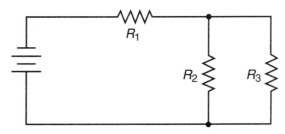

**Figure 2-12.** A series-parallel circuit.

Total current flows through the first series resistor but divides according to the branch resistances after that. There are definite relationships which must be explored here before that type of circuit can be fully understood. This will be done in a later part of this book.

## Open Circuits

An open circuit is an incomplete circuit. Figure 2-13 shows an open circuit that will become a closed circuit once the switch is closed. A circuit can also become open when one of the

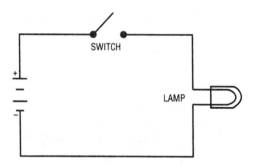

**Figure 2-13.** An open circuit produced by an open switch.

leads is cut or when one of the terminals has the wire removed. A loose connection can cause an open circuit.

## Short Circuits

The short circuit is something to be avoided because it can cause a fire or overheating.

A short circuit has a path of low resistance to electron flow. This is usually created when a low-resistance wire is placed across the consuming device (see Figure 2-14). The greater number of electrons will flow through the path of least resistance rather than through the consuming device. A short usually generates an excess current flow that can result in damage to a number of parts of the circuit. If you wish to prevent the damage caused by short circuits, use a fuse.

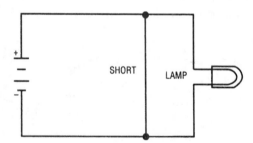

**Figure 2-14.** A short circuit. The wire has less resistance than the lamp.

## Fuses

Fuses are available in a number of sizes and shapes. They are used to prevent the damage done by excess current flowing in a circuit. They are placed in series with the consuming devices. Once too much current flows, it causes the fuse wire inside the fuse case to melt. This opens the circuit and stops the flow of current and prevents the overheating that occurs when too much current is present in a circuit.

The symbol for a fuse is ⌁. It fits into a circuit as shown in Figure 2-15.

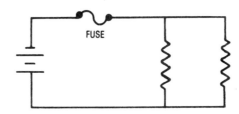

**Figure 2-15.** The location of the fuse in a circuit.

# Summary

Electricity is defined as the flow of electrons along a conductor. Electrons are the smallest part of an atom. An atom is the smallest part of an element which contains all its properties. Electrons can be directed along a given path called a circuit by means of magnetism, heat, light, or pressure. Electrons that move in a random motion are called free electrons. These free electrons when directed in a given direction make up what is called electricity.

Current flow is the flow of electrons along a conductor. Current flow is from negative to positive.

A material through which electricity passes easily is called a conductor. An insulator is a material with very few, if any, free electrons. No known materials are perfect insulators.

The diameter of an electron is about $0.000\,000\,000\,000\,22$ (or $2.2 \times 10^{-13}$) in. Materials used in the manufacture of transistors and diodes have a conductivity between that of a good conductor and a good insulator.

Electricity is measured in terms of volts, amps, and ohms. Meters are used to measure the flow of electrons, the voltage drop across resistors, and the opposition put up to the flow of electrons by certain materials.

Resistance is the opposition to the movement of electrons. Resistance is measured in ohms. Resistors are devices which provide measured amounts of resistance. There are two types of resistors: wirewound and carbon-composition.

Power is defined as the rate at which work is done. Power measured in terms of electrical energy is designated as watts. The watt is one volt times one ampere for one second. The kilowatt is one thousand watts. The kilowatthour is one thousand watts for one hour.

It takes 746 watts to produce one horsepower. It takes a mechanical horsepower defined in terms of 33,000 pounds lifted one foot in one minute to equal one electrical horsepower defined in terms of watts or 746 watts equal one horsepower.

Ohm's law states that the current in any circuit is equal to the voltage divided by the resistance. It can also be substituted so that the voltage is equal to the current times the resistance, or the resistance is equal to the voltage divided by the current.

There are a number of types of circuits. The open circuit does not have a complete path for electron flow from one terminal of the voltage source to the other. A short circuit has a resistance that is too small and therefore takes all the current and bypasses it from the intended load. The series circuit consists of resistors placed end to end. The parallel circuit consists of resistors placed across the power source. The current in a series circuit flows the same through all resistors. The current in a parallel circuit divides according to the branch resistance.

A combination of series and parallel circuits can be made with the use of at least three resistors.

Fuses are safety devices which protect circuits from overloads and overheating. They open the circuit when overheated.

# Review Questions

1. Where does the word electricity come from?
2. What is an atom?
3. How does an electron fit into an atom?
4. Where do you find free electrons?
5. How big is an electron?
6. List five insulators.
7. What are the two most common semiconductor materials?
8. Define voltage and current.
9. What is the unit for measuring electric power?
10. How much is a kilowatt?
11. How many watts are there in one horsepower?
12. What is the symbol for ohms?
13. What is the symbol for a resistor?
14. What type of resistor uses a color code?
15. How much is a kilo?
16. What is the term used to designate one million?
17. State Ohm's law.
18. State Kirchhoff's voltage law.
19. What is the formula used to find total resistance in a parallel circuit?
20. What is the purpose of a fuse?

# Chapter 3

# INDUCTANCE, CAPACITANCE, AND ALTERNATING CURRENT

**R**esistors and resistance are very important in the study of electricity and electronics. However, when two other devices are introduced into the circuits, there is the possibility of various combinations that can produce rather interesting final results. One of these devices is the inductor, which produces inductance. The other is the capacitor, which provides capacitance.

## Inductance

Inductance is the ability of a coil, or choke, or inductor (all three mean the same thing and are interchangeable) to oppose any change in circuit current. This is not so important in a direct current (dc) circuit because the current flows in only one direction, from negative to positive, when it is first turned on, and then it stops when it is turned off. The collapsing magnetic field that was produced by the coil of wire with a current through it produces an emf in the coil when it decays or collapses. This emf is in the opposite polarity to that which caused it to be produced. This emf is called a *counter emf*, abbreviated as *cemf*.

## Induction Coil

Michael Faraday was an Englishman who performed early experiments with coils of wire and electric current. He was born in London in 1791. His work was important in laying the foundation of the growing science of electricity. Faraday started to experiment with electricity about 1805. It was not until 1831 that he performed experiments on magnetically coupled coils. A voltage was induced in one of the coils by means of a magnetic field created by current flow in the other coil. From this experiment came the induction coil. Faraday's experiment and discovery made possible many of our modern conveniences. The automobile, doorbell, automobile radio, and television are all possible because of inductance.

Faraday also invented the first transformer. A transformer changes electricity into a higher or lower voltage. At that time it had very few practical uses. He made the first dc generator. At the time Faraday was working in England, Joseph Henry was making almost the same discoveries in the United States. Henry worked in New York and discovered the property of self-inductance before Faraday. The unit of measurement for inductance is the henry (H). The symbol for inductance is $L$.

An inductor has an inductance of one *henry* if an emf of one volt is induced in the inductor when the current through the coil is changing at the rate of one ampere per second. Keep in mind that the one-volt, one-ampere, and one-henry relationship deals with the basic units of measurement of voltage, current, and inductance.

# Inductor

Inductors come in many sizes and shapes (see Figure 3-1). Air-core inductors are coils that are wound without a core. They are used in circuits where the frequencies cannot be heard, such as radio frequencies. Radio frequencies are above the human hearing range. The symbol for a radio frequency coil is ⌒⌒⌒. Note how the radio frequency air core inductor in Figure 3-1B is very small and resembles a resistor. The size of the color band to the left is larger than the others. This indicates that it is an inductor instead of a resistor. The color code tells the size in microhenrys ($\mu$H).

**SUBMINIATURE RF CHOKES**

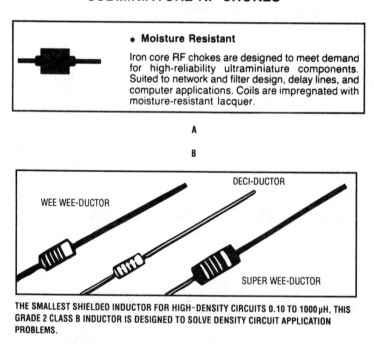

- **Moisture Resistant**

Iron core RF chokes are designed to meet demand for high-reliability ultraminiature components. Suited to network and filter design, delay lines, and computer applications. Coils are impregnated with moisture-resistant lacquer.

A

B

WEE WEE-DUCTOR

DECI-DUCTOR

SUPER WEE-DUCTOR

THE SMALLEST SHIELDED INDUCTOR FOR HIGH-DENSITY CIRCUITS 0.10 TO 1000 μH, THIS GRADE 2 CLASS B INDUCTOR IS DESIGNED TO SOLVE DENSITY CIRCUIT APPLICATION PROBLEMS.

**Figure 3-1.** Radio frequency chokes.

Inductors are also called *chokes* because of the way they hold back current or choke it. They are also called *coils* for the simple construction technique used to make them. They are nothing more than a coil of wire.

Inductors with iron cores are used in circuits where the frequencies can be heard. These are called *audio frequencies* and they are referred to as audio chokes or audio inductors.

The iron core is usually laminated sheets of iron. The iron is specially made silicon steel. Silicon steel is used because it can change its magnetic orientation rapidly without causing too much opposition to the changing field or polarity reversals. The symbol for an iron core choke is $\overline{\phantom{mmmm}}$ (see Figure 3-2).

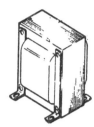 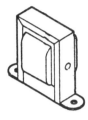

**Figure 3-2.** Audio frequency chokes.

# Mutual Inductance

When two coils are placed near one another *mutual induction* occurs. A change in the flux or magnetic field in one coil will cause an emf to be induced in the other coil. The two coils have mutual inductance. The amount of mutual inductance depends on the distance between the two coils. If the coils are separated a considerable distance, the amount of flux common to both coils is small and the mutual inductance is low. If the coils are close together, nearly all the flux on one coil will link the turns of the other. The mutual inductance can be increased greatly by mounting both coils on the same iron core.

## Factors That Influence Mutual Inductance

Mutual inductance of two adjacent coils depends upon five factors:

1. Physical size of the two coils
2. Number of turns in each coil
3. Distance between the two coils
4. Distance between the axes of the two coils
5. Permeability of the cores

*Permeability* is the ease with which magnetic lines of force distribute themselves throughout a material.

# Self-Inductance

Even a straight conductor or piece of wire has some inductance. When current flows through a conductor, a magnetic field is established around that conductor. As the current changes, so does the magnetic field. An emf is induced in the conductor when the field changes. This emf is called a *self-induced emf*. The direction of the emf, or current, has a definite relation to the direction of the field that causes the induced emf.

In a dc circuit, where the current varies only when the circuit is turned on and off, it produces a curve which resembles Figure 3-3. However, when alternating current (ac) is introduced in a coil, a whole new set of rules apply (to be discussed later in this chapter).

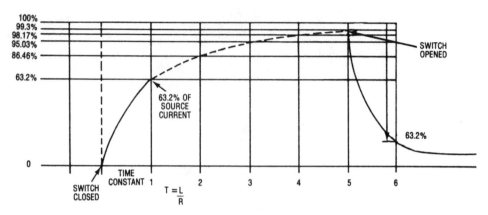

**Figure 3-3.** Time constant of a coil depends upon its inductance and resistance.

## Lenz's Law
The direction of the induced emf is always opposite to that which induced it. That is the basis for Lenz's law.

## Factors That Affect the Inductance of a Coil
Among the factors are:

1. The inductance of a coil is proportional to the square of the number of turns.
2. The inductance of a coil increases directly as the permeability of the core material increases. Coils with iron cores have higher permeability caused by the core material.
3. The inductance of a coil increases directly as the cross-sectional area increases.
4. The inductance of a coil decreases as its length increases.

Figure 3-4A shows two coils with a fixed number of turns. Each has a different cross-sectional area. The larger coil has a greater flux, or less reluctance. Figure 3-4B shows two coils, each with a fixed number of turns. They have the same cross-sectional area. They have different lengths. The longer coil has less total flux, or greater reluctance. Therefore, it has less inductance.

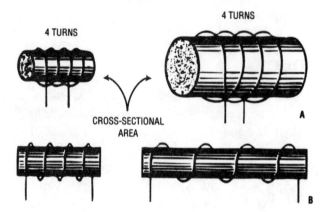

**Figure 3-4.** Factors which influence inductance of a coil.

# Alternating Current

Most of us have grown up with alternating current (ac). We are used to the 60-hertz (Hz) line current that is furnished to every house in the United States. It is much better than dc when it comes to transporting power over long distances without huge losses. It is also important since electronics relies so heavily on ac. The radio frequencies which bring us radio and television are also ac. In order to get a better understanding of ac, we should look at how it is generated and distributed.

Alternating current was developed by Nicholas Tesla. Its use was introduced around 1900. Niagara Falls, New York, had one of the first commercial ac generators, which was in operation until 1948 when a rock slide ruined it.

Alternating current has a distinct advantage over direct current. Alternating current can be stepped up to obtain higher voltages and lower currents and still produce the same amount of power at the other end of the line. It can be transported over long distances through small wires because of the higher voltages and lower currents. Current determines the size of the wire. Then it is stepped down for local distribution. Since transformers are very efficient machines, very low losses are experienced with ac.

Usually, ac is generated at 13,800 volts (V). This is then stepped up to at least 138,000 V for distribution. The most commonly used high voltages for long-distance transmission are 138,000, 250,000, and 750,000 V. Once the power reaches its destination, it is reduced to as low as 240 V for home use. This is further split for home circuits of 120 V.

## Nature of Alternating Current

Alternating current changes its direction of flow as it moves along a wire. It flows in one direction and then the other. Figure 3-5 shows a simple ac generator; Figure 3-6 shows the output of the generator.

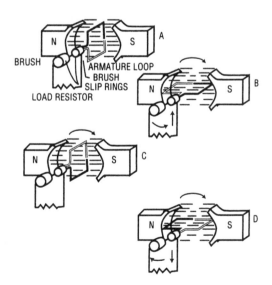

**Figure 3-5.** Simple ac generator called an alternator.

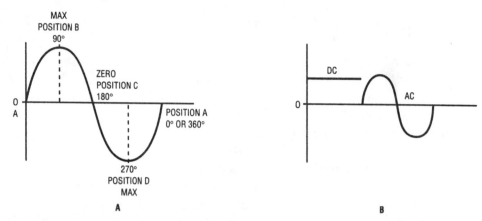

**Figure 3-6.** A. The sine wave. B. Comparison of ac to dc.

# Waveforms

Alternating current is constantly changing. It changes in magnitude and direction. The ac pattern, or *waveform*, is shown in Figure 3-6. The time base of one second (s) is standard in the electrical field. Whenever you see 60 Hz, you know it means 60 complete sine waves are generated in 1 s.

# Frequency

The term *frequency* is used to indicate how many times the alternating current changes direction. The ac you use at home changes direction 60 times per second. This means it moves back and forth 60 times per second. This current is described as having a frequency of 60 Hz. It is said to be 60-Hz ac. Frequency can be expressed in megahertz, or MHz. This is 1 million Hz/s. It can also be expressed in kilohertz, which means 1000 times or hertz per second. Kilohertz is abbreviated as kHz. Note that k is a small letter. However, MHz uses M since it stands for *mega*, or one million.

Figure 3-7 shows the difference between 4 Hz and 1 Hz. Note that both of them occur in 1 s. Note how the waveform becomes closer together as it becomes part of a higher frequency.

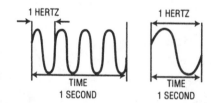

**Figure 3-7.** 4 Hz and 1 Hz compared.

# Maximum and Peak Values of AC

Three values are used to describe ac:

- Peak
- Average
- Root-mean-square (rms)

# Peak

The maximum point on a sine wave is the peak value. Both peaks of a single hertz may be included in a reference. If so, it becomes a peak-to-peak value. A peak value of 100 V means that the peak-to-peak value is 200 V (see Figure 3-8).

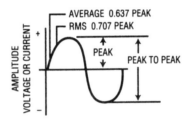

Figure 3-8. Ac sine wave values.

# Average

The average of all instantaneous values of a generator is measured at regular intervals. The values are taken at selected points in the generating process. The average of these is the average value of ac current (see Figure 3-8). The average value is 0.637 times the peak. This means that a peak of 100 V is equal to 63.7 V average. However, average is not often used in reference to ac. Instead, root-mean-square (rms) is used more often.

# Root-mean-square (rms)

This is a method used for equating ac to dc. Since ac is constantly changing, it forms a sine wave (see Figure 3-6). The values used for rms figuring are taken from selected points in the sine wave generating process. The ac is constantly changing and does not have the heating value of dc. That is because dc comes up to its peak and stays there until turned off. Therefore, if you take the ac sine wave and break it into four parts, each containing 90° of the complete cycle needed to generate 1 Hz, you will find that the instantaneous voltage and current when taken 90 times (once for each degree) and squared, then averaged (*mean* means average) and the square root taken of the average, you will have 0.7071 times the peak value of the sine wave. This shows that rms, effective heat, and heating effect all mean the same thing. You could also get the rms, or 0.7071 value, by taking the *sine* of 45°. Since 45° is one-half of the 90° to the peak of the waveform, it makes more sense mathematically. It also makes sense because the shape of the waveform is not a semicircle, but is shaped more like the mathematical equivalent to a sine value.

# Transformers

A practical application of alternating current is its use with the transformer. It has the ability to step up and step down voltages. But, only ac. Direct current does not work with a transformer.

A *transformer* is a device consisting of two coils that can change voltages. The voltage put into a transformer is either stepped up or stepped down. (In some instances, however, isolation transformers are used so that they have the same output as input voltage. They are used to eliminate the ground connection in conventional line current. That way you must come across both terminals to receive a shock instead of any ground and the hot side of the line.)

# Ground

A ground in electronics is defined as a common connection such as the chassis or a copper strip on a printed circuit board. The common ground usually has a negative polarity. The chassis of an electrical or electronic device serves as a common return to the power supply and is referred to as the ground. An *earth ground* has a potential of zero. To maintain a good earth ground, a metallic rod 4 to 6 feet long is driven into the ground. Any wire connected to that rod is *grounded*. The power company installs a rod like this when they connect power to your home. The telephone and cable-TV companies wire their network interfaces (lightning protection) to the power company's ground rod. The alternative to earth ground is a *floating ground*. A floating ground is simply a reference point that is not earth grounded. The negative terminal of a car's battery is a floating ground, and any home appliance that has a two-prong electrical plug also has a floating ground. In newer homes a third ground wire is provided for wall outlets for appliances. The symbol for a ground is shown in Figure 3-9.

**Figure 3-9.** Symbols used for a ground or earth.

# Types of Transformers

Figure 3-10 shows how transformers are used to distribute electric power from the generator to the consumer. Note that the symbols used for the transformer are two coils of wire with straight lines in between to designate the iron core.

A transformer has a primary winding and a secondary winding. A transformer is simply a coil when it has no load on the secondary. When a load is placed on the output side (the secondary winding), the device actually becomes a transformer.

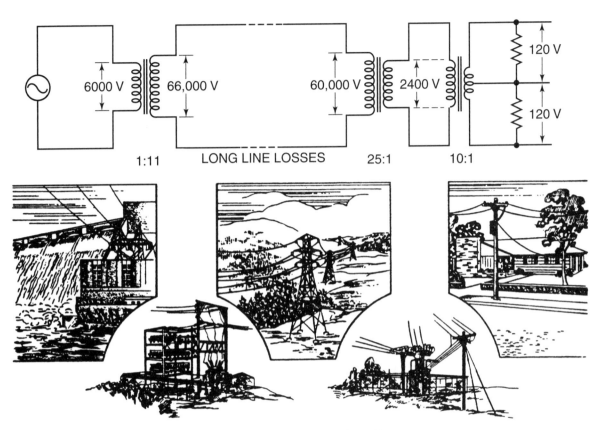

**Figure 3-10.** Using transformers to distribute ac over long distances.

Transformers come in many sizes and shapes (see Figure 3-11). They may be used on alternating currents at power line frequencies of 25, 40, or 60 Hz and also on frequencies of more than 1 million Hz. The transformers used on radio frequencies do not have cores. They have air for a core, and their physical size is much smaller than power frequency transformers.

Since there are no moving parts in a transformer, it can be up to 99 percent efficient. The only moving part is the current. Losses are eliminated by using silicon steel for the core laminations. The silicon steel reduces the losses due to hysteresis that are caused by changing the polarity many times per second. Eddy currents are small currents induced in the metal by the changing magnetic field. Laminations eliminate the losses caused by eddy currents. Copper losses are reduced to a minimum by using the proper size wire for the amount of current being handled.

# Power Transformers

A power transformer can be both a step-up and a step-down unit (see Figure 3-12). The secondary windings can furnish a number of different voltages. These voltages may be either higher or lower than the primary voltage. This type of transformer is used in electronics equipment where a number of different voltages are needed.

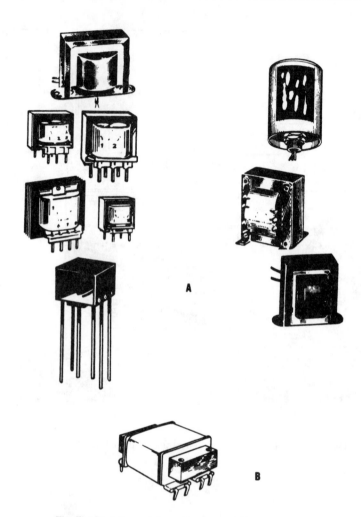

The Flat Pack™ power transformer meets the needs of low-clearance printed circuit boards and solid state power designs. The units can also be used for control and instrumentation applications. Voltages and current are chosen for widely used power applications and can be used in single, or dual-output supplies. The Flat Pack™ is designed to pass a 2000-V HiPot test.

**Figure 3-11.** A. Various sizes and types of transformers. B. Power transformer.

# Inductive Reactance

*Inductive reactance* is the opposition put up to alternating current by a coil. The coil has a definite time or delay built in due to the ratio of inductance to resistance. This built-in delay of current comes in conflict with the ac since the current is constantly changing and not necessarily at the same rate as the natural tendency of the coil. Therefore, the reaction or reactance is in the form of an opposition. Since it is an opposition, it is measured in ohms. Reactance is represented by the symbol $X$. Inductance is represented by its symbol $L$. When the inductive reactance is represented, it is written as $X_L$. It is measured in ohms.

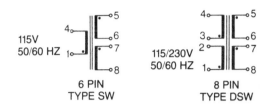

SIDE WINDER SPLIT BOBBIN PC MOUNT TRANSFORMERS

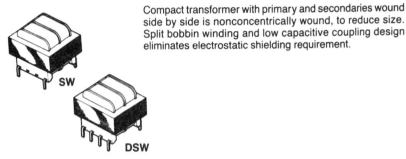

Compact transformer with primary and secondaries wound side by side is nonconcentrically wound, to reduce size. Split bobbin winding and low capacitive coupling design eliminates electrostatic shielding requirement.

**Figure 3-12.** Power transformers and schematics.

A number of factors determine $X_L$. One is the frequency of ac, which affects reactance. Another is the size of the inductor. The formula used to calculate $X_L$ is

$$X_L = 2\pi f L$$

In this equation, $f$ is the frequency, measured in hertz. $L$ is inductance, measured in henrys (H). Pi ($\pi$) is a standard mathematical term with a value of 3.141 592 654. Thus, $2\pi$ equals 6.28 when rounded off for quick answers.

By increasing the $f$, the $X_L$ increases. If the $f$ is decreased, the $X_L$ decreases. The same is true for $L$, or inductance. Since the $2\pi$ is a constant, it does not change. Therefore, the only two variables for $X_L$ are frequency and inductance.

Keep in mind that the voltage and current are 90 degrees out of phase whenever there is an inductor in the circuit. The voltage leads the current by 90 degrees in a purely inductive circuit. This property of an inductor can be useful in many applications as you will discover as you continue with the electronics part of this book. (See Figure 3-13.)

# Phase Shift

Figure 3-13 shows voltage and current out of phase by 90°. Wave A indicates the voltage waveform and wave B indicates the 90° lag in the current caused by an inductor. There is always some resistance in an inductor since it is made of copper wire and it has resistance, so there is never a full 90° phase lag in the current. But, for most practical purposes and theoretical purposes it is assumed to be 90° if not specifically mentioned otherwise. The waveform in Figure 3-13 can be displayed on an oscilloscope.

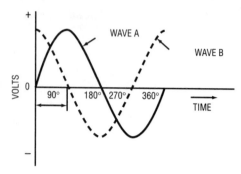

**Figure 3-13.** Voltage and current out of phase by 90°.

## $X_L$ in Series

Resistors and inductive reactances both produce opposition, and both are measured in ohms. This means that inductive reactances which are in series are simply added to obtain the total inductive reactance in a circuit:

$$X_{L_T} = X_{L_1} + X_{L_2} + X_{L_3} + \cdots$$

## $X_L$ in Parallel

Inductive reactances which are in parallel are treated the same as resistors in parallel. You can use the product divided by the sum formula or the reciprocal formula:

$$X_{L_T} = \frac{X_{L_1} \times X_{L_2}}{X_{L_1} + X_{L_2}}$$

$$\frac{1}{X_{L_T}} = \frac{1}{X_{L_1}} + \frac{1}{X_{L_2}} + \frac{1}{X_{L_3}} + \cdots$$

## Uses of $X_L$

The major use of inductance is to provide a minimum reactance for low frequencies. Inductors produce high opposition to higher frequencies.

One specific use of inductance is in filters. Filters are used when certain frequencies are desired and others are to be avoided. An inductor that has an $X_L$ that passes certain frequencies and opposes others is used.

The main use for inductive reactance is in electronic circuits. Such circuits, along with capacitors, tune in certain frequencies and reject others. An example is the tuner of your radio or television.

# Capacitors

Capacitors play an important role in the building of circuits. A *capacitor* is a device that opposes any change in circuit voltage. That property of a capacitor which opposes voltage change is called *capacitance*.

Capacitors make it possible to store electric energy. Electrons are held within a capacitor. This, in effect, is stored electricity. It is also known as an *electric potential*, or an *electrostatic field*. Electrostatic fields hold electrons. When the buildup of electrons becomes great enough, the electric potential is discharged. This process takes place in nature: clouds build up electrostatic fields. Their discharge is seen as lightning.

Figure 3-14 shows a simple capacitor. Two plates of a conductor material are isolated from one another. Between the two plates is a dielectric material. The dielectric does not conduct electrons easily. Electrons are stored on the plate surfaces. The larger the surface, the more area is available for stored electrons. Increasing the size of the plate therefore increases the capacitance.

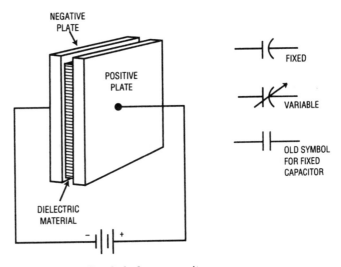

Figure 3-14. Design of a capacitor. Symbols for a capacitor.

## Operation of the Capacitor

If a capacitor has no charge of electrons, it is uncharged. This happens when there is no voltage applied to the plates. An uncharged capacitor is shown in Figure 3-15A. Note the symbol for a capacitor in this drawing. This is the preferred way to show a capacitor: a

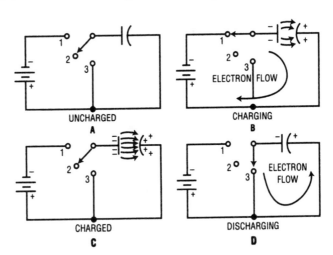

Figure 3-15. Charge and discharge of a capacitor.

straight line and a curved line facing each other. Note that the circuit has a dc source and a three-position switch that is in the open position.

In Figure 3-15B, the switch has been closed to position 1. This causes current to flow. A difference in potential is created by the voltage source. This causes electrons to be transferred from the positive to the negative plate. This transfer continues as long as the voltage source is connected to the two plates and until the accumulated charge becomes equal to the potential difference of the applied voltage. That is, charging takes place until the capacitor is charged.

In Figure 3-15C, the voltage has been removed. The switch is open. At this point, the potential difference, or charge, across the capacitor remains. That is, there is still a surplus of electrons on the negative plate of the capacitor. This charge remains in place until a path is provided for discharging the excess electrons.

In Figure 3-15D, the switch is moved to position 3. This opens the path for discharging the surplus electrons. Notice that the discharge path is in the opposite direction from the charge path. This shows how a change in circuit voltage results in a change in the capacitor charge. Some electrons leave the excess (negative) plate. They do this in an attempt to keep the voltage in the circuit constant.

As you can see from the foregoing, the ability of a capacitor to charge and discharge can be useful in many types of circuits. Its ability to oppose any change in the circuit voltage can also be helpful. All this will be put to work later in electronic circuits.

# Capacity of a Capacitor

The two plates of the capacitor may be made of almost any material. The only criterion is that the material will allow electrons to collect on it. The dielectric may be air, vacuum, plastics, wood, or mica.

Three factors determine the capacity of a capacitor:

- Area of the plates
- Distance between the plates
- Material used as a dielectric

**Area of the Plates.** Area of the plates determines the ability of a capacitor to hold electrons. The larger the plate area, the greater the capacity, or capacitance.

**Distance Between the Plates.** Distance between the plates of a capacitor determines the effect that electrons have upon one another. That is because electrons possess a charge, or field, around them that can react with those close by. Capacitance increases when the plates are brought close together.

Variable capacitors bring plates in and out of mesh and, in so doing, cause the plates to become closer or more distant accordingly (see Figure 3-16 for variable capacitors).

**Dielectric Materials.** One of the effects of the dielectric materials is determined by its thickness. The thinner the dielectric, the closer the plates will be. A thin dielectric can thus increase capacitance. Some dielectrics have better insulating qualities than others and will allow greater voltages to be applied between the plates before breaking down. Take a

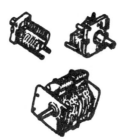

**Figure 3-16.** Variable capacitors.

**TABLE 3-1**

| Material | Dielectric Constant ($K$) |
|----------|---------------------------|
| Air or Vacuum | 1 |
| Rubber | 2–4 |
| Oil | 2–5 |
| Paper | 2–6 |
| Mica | 3–8 |
| Glass | 8 |
| Ceramics | 80–1200 |

look at the dielectric materials listed in Table 3-1 to see how various materials affect the capacitance of a capacitor.

# Working Voltage DC

The maximum safe working voltage of a capacitor in a dc circuit is identified as the working voltage dc, or WVDC. Above this voltage, a capacitor is expected to puncture or develop a short circuit.

If the temperature in which a circuit operates reaches 60°C or higher, the voltage rating is lowered.

Voltage ratings for mica, paper, and ceramic capacitors are usually 200, 400, and 600 V dc. Oil-filled capacitors have voltage ratings ranging up to 7500 V. As the voltage ratings become higher, the physical size of the capacitors becomes greater.

## Electrolytic Capacitors

The electrolytic capacitor has polarity. That is, it is marked with a + or −. It is polarized in order to be able to get large capacitances into small containers. Electrolytics are very much in evidence in transistor circuits since they need large capacitances in small packages.

Keep in mind that the electrolytic capacitor has polarity. Be sure to connect it properly in the circuit, or it can explode. If ac is applied to electrolytics, it can cause an explosion, and the capacitor will pop like a firecracker scattering its debris over a wide area.

Never operate a capacitor above its rated WVDC. It is customary to use a capacitor in a circuit with about 50 to 75 percent of its rated voltage.

Nonpolarized electrolytics are made by placing two electrolytics in series connected with + to + and/or − to −. This reduces the capacitance, but it eliminates the polarity. Nonpolarized electrolytics are usually made for crossover networks or for electric motors.

Electrolytics can dry up or open if not used for some time. They will also deteriorate somewhat even when used. However, electrolytics can be self-healing. They can change from an open condition back to proper operation with the passage of time, with no help from anyone or anything. The open condition is one that occurs sometimes spontaneously and is very frustrating when troubleshooting a circuit.

## Capacitors in Series

Capacitors can be connected in series, but the series reduces the capacitance. The formula used for finding capacitance for two capacitors in series is

$$C_T = \frac{C_1 \times C_2}{C_1 + C_2} \qquad \text{for two only}$$

$$\frac{1}{C_T} = \frac{1}{C_1} + \frac{1}{C_2} + \frac{1}{C_3} + \cdots$$

As you increase the distance between the plates, effectively, when placing them in series, you also increase the WVDC rating. Just add the WVDC ratings of the capacitors to obtain the higher value created with the placement in series (see Figure 3-17).

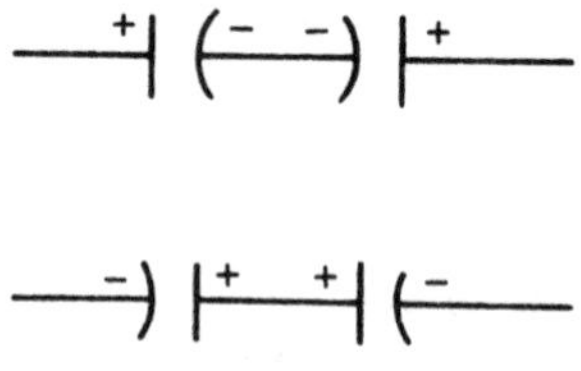

**Figure 3-17.** Capacitors connected back to back and in series to produce a nonpolarized electrolytic.

## Capacitors in Parallel

Capacitors can be connected in parallel if their polarity is observed in the case of electrolytics. For standard, nonpolarized types, it is not necessary to observe any particular connection procedure except to place the leads together in order to produce a parallel connection. Placing capacitors in parallel *increases* the capacitance. Just add the individual capacitances to obtain the total capacitance. However, keep in mind that the WVDC will be the rating of the *smallest* value of voltage in the WVDC ratings (see Figure 3-18).

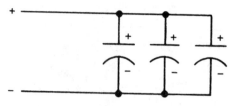

**Figure 3-18.** Capacitors in parallel. Note how the polarities of electrolytics are connected.

# Types of Capacitors

There are at least six different types of capacitors:

1. Air
2. Mica
3. Paper
4. Ceramic
5. Electrolytic
6. Tantalum

The electrolytic and the tantalum have polarity. The others do not need polarity stamped on them since they will operate in both dc and ac circuits.

# Units of Measurement

Capacitance is measured in farads (F). The *farad* is defined as having the ability to store enough electrons to produce a voltage difference of one volt across the terminals while producing one ampere of current for one second.

The farad is a very large unit of capacitance. The capacitors we use in electricity and in electronics are much, much smaller. They are measured in *microfarads* (0.000 001 F) and in microfarads, now called *picofarads* (0.000 000 000 001 F).

It is often necessary to interpret or change values. This occurs as you read circuit drawings or markings on capacitors. You may, for example, find yourself working with capacitors marked in terms of pF and drawings indicated in $\mu$F.

Many formulas are stated in terms of the farad. To convert from microfarads or picofarads to farads, just follow these simple rules:

- To convert from microfarads, move the decimal point six places to the left: 1 $\mu$F = 0.000 001 F.
- To convert from picofarads, move the decimal point 12 places to the left: 1 pF = 0.000 000 000 001 F.
- To change microfarads to picofarads, move the decimal point six places to the right: 0.1 $\mu$F = 100,000 pF, or 100 kpF.

## Conversion Methods

Keep in mind the following to make sure you get the conversion correct and that you are tuned in to the correct formulas.

- pF to $\mu$F, move the decimal six places to the left.
- $\mu$F to F, move the decimal six places to the left.
- F to $\mu$F, move the decimal six places to the right.
- $\mu$F to pF, move the decimal six places to the right.
- pF to F, move the decimal 12 places to the left.
- F to pF, move the decimal 12 places to the right.

# Capacitor Tolerances

Capacitors also have tolerances with a ± value that varies primarily with the price of the capacitor. The closer the + and – values, the more expensive the capacitor.

General-purpose ceramic-disk capacitors usually have tolerances of ±20 percent.

Paper capacitors usually have a tolerance of ±10 percent although some may be bought at ±5 percent.

Mica and ceramic tubular capacitors are used when closer tolerances are called for. Their tolerances range from ±2 to ±20 percent.

If very close tolerances are needed, silver-plated mica units are used. These have tolerances of ±1 percent.

Electrolytics usually have a wide range of tolerance. For instance, a 40-$\mu$F capacitor may have a tolerance of –10 percent as well as a +75 percent tolerance. This means that the 40-$\mu$F capacitor may be between 36 and 70 $\mu$F. The lower minus tolerance helps to ensure that there is enough capacitance in a circuit to prevent damage.

# Special-Purpose Capacitor

People in electronics have long dreamed of having a source of power, other than a battery, to power devices that may require small currents for long periods of storage time, such as the case with the memory in computers or calculators. This is now possible with the production of a capacitor of sufficient size (1 F) and both electrically and physically. The 1-F capacitor has long been thought of as a desirable thing for many uses. It is now possible to package a capacitor in the 1-F size in a 1.1-in. diameter unit only 0.55-in. high (see Figure 3-19).

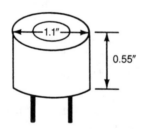

**Figure 3-19.** Electrolytic capacitor of 1 Farad fits a package of 1.1-in. diameter and 0.55-in. high.

This energy source makes it possible to support digital system backup applications without batteries. It has fast recharge time, easy interface, and virtually unlimited life. It is especially well suited for applications where the energy of a battery is not required, and reliability, long life, low cost, and simple design and implementation are of primary importance. The 64k bit CMOS memory can retain data while dissipating typically miniscule power of 0.1 $\mu$W. This low power consumption can be supported by the 1-F capacitor for several weeks.

Other applications for this type of capacitor are: relays, solenoids for starters, igniters, and actuators. Small motors and alarms for disc drives, coin metering devices, and security systems as well as toys can make use of a capacitor of this size. There are home appliances such as TVs, microwave ovens, dishwashers, refrigerators, energy management controls, personal computers, thermostats, vending machines, point-of-sale terminals, telephone autodialers, and programmable controllers that can also use the capacitor with characteristics of this nature.

## DIP Capacitors

Capacitors can now be purchased in dual in-line packages. They resemble the standard IC chip packaging. However, they contain capacitors connected between the pins as shown in Figure 3-20.

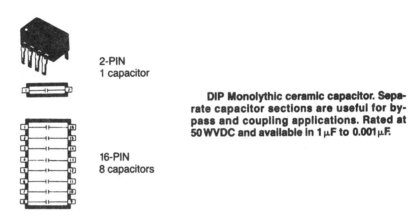

2-PIN
1 capacitor

**DIP Monolythic ceramic capacitor. Separate capacitor sections are useful for bypass and coupling applications. Rated at 50 WVDC and available in 1 μF to 0.001 μF.**

16-PIN
8 capacitors

**Figure 3-20.** DIP monolythic ceramic capacitor. Separate capacitor sections are useful for by-pass and coupling applications.

# Capacitive Reactance

Capacitive reactance is that opposition that a capacitor presents to alternating current (ac). A capacitor has a definite time period for charging: $T = R \times C$. The time $T$ in seconds (s) is equal to the resistance ($\Omega$) times the capacitance (F). This produces a time constant which is 63.2 percent of the maximum voltage presented to the capacitor. It takes five time constants for a capacitor to charge to its full, or 99.3 percent level. It also takes the same amount of time to discharge when presented with a resistance across its terminals.

When ac is present across the terminals of a capacitor, it changes faster than the capacitor can charge and discharge. This reaction or reactance is determined by the frequency of the ac and the capacity of the capacitor. A formula used to express capacitive reactance is

$$X_C = \frac{1}{2\pi f C}$$

where   $f$ = frequency, expressed in hertz

C = capacitance, expressed in farads

$X_c$ = capacitive reactance, expressed in ohms since it is in opposition to current flow

The following are conditions which occur when a capacitor is introduced into an ac circuit.

1. If the capacitance decreases, the capacitive reactance will increase for the same frequency.
2. If the capacitance increases, the capacitive reactance will decrease as long as the same frequency is presented to the capacitor.
3. If the frequency is decreased and the capacitor is the same, then the capacitive reactance will increase.
4. If the frequency is increased, then the capacitive reactance will decrease provided the capacitance stays the same.

As you can see from these statements and observations, the increase or decrease of the frequency or capacitance will cause the reverse reaction with the $X_C$.

## Capacitive Reactances in Series

The capacitive reactances placed in series, as in Figure 3-21, are similar to resistors placed in series. The formula is the same. Just add the individual reactances to obtain the total opposition. Or,

$$X_{C_T} = X_{C_1} + X_{C_2} + X_{C_3} + \cdots$$

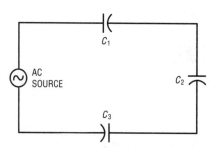

**Figure 3-21.** Three capacitors connected in series.

## Capacitive Reactances in Parallel

Once again, use the same approach for capacitive reactances as you would for resistors in parallel (see Figure 3-22):

$$\frac{1}{X_{C_T}} = \frac{1}{X_{C_1}} + \frac{1}{X_{C_2}} + \frac{1}{X_{C_3}} + \cdots$$

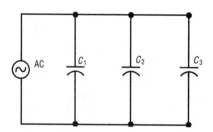

**Figure 3-22.** Three capacitors connected in parallel.

The other formula for finding resistors in parallel is modified here with $C$ instead of $R$, and the results are the same for *two only*.

$$X_{C_T} = \frac{X_{C_1} \times X_{C_2}}{X_{C_1} + X_{C_2}}$$

$X_C$ can be substituted in Ohm's law to produce another commonly used formula:

$$X_C = \frac{E_C}{I_C} \quad \text{or} \quad I_C = \frac{E_C}{X_C} \quad \text{or} \quad E_C = I_C \times X_C$$

# Uses for Capacitors and Capacitive Reactance

Capacitive reactance is useful in electronic circuits with both dc and ac. $X_C$ offers low opposition to higher frequencies of ac. In a dc circuit the capacitor stops or holds the flow of current (this is referred to as blocking) while it allows the ac to flow or to appear to flow through the capacitor. If you have a circuit with both ac and dc, it is possible to separate them by using a capacitor. This is done when signals are coupled from one amplifier stage to the other in an amplifier.

Capacitors can also be used for their ability to shift the phase of ac. In a capacitive circuit the voltage and current are out of phase by 90°. The voltage lags the current by 90°. This is useful when starting electric motors. Capacitor-start motors have high starting torque and put-out torque when overloaded. Capacitor-start motors are used on compressors, refrigerators, air conditioners, and wherever the motor must start under load or pull out of a suddenly applied load.

Capacitors used with inductors are used to establish frequencies that can be used to make music or broadcast television pictures and sound. When properly utilized, capacitors can be a very useful tool in electronic circuits. They are also useful in filters.

# Summary

Inductance is the ability of a coil, choke, or inductor to oppose any change in circuit current. An inductor is a device made up of a coil of wire and, in some cases, a core. The words inductor, choke, and coil are used interchangeably to mean the same. The unit of measure for inductance is the henry (H). The symbol for inductance is $L$. Five factors influence the mutual inductance of coils:

1. Physical size of the two coils
2. Number of turns in each coil
3. Distance between the two coils
4. Distance between the axes of the two coils
5. Permeability of the cores

Self-inductance occurs when a collapsing magnetic field around a coil induces an emf in the coil. Self-inductance and mutual inductance are both properties of a coil or choke.

Lenz's law says the induced current, or emf, in a coil is in the opposite direction of that which caused it.

Alternating current is constantly changing in magnitude and direction. Alternating current (ac) is generated mostly in power line frequencies of 25, 40, and 60 hertz (Hz). It is also generated in three phase by large generators and separated into single phase when used by homes and stores.

The term *frequency* is used to indicate how many times the alternating current changes direction. The unit of measurement for frequency is the hertz (Hz).

Root-mean-square is abbreviated rms, and refers to ac being converted to its equivalent of dc in terms of heating effect. The transformer has the ability to step up or step down ac voltages and currents.

Inductive reactance is the opposition generated to the changing of direction of the ac by a coil or choke.

Capacitors oppose any change in circuit voltage. They have the capacitance measured in farads (F). But in most instances they are made in microfarad ($\mu$F) (0.000 001 F) and in picofarad (pF) (0.000 000 000 001 F) size.

Capacitors have capacitive reactance when they are used in an ac circuit. Capacitive reactance is measured in ohms as is inductive reactance. Both are units of opposition to the flow of ac in a circuit.

A number of different types of capacitors and inductors are available to power line distributors and to those who work with electronics. They are basically the same, only the size changes.

Capacitors, inductors, and transformers are very valuable items when it comes to electronic circuits. They are the heart of the circuits in most instances.

# Review Questions

1. What is inductance?
2. What are three names given to devices which have inductance?
3. What is the unit of measure for inductance?
4. What is mutual inductance?
5. List five factors that influence mutual inductance.
6. Which four factors influence the inductance of a coil?
7. What are the advantages of ac over dc?
8. Who is responsible for ac in the United States?
9. At what voltages is ac generated commercially?
10. What is a waveform?
11. Describe peak, average, and root-mean-square as they apply to ac.
12. Why is a transformer so efficient?
13. What is eddy current?
14. What is hysteresis?
15. What are the causes of copper losses in a transformer?
16. What is a capacitor?
17. Which three factors determine the capacity of a capacitor?
18. What is meant by WVDC?
19. How are electrolytics made nonpolarized?

20. What happens to capacitance when capacitors are connected in parallel?
21. How much is a microfarad?
22. What is meant by $X_C$?
23. How are capacitors and inductors used together?
24. What is the formula for finding total $X_C$ when the capacitive reactances are known for capacitors connected in series?
25. What is the formula for finding total $X_C$ when the capacitive reactances are known for capacitors connected in parallel?

# Chapter 4

# RESONANCE

esonance is a very important part of electronics. It is necessary for the operation of the many types of television receivers and FM and AM radios. Resonance is created through the proper arrangement of a coil and capacitor.

Circuits with resistance, capacitance, and inductance behave differently from those with only one or two of these factors. For instance, a circuit with resistance reacts to alternating current (ac) and direct current (dc) the same way. However, when both a resistor and inductor are in a circuit, another factor is introduced because ac is applied to the combination. It behaves completely different from the dc circuit consisting of only a resistor and coil. The same is true with a resistor and capacitor combination. The ac introduces the capacitive reactance, but dc causes only the charging of the capacitor at a time determined by the values of the resistor and the capacitor.

It becomes important, then, for us to look closely at the combination of devices connected to a circuit. It is very evident that ac and dc cause different things to happen in an electric circuit. These things are going to be the subject of this chapter.

You will learn how circuits with resistors, capacitors, and inductors behave with both alternating and direct currents. The use of vectors will aid in the understanding of phase angle introduced by various combinations of these three devices (inductors, capacitors, and resistors).

# Resistance, Capacitance, and Inductance

Resistance produces an opposition to current flow in a circuit. The resistor is the device that produces the opposition. It behaves the same with either ac or dc.

Capacitive reactance produces an opposition to current flow in a circuit. The capacitor is the device that produces the opposition. The capacitor behaves differently with ac than with dc. Remember (just as in resistance) the opposition, or capacitive reactance, is measured in ohms.

Inductive reactance produces an opposition to current flow in a circuit. The inductor is the device that produces the opposition. The inductor behaves differently with ac than with dc. Inductive reactance is also measured in ohms.

Bear in mind, also, that the capacitor opposes any *change* in circuit *voltage*. The inductor opposes any *change* in *circuit current*. These two simple statements make a great deal of difference between understanding resonance and not being able to visualize it. So reread them to make sure you have them clearly in mind.

The relationship between current and voltage is vital to an understanding of electronics and electric circuits.

# Impedance

The *total* opposition to current flow within a circuit is impedance. The symbol for impedance is $Z$. Impedance impedes or opposes current flow. It is a term used when either resistance, capacitive reactance, inductive reactance, or any combination of the three is used. In dc circuits, opposition to voltage and current is resistance only, since capacitors and inductors do not have reactance with dc — only ac. $Z$ is measured in ohms.

Impedance can be $R$ and $X_L$. It can be $R$ and $X_C$, or it can be $R$ and $X_C$ and $X_L$. Impedance ($Z$) can also be used when there is $X_L$ and $X_C$. Any combination of these oppositions can be referred to as an impedance.

# Leading and Lagging

Remember Chapter 3, where you learned that the voltage lagged the current in a capacitive circuit? You also learned that the current lagged the voltage in an inductive circuit. Voltage lagged the current in a capacitor by 90° in a purely capacitive circuit. Voltage led the current in a purely inductive circuit. That is another way of saying that the current lagged the voltage by 90°. This leading and lagging is very important in any understanding of impedance for it takes into account the *phase angle*.

# Phase Angle

The phase angle is the difference between the voltage and the current in a circuit caused by either capacitance or inductance in an ac circuit. If we want to combine these phase angles such as when a capacitor and inductor are in a circuit, we have to do it vectorially (see Figure 4-1). Figure 4-1 is a diagram of how impedance is represented by $Z$, resistance by $R$, and reactance by $X_L$. Note the angle formed by lines $BA$ and $AD$. $Z$ is shown halfway between the two lines and as a result shows a phase angle of 45°. This happens when the

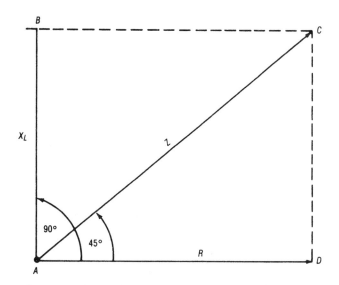

**Figure 4-1.** Impedance shown in vector form.

resistance ($R$) and inductive reactance ($X_L$) are equal. As you can see, impedance is the *vector sum* of resistance and reactance. A *vector* is a line segment used to represent a quantity that has both direction and magnitude. Vectors are used to represent current, voltage, or any combination of the electrical quantities encountered.

A vector can show direction of current flow. It can also show the magnitude, or amount, of current flowing. A vector sum is a line representing the total of two or more vectors. Impedance is stated in terms of a vector sum.

# Finding a Vector Sum

Inductive reactance causes the current in an inductor to lag 90° behind the voltage. Therefore, a graphic way of presenting the impedance of current can be drawn as shown in Figure 4-1. In this illustration, resistance is plotted on the horizontal line $AD$. The length of the line $AD$ is proportional to the amount of resistance in the circuit. Proportional means that the quantity of resistance within the circuit is represented by line $AD$. Zero resistance is indicated by point $A$. The value of resistance in the circuit is indicated at point $D$. Using the same scale, the amount of inductive reactance is plotted on a line 90° from the resistance line. This is because the voltage and current in the resistor are in phase with one another. This means that the resistance line $AD$ can be used as the horizontal reference and everything else will be plotted up or down in reference to this horizontal reference.

The vertical line $AB$ represents the inductive reactance. This is also proportional. Zero inductive reactance is shown at point $A$. The value of inductive reactance is indicated at point $B$.

The impedance $Z$ is the vector sum of the two lines. It is represented by line $AC$. To find the value of $C$, begin by constructing a parallelogram. This is shown in Figure 4-1 by the dotted lines finishing up the figure. A *parallelogram* is a four-sided figure whose opposite sides are parallel and equal. In Figure 4-1, the dotted line $CD$ is parallel to $AB$, and $BC$ is parallel to $AD$.

$C$ is the point where the parallelogram is completed. The value of $Z$ is found by drawing a straight line between $C$ and $A$. This line can be measured and the value of $Z$ found by equating it to the units used in $X_L$ and $R$.

The value of this graphic method is that it helps you to visualize the procedure. In practice there are faster methods for calculating impedance. This is a simple operation on most calculators. Using a calculator becomes even easier once you can visualize and understand the values involved.

# Finding the Impedance of an $RL$ Circuit

Figure 4-2 shows an ac circuit with a resistor and an inductor connected in series. The resistor has a value of 6 ohms ($\Omega$). The inductor has a reactance of 8 $\Omega$.

These values are indicated by the solid horizontal and vertical lines in Figure 4-3. In

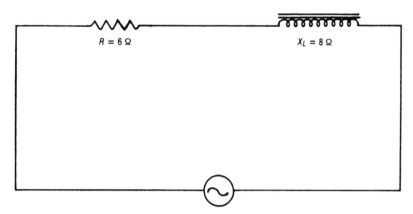

Figure 4-2. A series *RL* circuit.

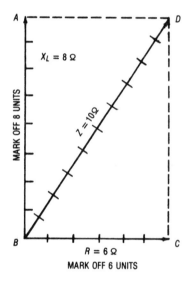

Figure 4-3. Drawing a parallelogram to find impedance.

this figure, a parallelogram has been formed, and the value of it has been plotted. The length of line *BD* is ten units. This is proportional with the six units in *BC* and the eight units in *BA*. Therefore, the impedance (Z) of the circuit is 10 Ω.

# Finding the Impedance of an *RL* Circuit Mathematically

You can also use the mathematical method of finding the impedance of a series *RL* circuit. *RL* means resistive-inductive. The equation used is known as the *Pythagorean theorem*. The Pythagorean theorem states that the square of the hypotenuse of a right triangle is equal to the sum of the squares of the other two sides.

The lines representing resistance and inductance in Figure 4-3 form a 90° angle, or a right triangle. The hypotenuse is the line in a right triangle that is opposite the right angle. Line *BD* is the hypotenuse of triangle *CBD*. The side opposite *AB* is *CD*, which means that *CD* is the same length, and that both lines are parallel. Since *CD* is the same length as *AB*, it is also equal to 8 Ω. So, *BC* = 6 Ω and represents the resistance; *CD* = 8 Ω and represents the inductive reactance. The resulting angle made at $CBD$ is the phase angle. It can be measured by a protractor to determine its value in degrees. If you have the unit-measuring device that you used to lay out the $X_L$ and $R$ on the graph, you can use the same measuring device to measure the length of $BD$ to obtain the hypotenuse length ($BD$). However, if you are going to use the mathematical method to obtain the value of $BD$, it can be done with a formula derived from the Pythagorean theorem:

$$Z^2 = X_L^2 + R^2$$

A simplified version becomes

$$Z = \sqrt{X_L^2 + R^2}$$

Substitute the values of the two known elements ($R$ and $X_L$) to find the value of $Z$. Or,

$$Z = \sqrt{8^2 + 6^2}$$
$$= \sqrt{64 + 36}$$
$$= \sqrt{100}$$
$$= 10\,\Omega$$

These steps show how the mathematical approach to the solution of the problem is handled. Use an electronic calculator to simplify the process by using the square and the square root keys to get an answer quickly.

# Impedance in an $RC$ Circuit

A similar method can be used to find $Z$ in circuits that have a resistor and capacitor in series. Such a circuit is shown in Figure 4-4.

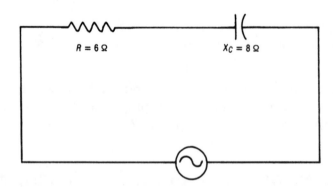

Figure 4-4. A series $RC$ circuit.

Note the difference in the structure of the parallelogram in this figure. In a capacitive circuit, the current leads the voltage by 90°. Expressed another way, the voltage lags the current by 90°. Because the voltage lags the current, reactance is considered to be at −90° with respect to resistance. This is indicated by having the $Z$ vector point downward. In a series circuit, remember, the current was the stable element and the voltage the varying factor. This is still true here since it is a series circuit. The current is the stable element and the voltage either leads or lags and is plotted accordingly: either up or down to show lead and lag.

Figure 4-5 shows the downward line representing the impedance. Note also that the

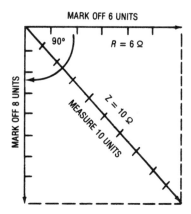

**Figure 4-5.** Parallelogram method of finding the impedance.

resistance line is still horizontal since voltage and current are in phase in a resistor. Line *AD* represents capacitive reactance. Line *AD* is downward to show the negative or lagging connotation assigned to the voltage lagging the current in a capacitor. The problem can be solved the same way the series *RL* circuit was solved:

$$Z^2 = X_C^2 + R^2$$
$$Z = \sqrt{X_C^2 + R^2}$$
$$= \sqrt{8^2 + 6^2}$$
$$= \sqrt{64 + 36}$$
$$= \sqrt{100}$$
$$= 10\,\Omega$$

The Pythagorean theorem is used in the same way as before to solve for the impedance where a capacitor and resistor are connected in series. Note that $X_C$ is substituted for $X_L$ and that the direction of the $X_C$ is downward from point *A* instead of upward as shown in Figure 4-3 for $X_L$.

In either of the impedance circuits shown, the phase angle is determined by the size relationship between the impedance line and the resistance line. The phase angle of any circuit can, therefore, be between +90° and −90°. The angle depends upon whether the reactance part of the impedance is inductive or capacitive. The angle also depends upon the relative values of the reactances and the resistance.

Another factor that affects the phase angle is the frequency of the ac being applied to the circuit. Frequency affects reactance in both the capacitor and inductor. By affecting reactance, the frequency can also affect the amount of impedance in the circuit.

# Series *RCL* Circuit

Take a look at a series circuit consisting of a resistor, a capacitor, and an inductor. The resistor has a resistance of 12 Ω, while the inductive reactance of the coil is 20 Ω and the capacitor presents a reactance of 4 Ω (see Figure 4-6).

A vector diagram can be used to express the three values in graphic form (see Figure 4-7).

Note that the vertical line in Figure 4-7 is longer than the horizontal line. The horizontal line represents resistance. The vertical line represents reactance, as usual.

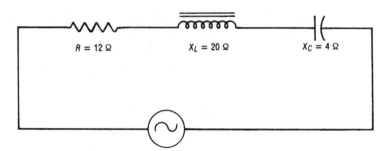

$R = 12\ \Omega$      $X_L = 20\ \Omega$      $X_C = 4\ \Omega$

**Figure 4-6.** A series *RCL* circuit.

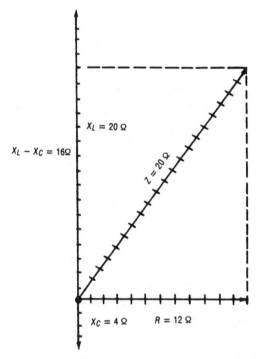

**Figure 4-7.** A parallelogram method of finding the impedance of an *RCL* circuit.

However, there are some small differences. The inductive reactance line is longer than the other two lines. The $X_L$ line runs from point $A$ upward 20 units. The $X_C$ line runs downward 4 units. In order to find the *resultant* reactance, you subtract the $X_C$, which is in the negative direction, from the $X_L$, which is in the upward direction. As a result, you obtain $20 - 4 = 16$ units of reactance. These 16 units are plotted as shown in the dotted lines of Figure 4-7. The parallelogram is constructed and bisected to produce the two right triangles. By using the information already learned, you can arrive at the impedance mathematically or measure it on the graph paper where it has been plotted.

The formula is

$$Z^2 = R^2 + (X_L - X_C)^2$$

$X_L - X_C$ is now one leg of the triangle. That means you can obtain an answer by further reducing the formula to

$$Z = \sqrt{R^2 + (X_L - X_C)^2}$$

and then substituting the values into the formula. This produces

$$Z = \sqrt{(12)^2 + (16)^2}$$
$$= \sqrt{144 + 256}$$
$$= \sqrt{400}$$
$$= 20 \ \Omega$$

# Phase Angle

In ac circuits, reactance and resistance change the phase differences between current and voltage. In a resistor, the voltage and current are in phase. In an inductor, the voltage leads the current by 90°. In a capacitor, the voltage lags the current by 90°. When these devices are placed in a circuit with ac, they produce a resulting phase angle somewhere between 0 and 90°. The relationship between the resistance, inductive reactance, and capacitive reactance determines the phase angle.

If you refer to Figure 4-6, you will see that a resistor, a capacitor, and an inductor are connected in series. The phase angle for this circuit is graphed into a vector diagram in Figure 4-8. The phase angle is between the lines for $Z$ and $R$. It is marked as angle *theta* ($\theta$).

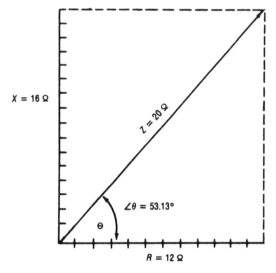

**Figure 4-8.** Series *RCL* circuit diagram of impedance.

As you can see from the drawing in Figure 4-8, the angle is less than 90°. The circuit represented in the diagram has more inductive reactance than capacitive reactance. Thus, the circuit tends to be inductive. That means the voltage is leading the current in the circuit. And this also means that the $Z$ or impedance line will be above the horizontal line used to represent resistance. In the case of capacitive circuits where there is more capacitive reactance than inductive reactance, the $Z$, or impedance line, will be below the horizontal line which represents resistance. Remember, the R line represents resistance and is horizontal simply because voltage and current are in phase in a resistor, and this means it can then serve as a reference line in respect to the voltages leading or lagging the current in the other devices in the circuit that cause a phase shift.

# Other Circuit Values in a Series $RCL$ Circuit

Figure 4-9 presents a schematic drawing with a series resistor, capacitor, and inductor. The values of the components are given as 5000 $\Omega$ of resistance, 10 henrys (H) of inductance, and 20 microfarads ($\mu$F) of capacitance.

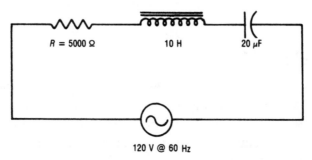

R = 5000 $\Omega$       10 H       20 $\mu$F

120 V @ 60 Hz

**Figure 4-9.** Series $RCL$ circuit.

Formulas can be used to determine several values of voltage and current in the circuit. We do know that 120 V at 60 hertz (Hz) is applied to the circuit. What we can do now is find how the voltage is distributed across each component and how much current flows through each.

We can also find the power factor and the phase angle of the circuit. Inductive reactance, capacitive reactance, and impedance can also be found using formulas and the information already at hand. For instance, the current in a series circuit is the same in all parts of the circuit. Thus, if we find the current, we know the value of current through each component. The formula used to find current is $I_T = E_A/Z$. We know the applied voltage, but we do not know the impedance. That means we must find the $Z$, or impedance, before we can obtain the total current. Keep in mind that $I_T = I_R = I_L = I_C$.

To find the impedance we must know the inductive reactance and the capacitive reactance. These we can find by using the formulas already developed in a previous chapter. $X_L = 2\pi fL$ and $X_C = 1/(2\pi fC)$. Since we know the frequency (60 Hz) and the inductance (10 H), and the value for $2\pi$ (= 6.28), we can begin here:

$$X_L = 6.28 \times 60 \times 10$$
$$= 3768 \ \Omega$$

The next step is to find $X_C$. Use the formula and substitute. Convert the 20 $\mu$F into farads (F) for the formula.

$$X_C = \frac{1}{6.28 \times 60 \times 0.00002}$$
$$= 132.696 \ \Omega$$

To calculate the impedance we use the two values just found and combine with the resistance in the $Z$ formula:

$$Z = \sqrt{R^2 + (X_L - X_C)^2}$$
$$= \sqrt{5000^2 + (3768 - 132.696)^2}$$
$$= \sqrt{25,000,000 + 13,215,435.23}$$
$$= \sqrt{38,215,435.23}$$
$$= 6181.86 \ \Omega$$

Now it is possible to find the total current in the circuit.

$$I_T = \frac{E_A}{Z}$$
$$= \frac{120}{6181.86}$$
$$= 0.0194 \text{ A}$$

## Finding the Voltage Across Each Component

It is now possible to find the voltage drop across each of the components in the circuit. Voltage $E$ = current $I$ times resistance $R$, or $E = I \times R$.

$$E_R = 0.0194 \times 5000 = 97 \text{ V}$$
$$E_L = 0.0194 \times 3768 = 73.1 \text{ V}$$
$$E_C = 0.0194 \times 132.696 = 2.57 \text{ V}$$

If you added the voltages you would get 172.67 V, which is more than was applied (120 V). This is because a phase angle was introduced, and the voltage and current were out of phase in this circuit.

# Power Factor

In a circuit with ac applied to a capacitor, a resistor, and an inductor, there is a difference between the actual power dissipated or used and that which appears to be used if the volts-times-amps formula is used. Remember that in a previous chapter power was measured in watts and was found by multiplying the voltage times the current? Well, that is still possible, but only *if* you have purely resistive circuits or dc applied to the circuit. Power is measured in watts. We now must further identify this power measured in watts. We call it true power (TP) to differentiate it from the apparent power (AP) that is obtained when the voltage is multiplied by the current in an ac circuit with an inductor or capacitor. Apparent power is measured in voltamperes (VA).

In order to find the true power, you must multiply the apparent power by a power factor (PF).

The power factor is found by utilizing the phase angle. In Figure 4-8, you saw that the angle $\theta$ was 53.13°. If you use a calculator with trigonometry tables or if you use a trigonometry table in a book, you will find that the angle is found by dividing the resistance by the impedance, or $R/Z$. This produces 0.6000 which, if located in the trig table, produces an angle of 53.13°. This 0.6000 is also the PF since the cosine of $R/Z$ indicates the percentage of power actually being consumed in the circuit. In this case it can be referred to as 0.6000, or 60 percent. Power factor, then, is used to determine the actual power used in a circuit. The apparent power times this power factor produces the true power: AP × PF = TP.

In this instance (see Figure 4-8), the total voltage was 120. The current was found by $I_T = 120/20 = 6$ A. This means that you take the voltage times the current times the power factor to produce the true power. For that circuit it would be $120 \times 6 \times 0.6000 = 432$ W consumed. If you wanted the apparent power, you would simply multiply the voltage times the current, or $120 \times 6 = 720$ VA.

As you can see, the difference between the amount of power that is paid for (in this case

720 VA) and the power actually consumed are quite appreciable. The total amount can become rather huge in large factories. This is why the synchronous motor is used as a rotating capacitor to bring the phase angle back close to zero in these plants. Of course, it would not be practical, yet, to do the same for home use. However, some motors are made to run as capacitor-start and capacitor-run types in order to take advantage of the phase angle correction. This is very helpful when you have a large home air conditioner which pulls a lot of current.

# Series Resonance

Resonance is observed when the inductive reactance and capacitive reactance are equal. Since the current in the circuit is the same, the current through the capacitor and the inductor is also the same. (Technically, it is what appears to be current flow through a capacitor. Actually, having current flow through a capacitor is the same as having no capacitor in the circuit. It is just easier to say or read current flow through a capacitor.) This means that the current times the inductive reactance and the current times the capacitive reactance would produce the same result, and the voltage across the inductor and the voltage across the capacitor would also be equal.

At resonance, then, we can say these things:

1.  Current through both components is the same.
2.  $X_L$ is equal to $X_C$.
3.  $E_C$ is equal to $E_L$.

The circuit in Figure 4-10 has an inductor and capacitor connected in series. The inductive reactance and the capacitive reactance are the same, or equal. When this occurs, the two reactances tend to cancel one another. That leaves only the resistance in the circuit for opposition to current flow. A resonant circuit occurs only when there is one frequency that will cause each component to produce the same reactance. Since there is no resistance in the circuit (theoretically), it presents a short circuit to the resonant frequency. Current in a series $LC$ circuit is theoretically infinite as the impedance is theoretically zero. Actually, there is always some resistance, and so you refer to the current as maximum and to the impedance as minimum.

Inductance of the circuit in Figure 4-10 is 0.704 H, or 704 millihenrys (mH). Internal resistance is 100 $\Omega$. This is the resistance of the windings or copper wire that make up the coil. The coil is connected in series with a 10-$\mu$F capacitor. Applied voltage is 120 at 60 Hz.

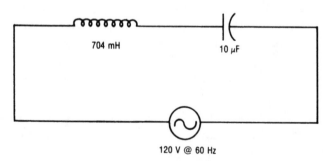

704 mH          10 μF

120 V @ 60 Hz

**Figure 4-10.** Series $LC$ circuit.

In order to solve this problem for inductive and capacitive reactances, as well as for impedance, the following procedure should be followed:

$$X_L = 2\pi fL$$
$$= 6.28 \times 60 \times 0.704$$
$$= 265.2672 \ \Omega$$

Then solve for the capacitive reactance:

$$X_C = \frac{1}{2\pi fC}$$
$$= \frac{1}{6.28 \times 60 \times 0.000\,01}$$
$$= 265.393\,781\,3 \ \Omega$$

The next step is to find the impedance:

$$Z = \sqrt{R^2(X_L - X_C)^2}$$
$$= \sqrt{100^2 + 0.126\,581\,3^2}$$
$$= \sqrt{10,000 + 0.016\,022\,825\,5}$$
$$= \sqrt{10,000.016\,022\,825\,5}$$
$$= 100.000\,081 \ \Omega \qquad (\text{for all practical purposes, } 100 \ \Omega)$$

From this answer you can see that the $X_L$ and $X_C$ were cancelled (only $0.126\,581\,3$ difference between the two), and the only thing actually left in the circuit was the resistance of $100 \ \Omega$.

## Plotting a Curve for Resonance

The impedance for this circuit is at its minimum point at the resonant frequency of 60 Hz. Look at Figure 4-11. The value of impedance at the frequency applied to the circuit in

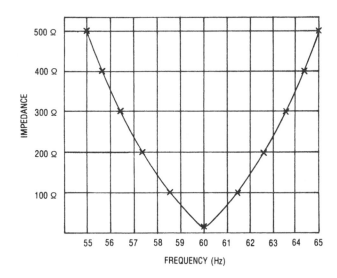

**Figure 4-11.** Series resonant circuit curve.

Figure 4-10 is shown in Figure 4-11. The x's indicate frequency readings in a range from 55 to 65 Hz. A curve is drawn through the frequency marking points to produce what is referred to as a resonant curve. The lowest point is located at 60 Hz.

With the impedance at its minimum point, the current in the circuit will be at its maximum. This characteristic of a series resonant circuit means you must carefully choose the capacitor. The voltage is very high, and the current is very high. This means that the capacitor must be able to withstand the current and voltage. The current in the circuit can be found by

$$I_T = \frac{E_A}{Z}$$
$$= \frac{120 \text{ V}}{100 \ \Omega}$$
$$= 1.2 \text{ A}$$

This current means that the inductor must be wound with wire that is capable of handling 1.2 A. When you refer to a wire table, you will find that the wire size should be no smaller than no. 20 (see Table 4-1).

## TABLE 4-1

### WIRE TABLE (COPPER)

| AWG (B&S) Gage Number | Diam. (in.) | Circular Mils | Ohms/ 1,000[1] | Current Capacity @ 700 CM/amp. | Single Formvar | | | Heavy Formvar | | |
|---|---|---|---|---|---|---|---|---|---|---|
| | | | | | DIAM. | TURNS/IN. | TURNS/IN.[2] | DIAM. | TURNS/IN. | TURNS/IN.[2] |
| No. 8 | 0.1285 | 16,509.65 | 0.6282 | 23.5859 | 0.1306 | 7 | 49 | 0.1323 | 7 | 49 |
| No. 9 | 0.1144 | 13,092.75 | 0.7921 | 18.7039 | 0.1165 | 8 | 64 | 0.1181 | 8 | 64 |
| No. 10 | 0.1019 | 10,383.02 | 0.9988 | 14.8329 | 0.1039 | 9 | 81 | 0.1055 | 9 | 81 |
| No. 11 | 0.0907 | 8,234.11 | 1.2595 | 11.7630 | 0.0927 | 10 | 100 | 0.0942 | 10 | 100 |
| No. 12 | 0.0808 | 6,529.95 | 1.5882 | 9.3285 | 0.0827 | 12 | 144 | 0.0842 | 11 | 121 |
| No. 13 | 0.0720 | 5,178.48 | 2.0027 | 7.3978 | 0.0738 | 13 | 169 | 0.0753 | 13 | 169 |
| No. 14 | 0.0641 | 4,106.72 | 2.5254 | 5.8667 | 0.0659 | 15 | 225 | 0.0673 | 14 | 196 |
| No. 15 | 0.0571 | 3,256.78 | 3.1844 | 4.6525 | 0.0588 | 17 | 289 | 0.0602 | 16 | 256 |
| No. 16 | 0.0508 | 2,582.74 | 4.0155 | 3.6896 | 0.0524 | 19 | 361 | 0.0538 | 18 | 324 |
| No. 17 | 0.0453 | 2,048.21 | 5.0634 | 2.9260 | 0.0469 | 21 | 441 | 0.0482 | 20 | 400 |
| No. 18 | 0.0403 | 1,624.30 | 6.3849 | 2.3204 | 0.0418 | 23 | 529 | 0.0431 | 23 | 529 |
| No. 19 | 0.0359 | 1,288.13 | 8.0512 | 1.8402 | 0.0374 | 26 | 676 | 0.0386 | 25 | 625 |
| No. 20 | 0.0320 | 1,021.53 | 10.1524 | 1.4593 | 0.0334 | 29 | 841 | 0.0346 | 28 | 784 |
| No. 21 | 0.0285 | 810.11 | 12.8019 | 1.1573 | 0.0299 | 33 | 1,089 | 0.0310 | 32 | 1,024 |
| No. 22 | 0.0253 | 642.45 | 16.1429 | 0.9178 | 0.0266 | 37 | 1,369 | 0.0277 | 36 | 1,296 |
| No. 23 | 0.0226 | 509.49 | 20.3558 | 0.7278 | 0.0238 | 42 | 1,764 | 0.0249 | 40 | 1,600 |
| No. 24 | 0.0201 | 404.04 | 25.6682 | 0.5772 | 0.0213 | 46 | 2,116 | 0.0223 | 44 | 1,936 |
| No. 25 | 0.0179 | 320.42 | 32.3670 | 0.4577 | 0.0190 | 52 | 2,704 | 0.0200 | 50 | 2,500 |
| No. 26 | 0.0159 | 254.10 | 40.8141 | 0.3630 | 0.0169 | 59 | 3,481 | 0.0179 | 55 | 3,025 |
| No. 27 | 0.0142 | 201.51 | 51.4656 | 0.2879 | 0.0152 | 65 | 4,225 | 0.0161 | 62 | 3,844 |
| No. 28 | 0.0126 | 159.81 | 64.8969 | 0.2283 | 0.0135 | 74 | 5,476 | 0.0145 | 68 | 4,624 |
| No. 29 | 0.0113 | 126.73 | 81.8335 | 0.1810 | 0.0122 | 81 | 6,561 | 0.0131 | 76 | 5,776 |
| No. 30 | 0.0100 | 100.50 | 103.1901 | 0.1436 | 0.0108 | 92 | 8,464 | 0.0116 | 85 | 7,396 |
| No. 31 | 0.0089 | 79.70 | 130.1204 | 0.1139 | 0.0097 | 103 | 10,609 | 0.0104 | 96 | 9,215 |

## Deriving the Resonant Frequency Formula

In order to obtain the resonant frequency of an inductor-capacitor combination (either in series or parallel combinations), it is best to have a formula where the values of the components can be plugged in and an answer arrived at using a calculator.

Remember, at resonance, $X_L$ is equal to $X_C$. This provides a basis for developing the formula for $f_r$, or resonant frequency. Start with $X_L = X_C$. That means

$$2\pi fL = \frac{1}{2\pi fC}$$

Cross-multiply to produce

$$1 = 4\pi^2 f^2 LC$$

Divide each side by $f^2$:

$$\frac{1}{f^2} = \frac{4\pi^2 f^2 LC}{f^2}$$

That produces

$$\frac{1}{f^2} = 4\pi^2 \, LC$$

Take the square root ($\sqrt{\phantom{x}}$) of each side of the equation. Invert both sides of the equation to obtain the formula

$$f = \frac{1}{2\pi\sqrt{LC}}$$

# Parallel Resonance

Figure 4-12 has a capacitor and inductor connected in parallel. In order to find the resonant frequency ($f_r$) of the inductor-capacitor combination, simply plug in the values of the components into the formula just derived. Or,

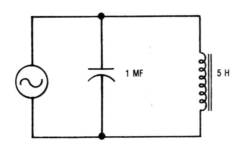

1 MF    5 H

Figure 4-12. A parallel $LC$ circuit, otherwise known as a tank circuit.

$$f_r = \frac{1}{2\pi\sqrt{5 \times 1 \times 10^{-6}}}$$

$$= \frac{1}{0.0140425069}$$

$$= 71.21235596 \, \text{Hz}$$

At the resonant frequency of a parallel resonant circuit, an action takes place that is known as the *flywheel effect*. The flywheel effect is a continuing oscillation in a resonant circuit between pulses of electric energy. These circuits are sometimes called *tank circuits*. They are used to tune in radio stations and television stations. Take a closer look at the tank circuit.

The operation of a tank circuit is shown in Figure 4-13. In Figure 4-13A, a current flows from the external circuit through the inductor and also onto the plates of the capacitor. The capacitor is charged to the supply voltage. The current flowing through the inductor causes it to have a magnetic field surrounding it.

The external source is effectively removed because the capacitor is charged to the source voltage. The capacitor then sees the inductor as a means of discharging and equalizing the charge on its plates (see Figure 4-13B).

At the same time, the inductor will keep the current flowing through its windings since its initial reactance to current flow has been overcome. The current that was supplied by the source overcame the opposition of the coil.

Once the capacitor is charged, the magnetic field of the capacitor starts to collapse for there is no longer a current flow in the circuit. This collapsing magnetic field cuts across the windings of the coil and induces an emf in the coil. This emf is in the opposite direc-

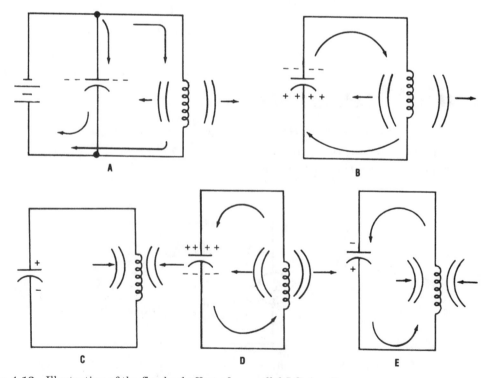

**Figure 4-13.** Illustration of the flywheel effect of a parallel *LC* circuit.

tion from that which caused it. This coil becoming a source of emf adds to the capacitor charge. As the magnetic field weakens, the capacitor is charged to its maximum point. Then the capacitor sees the inductor windings as a low-resistance path to discharge or try to equalize the charge on both its plates. That means the discharging of the capacitor produces a current flow through the coil windings which produces a magnetic field, and the whole process is repeated. This charge-discharge continues until the resistance of the circuit (made up mainly of the windings of the coil and the connections) dissipates the energy. If the source is still active and putting the same frequency into the circuit, it will add whatever is needed to keep the circuit going back and forth, or oscillating.

The external source is effectively removed as no more current flows once the circuit gets started, that is, of course, if there is no resistance whatever in the circuit. Zero resistance occurs only theoretically since there is always *some* resistance (the coil has resistance and so do the connections and connecting wires). Figures 4-13C, D, and E show the steps followed by the circuit in producing the flywheel effect. Makeup current (to overcome the inherent resistance) is drawn from the source. This makeup current keeps the circuit going at its resonant frequency. That also means the resonant circuit presents a maximum circuit impedance to the source voltage. A maximum impedance draws a minimum of current. So, at resonance (in a parallel *LC* circuit), the impedance is maximum (theoretically infinite), and the current is minimum (theoretically zero). The *circulating* current in the tank circuit is at its maximum.

The flywheel action produces a pure sine wave as its output if energized and it the source of voltage is removed immediately (see Figure 4-14). If the source is used to charge the capacitor and is then removed, the circuit will produce a damped oscillation such as shown in Figure 4-15. A *damped* oscillation is one that has the waveform gradually grow

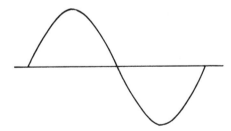

**Figure 4-14.** Sine wave.

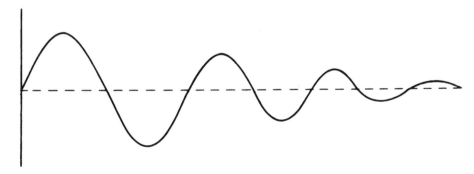

**Figure 4-15.** Damped oscillation.

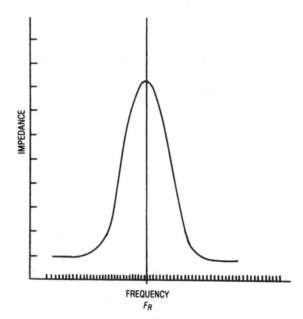

**Figure 4-16.** Parallel resonant circuit impedance curve.

smaller until it is dissipated over the resistance of the circuit. Finally, the energy is dissipated and the oscillations cease.

Figure 4-16 shows the impedance curve for a parallel resonant circuit.

# Series Resonance Versus Parallel Resonance

Note some of the characteristics of each of the series and the parallel resonant circuits. Each has value in electronic circuits. Each has its own way of doing a job. These specific characteristics lend themselves to making a radio or television set operate properly.

The series resonant circuit offers a *high* impedance to frequencies other than the resonant frequency. The parallel resonant circuit offers a low impedance to all frequencies other than its resonant frequency. This allows for most of the uses of the parallel resonant circuit.

Parallel resonant circuits may be used for filters. Filters separate one frequency from another.

Figure 4-17 shows a diagram of a simple parallel resonant circuit with an inductor, a capacitor, an antenna, and a ground connection. The antenna receives radio waves of sufficient amplitude and conducts them to the parallel resonant circuit. Frequencies other than the resonant frequency find a low-frequency path to ground by way of the inductor. Frequencies higher than the resonant frequency find a low-impedance path to ground by way of the capacitor. Only the resonant frequency meets with a high impedance. It is conducted on to the detector circuit consisting of a diode and a headset as shown in Figure 4-18. At *series* resonance, the circuit:

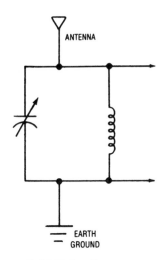

**Figure 4-17.** Antenna attached to a parallel *LC* circuit.

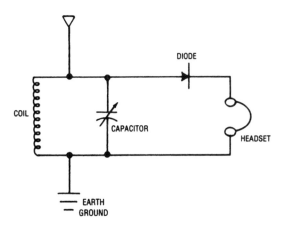

**Figure 4-18.** Crystal set radio receiver.

Has *low* impedance
Resembles a resistance of *low* value
Has high *voltages* developed across each element

At *parallel* resonance, the circuit:

Has *high* impedance
Resembles a resistance of *high* value
Has high *currents* through each element

# Ground

Ground is usually thought of as earth. In electronics and electrical work the ground is a common connection to which one side of the power supply is connected. The common ground may be a copper rod 4 to 6 feet into the soil, or it may be the chassis of a radio, TV set, or other electronics device.

Earth ground is used by power companies and telephone companies to protect their equipment from lightning. The alternative to the earth ground is the floating ground, which is simply a reference point that is not earth grounded. The negative terminal of the automobile is connected to the chassis, which then becomes ground, but it does not touch the earth because of the rubber tires. Newer homes have a third connection point — usually a round hole in the wall receptacle — that is actually connected to the earth, most often through the copper cold water pipes.

# Practical Uses for This Information

In making a choice of radio receivers, whether it is AM, FM, or television, do not buy more than you need or want. A communications receiver is rated according to its sensitivity. The rating indicates how weak the signal can be and still be heard. If you want to use your receiver for local stations only, sensitivity is not critical. Do not pay for more sensitivity than you need.

If the receiver you are buying has a squelch control for easier listening on weak stations, you may miss some of the weaker ones if the squelch is not disabled. Make sure you can turn it off when tuning so that you can pick up the weak stations and then put it back on once the station has been located. Just make sure the squelch control is adjustable.

The more tuned circuits the radio has, the better its selectivity. Selectivity is the ability to separate stations that are close together. Selectivity is usually easily identified by counting the number of little IF cans (small shiny metal containers) on the printed circuit board. The more IF cans, the better the selectivity of the TV set, FM receiver, or AM receiver. Do not settle for fewer than three in an AM receiver or for fewer than four in an FM set. There are many more in the TV set. Usually the more it has, the more expensive the set is. This is very important when you live in an area that has many TV stations to choose from or when you use cable with its many channels that need separating.

# Summary

In this chapter you have learned about the performance and values of circuits with combined properties. You have learned about resistance and capacitive reactance in series plus resistance and inductive reactance in series. You have also learned about the combination of all three in both series and parallel. Some of the things you should know now:

- Impedance is the total opposition to the flow of current in a circuit.
- Impedance is the vector sum of resistance and reactance. Vectors are used to represent two dimensions of current — direction and magnitude.
- What the electrical properties of series and parallel resonant circuits are.
- What the effect on phase angles of both resistance and reactance is.
- What the properties of a tank circuit are.
- How to figure phase angle and power factor.
- How resonant circuits are used to pick one frequency from a number present at a given point.

# Review Questions

1. What is the symbol for impedance?
2. What is a vector sum?
3. In what unit is impedance measured?
4. Why do capacitors and inductors present opposite reactance values?
5. What is a phase angle?
6. What is the formula for finding the impedance of a series $RCL$ circuit?
7. What is the formula used for determining the current in an $RCL$ circuit?
8. How is Ohm's law for ac circuits useful in $RCL$ circuits?
9. What condition exists when $X_L$ is equal to $X_C$ in a circuit?
10. What conditions must be present for a resonant circuit?
11. What is the impedance of a series $LC$ circuit when the voltages across each component are equal?
12. What is the impedance of a parallel $LC$ circuit when $X_L = X_C$?
13. What is the formula used to find the resonant frequency?
14. When does the flywheel effect take place?
15. What is a tank circuit?
16. What is a damped oscillation?
17. At what point is impedance the highest in a tank circuit?
18. What is the phase angle of a series resonant circuit?
19. In a parallel resonant circuit, what is the impedance presented to frequencies higher than the resonant one?

# Chapter 5

# CHANGING ALTERNATING CURRENT TO DIRECT CURRENT

Since alternating current (ac) is inexpensive, it is used in thousands of devices. It can be stepped up or stepped down by using a transformer. It is a versatile type of power that can easily be changed to fit the voltage or current needs of particular circuits. However, direct current (dc) is also useful for many devices. Electronics depends upon dc for many of its circuit components. This dependence upon dc requires a source of inexpensive direct current for a variety of voltages and currents.

Historically, the changing of ac to dc began around the turn of the century when ac became available at Niagara Falls, New York, but was not easily transported to Buffalo (26 miles away) where it was needed by the milling industry. The demand for soap products became rather important in our country around 1900 when the newly "arrived" middle class was demanding the cleanliness of everything, and it became apparent that dc current could be used to produce any number of soap products from various cheap chemicals. Niagara Falls with its inexpensive source of falling water–generated ac would easily become one of the chemical centers of the country if only the ac could be converted to dc. The change was vital because dc was much more easily used in the chemical processes of the time.

Large copper oxide rectifiers were designed to change the ac to dc. One side of the rectifier disk was copper, and the other was copper oxide. The copper allowed the current to flow easily, but the copper oxide side put up a great opposition to the flow of current in the other direction. This meant the ac could be rectified since the copper oxide rectifier allowed the current to flow in only one direction. This produced pulsating dc which was useful for a number of chemical processes and for driving motors which were being developed at the time.

In the copper oxide rectifier (see Figure 5-1) the oxide is formed on the copper by partial oxidation of the copper by a high temperature. In this type of rectifier the electrons flow more readily from the copper side to the oxide than from the oxide to the copper.

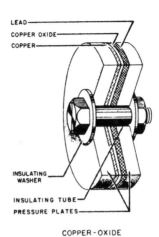

Figure 5-1. Construction of the metallic rectifier.

External electrical connection may be made by connecting terminal lugs between the left pressure plate and the copper and between the right pressure plate and the lead washer.

For the rectifier to function properly, the oxide coating must be very thin. Thus, each individual unit can stand only a low inverse voltage. Rectifiers designed for moderate- and high-power applications consist of many of these individual units mounted in series on a single support. The lead washer enables uniform pressure to be applied to the units so that the internal resistance may be reduced. When the units are connected in series, they normally present a relatively high resistance to the current flow. The resulting heat developed in the resistance must be removed if the rectifier is to operate satisfactorily. Many commercial rectifiers have copper fins between each unit to dissipate the excess heat. The useful life of the unit is extended by keeping the temperature low, below 140°F (60°C). The efficiency of this type of rectifier is generally between 60 and 70 percent. They are found in older types of equipment. No one uses this type of rectifier today in newly designed equipment with inexpensive present-day semiconductor devices with over 90 percent efficiency readily available.

# Selenium Rectifiers

Selenium rectifiers are similar to and function much the same as copper oxide rectifiers. Figure 5-2 indicates the similarities between the two rectifiers. The selenium rectifier is made up of an iron disk coated with a thin layer of selenium. In this type of rectifier, the electrons flow from the selenium to the iron.

Commercial types of selenium rectifiers were used in early models of television sets. They were designed to pass 50 milliamperes per square centimeter (mA/cm$^2$) of plate area. This type of rectifier may be operated at a somewhat higher temperature than the copper oxide type. The efficiency is between 65 and 85 percent. Many units may be bolted together to increase the voltage rating when connected in series. Larger element disks and larger cooling fins must be used if higher currents are drawn. Forced-air cooling is required in some instances to keep them cool.

Metallic rectifiers may be used in battery chargers, instrument rectifiers, and many

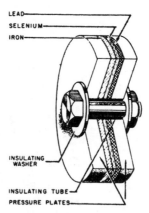

**Figure 5-2.** Construction of the selenium rectifier.

other applications including welding and electroplating. Commercial radios and television sets have been designed to utilize the selenium type of metallic rectifier.

*One caution:* If you are going to reuse one of these units, and you overheat the selenium type, a pungent odor will be quickly detected. It smells something like rotten eggs, or hydrogen sulfate, or like molten sulfur. Thus, one easy way to tell if the selenium rectifier is "gone" is by the odor. It can be replaced in most instances with a newer type of semiconductor diode.

# Solid State Rectifiers

Semiconductor materials are used to make a diode. A *diode* is a device which allows current to flow in one direction and not the other. Germanium and silicon are used as the materials for semiconductor diodes.

Crystals of germanium or silicon are grown from a *melt* which includes small quantities of impure substances such as phosphorus, indium, boron, and other *impurity* atoms. The crystal structure of the resulting metallic chips or wafers permits current flow in one direction only.

Historically, the crystal diode dates to the crystal set used by the first amateurs who worked on making their own receivers of radio signals. Figure 5-3 shows how a crystal set may have looked in the early days of radio. Note particularly the crystal and cat whisker. This was the forerunner of today's crystal diodes.

The first use of the crystal semiconductor as a rectifier (detector) was in the early days of radio. A crystal was clamped in a small cup or receptacle and a flexible wire (cat whisker) made light contact with the crystal. Tuning the receiver was accomplished by operating the adjusting arm until the cat whisker was positioned on a spot of the crystal that resulted in a sound in the headset. Tuning the variable capacitor provided maximum signal in the headset. Trying to find the correct or loudest point on the crystal was quite time-consuming. Today's point contact diode is identical to the crystal diode of yesteryear (see Figure 5-4).

The development of the point contact transistor was announced in 1948. The physical construction of the point contact transistor is similar to that of the point contact diode

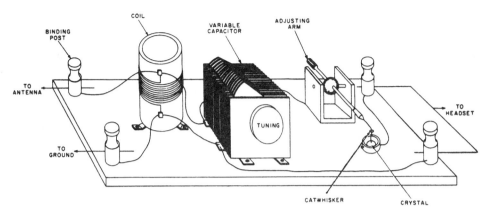

**Figure 5-3.** Early receiver consisting of a crystal, cat whisker, coil, and capacitor, known as the *crystal set* to early radio buffs.

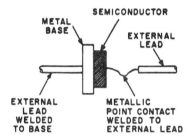

**Figure 5-4.** Physical construction of the point contact diode.

except that a third lead with a metallic point contact is placed near the other metallic point contact on the semiconductor.

The junction diode was first announced in 1949. The junction diode consists of a junction between two dissimilar sections of semiconductor material (see Figure 5-5). One section, because of its characteristics, is called a P semiconductor. The connections to the junction diode consist of a lead to the P semiconductor material and a lead to the N semiconductor. The P material has a deficiency of one electron for every covalent bond of the material. The N material has an extra electron [therefore the (–) or N designation] for every covalent bond of the material. Covalent means that the atoms share electron orbits with adjacent atoms.

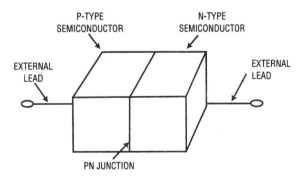

**Figure 5-5.** Physical construction of the junction diode.

The junction diode handles larger power than the point contact diode, but the junction diode has a larger shunt capacitance.

More about the exact construction of the diode and its different types will be presented in Chapter 6.

Many types of semiconductor diodes are available. They vary in size from so small that they are hard to see to as large as 2 in. in diameter. They can withstand high voltages and carry large currents. The improvement of the semiconductor material creates better-quality diodes.

# Power Supply Rectifiers

There are three types of rectifier circuits discussed in this chapter: the half-wave, the full-wave, and the bridge rectifier.

## Half-Wave Rectification

A half-wave rectifier is a device by means of which ac is changed to pulsating direct current (pdc) by permitting current to flow through the device during one-half of the power supply hertz. This one-way current is controlled by a semiconductor device known as a diode. Figure 5-6 shows a number of different packages for diodes. These are general purpose rectifiers.

Figure 5-7 shows how a diode is inserted in a circuit with a transformer. This is a simple half-wave rectifier circuit. The main reason for the transformer is to increase or decrease the 120 V to a higher or lower level to be rectified.

Note that the resistor ($R$) is connected in series with the cathode (+) of the diode. The load resistor ($R$) is there for a purpose.

The operation of the half-wave rectifier circuit is shown in Figure 5-7. The alternations of the input voltage $e_1$ are reproduced by the transformer with an increase in voltage $e_2$ in the secondary windings. The waveforms indicate a 180° difference in phase between $e_1$ and $e_2$. The difference is characteristic of induced voltages. The induced secondary voltage $e_2$ is impressed across the diode and its series load resistance. This voltage causes current $i$ to flow through the diode and its series resistor on the positive half of the hertz. The resultant voltage $e_3$ across the load resistor has a pulsating waveform, as shown in Figure 5-7. This pulsating waveform is referred to as a *ripple voltage*.

Refer to Figure 5-7 and locate points 1 and 2 on the drawing. When point 2 is positive and point 1 is negative, electrons will flow from the ground (point 1) junction, through the load resistor ($R$), to the cathode and thus to the upper terminal, point 2, of the transformer. The secondary winding acts as the immediate source of voltage for the current flow.

Because current flows in only one direction (point 1 to point 2) through the diode and its load, the polarity of the voltage across the load resistance is always as shown. The voltage drop across the diode is quite small when compared with that across the load resistance. The diode has very little forward resistance so there would be a very small voltage drop across it since $E = I \times R$.

The half-wave rectifier uses the transformer during only one-half of the hertz, and therefore for a given size transformer, less power can be developed if the transformer were

## GENERAL PURPOSE RECTIFIERS

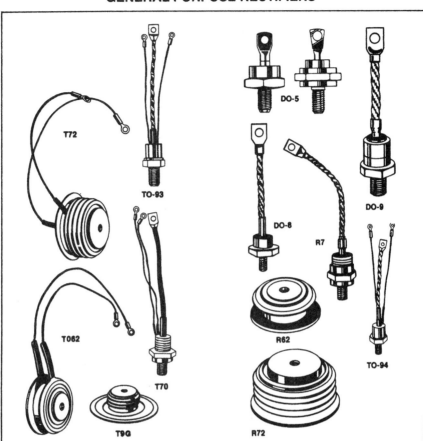

**Figure 5-6.** General-purpose diodes used in high-current rectification circuits. The T70, TO-93, R72, DC-8 are package designations.

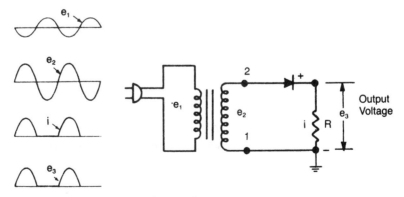

**Figure 5-7.** Half-wave rectifier circuit with waveforms.

utilized on both halves of the hertz. If any considerable amount of power is to be developed in the load, the half-wave rectifier's transformer must be relatively large compared with what it would have to be if both halves of the hertz were utilized. This disadvantage limits the use of the half-wave rectifier to applications that require a very small current drain. The half-wave rectifier is widely used in small commercial receivers and in some early television receivers and oscilloscopes. It is also used in some battery charging circuits since the battery has a tendency to act as a filter and smooth out the pulses.

# Full-Wave Rectification

A full-wave rectifier is a device that has two or more elements so arranged that the current output flows in the same direction during each half-hertz of the ac power supply.

Full-wave rectification may be accomplished by using two diodes, such as shown in Figure 5-8. The end of the load resistor is connected to the center tap (Figure 5-8) of the transformer. The diode is a device that allows current to flow in only one direction.

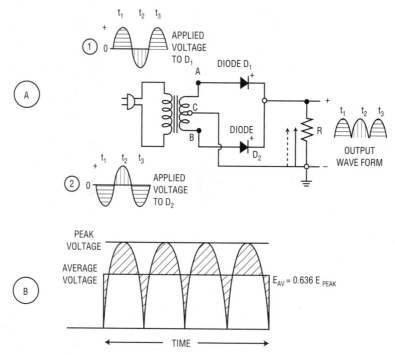

**Figure 5-8.** Full-wave rectifier circuit with waveforms.

The two halves of the secondary winding, $AC$ and $BC$, may be a center-tapped winding as shown, or they may be separated windings. In either case the load circuit is returned to a point midway in potential between $A$ and $B$ so that the load current is divided equally between the two diodes.

The part of the secondary winding between $A$ and $C$ may be considered a voltage source that produces a voltage of the shape shown in ① in Figure 5-8. This voltage is impressed across $D_1$ in series with the load resistor $R$. During the half-hertz marked $t_1$, the electrons flow in the direction indicated by the arrow. This flow of electrons from the ground up through the load resistor $R$ makes the cathode positive with respect to ground. Thus, the load voltage is developed across $R$ between cathode and ground. During this same half-hertz the voltage across $BC$ is negative, as shown in Figure 5-8A and $D_2$ is nonconducting.

A half-hertz later, during interval $t_2$, the polarity of the voltages on the diodes is reversed. $D_2$ now conducts, and $D_1$ is nonconducting. The electron flow through $D_2$ is in the direction indicated by the dotted arrow. This current flow also flows from the ground up through $R$ and makes the cathode positive with respect to ground. Thus another half-hertz of load voltage is developed across $R$. Close examination of the drawing shows that only one diode is conducting at any given instant.

Because there are two pulsations of current in the output for each hertz of the applied ac voltage, the full-wave rectifier uses the power supply transformer for a greater percentage of the input hertz. Therefore, the full-wave rectifier is more efficient than the half-wave rectifier. The full-wave rectifier has less ripple effect, and it may be used for a greater range of electronic devices and circuits.

The dc load current flows equally through the two halves of the transformer secondary and in opposite directions. Thus the ampere turns are equal and act in opposite directions so that there is no tendency to orient the molecules of the iron core in any one direction. Therefore, the transformer inductance is not reduced as it is in the half-wave rectifier and both the voltage regulation and efficiency are improved. The full-wave rectifier is widely used in radio transmitters, receivers, and television receivers.

# Bridge Rectification

Take a look at Figure 5-9 for a four-diode arrangement referred to as a bridge rectifier circuit. This is a full-wave rectifier; the output is full-wave as shown in the previous two-diode circuit. In Figure 5-9A the diodes, $D_1$, $D_2$, $D_3$, and $D_4$ are a semiconductor bridge arrangement. Notice how the negatives and the positives are tied together to produce a (−) and a (+) output from the arrangement. Where the − and + are connected indicates there is no polarity, and the ac can be input there. In reality, the diodes will have a black body with a white ring around one end. The white ring indicates the + end (called cathode in semiconductor parlance). In some instances the body of the diode may be white and a black ring around one end to denote the +, or cathode connection.

During one-half of the hertz of the applied ac voltage, point $A$ becomes positive with respect to point $B$ by the amount of the voltage induced in the secondary of the transformer. During this time, the voltage across $AB$ may be considered to be impressed across a load consisting of $D_1$, load resistor $R$, and $D_3$ in series. The voltage applied across these diodes allows them to conduct in one direction. See the solid arrows. The waveform of this current is shown in ② and ③. One half-hertz later, $D_1$ and $D_3$ are nonconducting, and an electron stream flows through $D_4$, $R$, and $D_2$ in the direction indicated by the dotted arrows. The waveform of this current is shown in ④ and ⑤. The current through the load $R$ is always in the same direction. As this current flows through $R$, it develops a voltage having the waveform shown at ⑥. The bridge rectifier is a full-wave rectifier because current flows in the load during both halves of each hertz of applied alternating voltage.

There are a few advantages to the bridge rectifier over the regular semiconductor full-wave rectifier. First, the bridge arrangement allows for twice the voltage output from the same power transformer. Second, a bridge rectifier is so designed that it has only half the inverse voltage as the full-wave rectifier with only two diodes. Peak inverse voltage (piv) is a negative voltage applied across the diode when there is no current flowing through the diode.

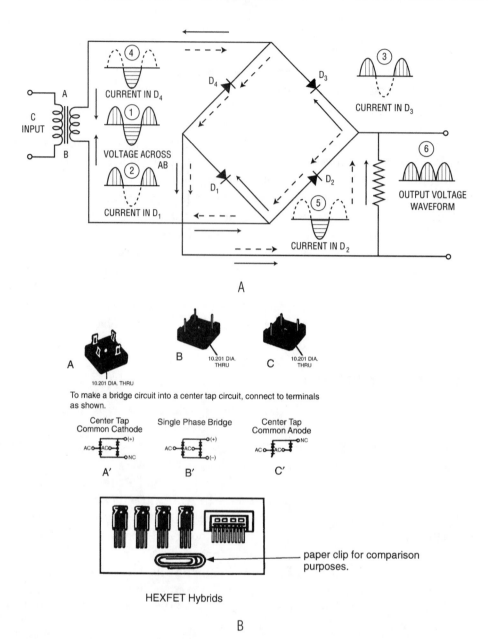

**Figure 5-9.** A. Bridge rectifier with waveforms. B. Single-phase molded diode bridges — 10 to 35 amps.

Another advantage of the bridge rectifier is that it can be packaged easily into molded units for ease in handling and replacement if necessary (see Figure 5-9B).

These **HEXFET** transistors feature all of the advantages of MOSFETs (See p. 161), such as voltage control, freedom from second breakdown, fast switching and temperature stability of electrical parameters. They are designed for a wide range of applications. The full bridge configuration makes hybrids suitable for motor drives, power supplies, and reversing switches among other applications. HEXFET design achieves low on-state resistance, high transconductance, and rugged, long-term usage. The copper film on alumina substances gives increased packaging density and improved field reliability. HEXFET transistors have the same standard configurations as found with modular plug-in units.

# Filters

Up to this point we have discussed the making of pulsating dc from ac. This is fine for some uses, but for most electronic circuits it is not pure enough dc for proper operation of the circuits. Too much pulsating will make a high level of hum, and in some — computer circuits, for instance — it will give unreliable results. That, then, calls for something a little purer in the way of making the dc usable and of the proper form to do the work. Filter circuits are the answer to smoothing out the ripple and pulsations of full-wave and half-wave rectifiers.

The unfiltered output of a full-wave rectifier is shown in Figure 5-10. The polarity of the output voltage does not reverse, but its magnitude fluctuates about an average value as the successive pulses of energy are delivered to the load. In Figure 5-10, the average voltage is shown as the line that divides the waveform so that area $A$ equals area $B$. The fluctuations of voltage above and below this average value is called the *ripple*. The frequency of the main component of the ripple for the full-wave rectifier shown in Figure 5-10 is twice the frequency of the voltage that is being rectified. In the case of the half-wave rectifier the ripple has the same frequency as the input ac voltage. Thus, if the input voltage is obtained from a 60-Hz source, the main component of ripple in the output of a half-wave rectifier is 60 Hz, and in a full-wave rectifier it is 120 Hz.

**Figure 5-10.** Unfiltered output voltage of a full-wave rectifier.

The output of any rectifier is composed of a direct voltage and an alternating or ripple voltage. For most uses, the ripple voltage must be reduced to a very low amplitude. The amount of ripple that can be tolerated varies with different uses of the semiconductors.

The *percentage of ripple* is 100 times that ratio of the rms value of the ripple voltage at the output of a rectifier filter to the average value, $E_O$, of the total output voltage. Figure 5-11 indicates graphically how the percentage of ripple may be determined. It is assumed that the ripple voltage is of sine waveform. The formula for determining the percentage of ripple is

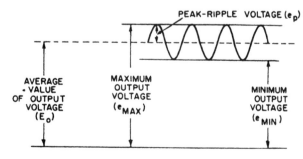

**Figure 5-11.** Percentage of ripple.

$$\text{Percent of ripple} = \frac{E_{rms}}{E_o} \times 100$$

where $E_{rms}$ = 0.707 of $e_p$

$e_p$ = peak value of ripple voltage

A circuit that eliminates the ripple voltage from the rectifier output is called a *filter*. Filter systems in general are composed of a combination of capacitors, inductors, and, in some cases, resistors.

# Capacitor Input Filter

The load in a rectifier circuit has the voltage supplied in pulses from the changing of ac to dc by the diode action. The fluctuations can be reduced if some of the energy can be stored in a capacitor while the rectifier is delivering its pulse and is allowed to discharge from the capacitor between pulses.

Figure 5-12A shows the output of a half-wave rectifier. This pulsating voltage is applied to a filter capacitor represented by $C$ in Figure 5-12B. Note that the capacitor is placed across the load. This means the capacitor will react to any change in circuit voltage. Because the rate of charge of the capacitor is limited only by the reactance of the transformer secondary and the plate resistance of the diode or tube in the rectifier, the voltage across the capacitor can rise nearly as fast as the half sine wave voltage output from the rectifier. In other words, the $RC$ charge time is relatively short. The capacitor, therefore, is charged to the peak voltage of the rectifier within a few hertz. The charge on the capacitor represents a storage of energy. When the rectifier output drops to zero, the voltage across the capacitor does not fall immediately. Instead, the energy stored in the capacitor is discharged through the load during the time that the rectifier is not supplying energy. This occurs when the voltage on the vacuum tube diode plate is negative or the cathode of the semiconductor is negative. Bear in mind that diodes have a (+) sign to indicate the cathode of a semiconductor.

The voltage across the capacitor (and the load since it is in parallel with the capacitor) falls off very slowly if it is assumed that a large capacitance and a relatively large value of load resistance are used. In other words, the $RC$ discharge time is relatively long. The amplitude of the ripple therefore is greatly decreased, as seen in Figure 5-12C.

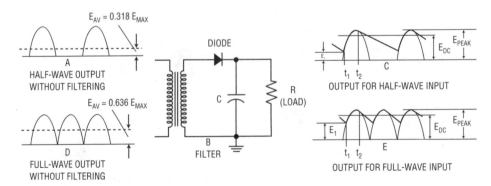

**Figure 5-12.** Capacitor-input filter with waveforms.

Figure 5-12D shows the input voltage to the filter when a full-wave rectifier is used, and Figure 5-12E shows the resulting output voltage waveform.

After the capacitor has been discharged (with either the half-wave or full-wave output), the rectifier does not begin to pass current until the output voltage of the rectifier exceeds the voltage across the capacitor. Thus, in Figure 5-12C and E, current begins to flow in the rectifier when the rectifier output reaches a voltage equal to the capacitor voltage. This occurs at some time, $t_1$, when the rectifier output voltage has a magnitude $E_1$. Current flows in the rectifier until slightly after the peak of the half sine wave, at time $t_2$. At this time the sine wave voltage is falling faster than the capacitor can discharge. A short pulse of current, beginning at $t_1$ and ending at $t_2$, is therefore supplied to the capacitor by the power source.

The average voltage of the rectifier output is shown in Figure 5-12A and D. Because the capacitor absorbs energy during the pulse and delivers this energy to the load between pulses, the output voltage can never fall to zero. Hence, the average voltage of the filter output (see Figure 5-12C and E) is greater than that of the unfilter output (see Figure 5-12A and D). However, if the resistance of the load is small, a heavy current is drawn by the load and the average or direct voltage falls. For this reason, the simple capacitor filter is not used with rectifiers that must supply a large load current. Also the input capacitor acts like a short circuit across the rectifier while the capacitor is charging. Because of this high peaked load on the rectifier tubes, the capacitor input filter is seldom used with gas tubes in high-current installation. This high peaked load can also ruin semiconductor diodes.

## Inductor Input Filter

The capacitor was placed in parallel across the load since it reacted to variations in voltage. The inductor is a device that reacts to changes in current. The inductor causes a delay in current. Since the current is the same in all parts of the series circuit, the inductor is placed in series with the load it is to filter (see Figure 5-13C).

The inductor is placed in series with the load to prevent abrupt changes in load current. An inductor filter is shown in Figure 5-13. The input waveforms from a half-wave rectifier and a full-wave rectifier are shown in the other parts of the figure. When no inductor is used in series with R, the output current waveforms are indicated by dotted lines. The solid lines indicate the output current waveforms when an inductor is placed in the circuit. The use of an inductor prevents the current from building up or dying down too quickly. If the inductance is made large enough, the current becomes nearly constant.

The inductance prevents the current from ever reaching the peak value that is reached without the inductance. Consequently, the output voltage never reaches the peak value of

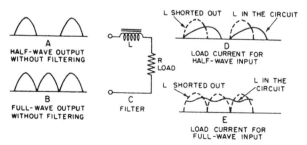

**Figure 5-13.** Inductor-input filter with waveforms.

the applied sine wave. Thus, a rectifier whose output is filtered by an inductor cannot produce as high a voltage as can one whose output is filtered by a capacitor. However, this disadvantage is partly compensated because the inductance filter permits a larger current drain without a serious change in output voltage.

# Pi Filter

The pi ($\pi$) filter is used to improve the filtering action of rectified current and voltage. The capacitor alone or the inductor alone cannot do the job properly. However, if both the capacitor and the inductor are combined, they have a better chance of producing a high-quality dc for use by electronics circuits.

The function of the capacitor is to smooth out the variations in voltage while the inductor is used to smooth out the variations in current produced by the pulsating dc output of a rectifier. The result is a voltage having a nearly constant magnitude.

Figure 5-14 is a pi filter used to smooth the full-wave output of a rectifier. This type is used with receivers and transmitters.

Besides being referred to as a pi filter, it is also called a capacitor-input filter. The first or input capacitor, $C_1$, in Figure 5-14, acts to bypass the greatest portion of the ripple component to ground. In all filters the major portion of the filtering action is accomplished in this first component. The series choke in the pi filter serves to maintain the current at a nearly constant level during the charging and discharging cycles of the input capacitor.

At the bottom of Figure 5-14 are shown the waveforms of current through the diodes and the voltage across $C_1$. The other capacitor, $C_2$, acts to bypass the residual fluctuations existing after filtering by the input capacitor and inductor. The current flow through the rectifier tubes is a series of sharp peaked pulses, because the input capacitor acts as a short circuit across the rectifier while the capacitor is charging.

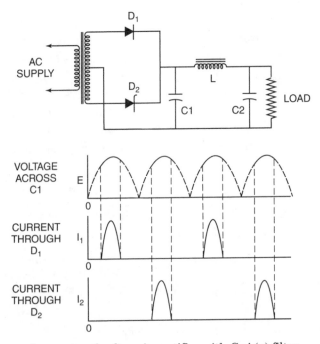

Figure 5-14. Waveforms of current and voltage in rectifier with $C$ pi ($\pi$) filter.

Because of this high peaked load on the rectifiers, the *pi* filter is used only in low-current installations such as radio receivers and television receivers.

You can hear the difference if the capacitors open while in a receiver. If the signal on the radio is very weak with a strong hum, it means $C_1$ is open. If the signal is very strong, but has a low-level hum that is annoying, it is $C_2$ that is open. Electrolytics do dry up and appear as open when they are not used for some time.

Most *pi* filters use a resistor instead of the inductor. That is because for tubes the voltage is more critical than the current. Filtering the voltage makes a difference in the hum level of the output of the receiver. The resistor keeps the two capacitors from causing a "motorboating" sound in the output of the receiver or amplifier. The filter resistor does, however, drop some voltage, and the output voltage is less than if an inductor were used instead. However, this is acceptable in most circuits that use the *pi* filter.

Once you learn about the power supplies and how they perform, you will be able to look at a schematic of a receiver or piece of electronics equipment and know if the circuit you are getting is worth the money you are spending. This type of information makes you a wiser consumer of electronics.

## *L* Filter

Another, not so often used, filter is the *L* type. It uses an inductor (in cheaper circuitry a resistor is used instead of the inductor) and a capacitor to filter the pulsating dc output of a rectifier. It is used primarily in high-current applications (see Figure 5-15).

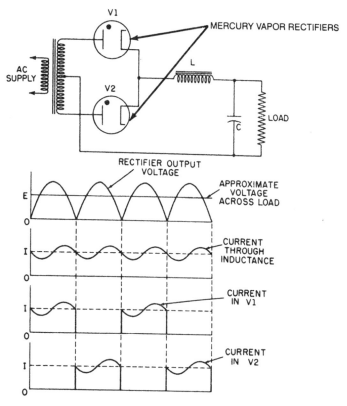

**Figure 5-15.** Current and voltage waveforms in full-wave rectifier with an *L* filter.

The components perform the same functions as they do in a *pi* filter, except that the inductor, or choke, input reduces the voltage output of the filter. This filter is also referred to as a *choke input filter*. The input choke allows a continuous flow of current from the diodes rather than the pulsating current flow demanded by the capacitor input filter. The *L* filter is seldom used with half-wave rectifiers because there is no device to maintain the current flow between the half-hertz.

Because of the uniform flow of current, the *L* filter has uses in most high-power circuits. It is used primarily with mercury-vapor rectifier tubes. It also has the advantage of voltage regulation. The inductive reactance ($X_L$) of the choke reduces the ripple voltage without reducing the dc output voltage.

Two *L* filters are sometimes used in series to produce a higher degree of filtering action.

# Voltage Multipliers

High voltage is needed in many circuits and devices. For instance, cathode ray tubes used in oscilloscopes, radar displays, and video displays call for a typical voltage of 2 kV up to 20 kV. They need only a few milliamperes of current and a regulation of ±1%.

Photomultiplier tubes are used in scintillation counters, flying spot scanners, as well as low-level photometry.

The typical voltages for this type of tube is 1 to 3 kV at about one-half a milliampere up to as much as 5 mA. Required regulation is within the 0.1% range.

The voltage supply to the tube must be held constant since the electron gain of a photomultiplier tube is dependent on the applied voltage. A change of only 1% in voltage results in about 10% change in the tube gain.

Other uses for high-voltage supplies include the photocopier. These lamps require a typical voltage of 5 to 10 kV with about 5 mA current and a regulation of less importance than photomultiplier tubes. These are but a few uses for high-voltage power supplies. Camera tubes, electron microscopes, and image intensifier tubes also require high voltage at comparatively low currents.

Voltage doublers, multipliers, and quadruplers are used to produce the high voltage needed for these devices. Inasmuch as the current requirements are low, it is possible to make an arrangement such as in Figure 5-16A. This is a doubler circuit that uses two diodes and two capacitors to increase the input voltage. It is also possible to increase the line voltage (120 V) without the use of a transformer if about 240 V is sufficient for the device. This type of "keep-alive" voltage is sometimes employed in a strobe circuit where the xenon flash tube is held near ionization by a voltage just below the flash point.

## Voltage Doublers

The simple doubler circuit is nothing more than two half-wave rectifiers with the outputs summed by using two capacitors. $D_1$ conducts on the positive half-cycle to charge $C_1$ to the peak value of the ac secondary, $V_P$, and then on the negative half-cycle, $D_2$ conducts to charge $C_2$ also to $V_P$. Since these two capacitors are in series, the voltage available across the two of them is additive and can be taken off from the top and bottom points of the two-capacitor string. By taking the output from *A-B* it produces the *sum* of the two charges. As

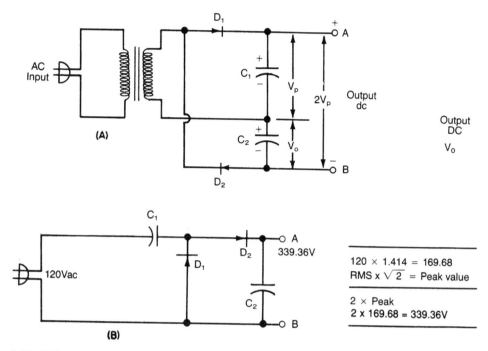

**Figure 5-16.** Voltage doublers. A. Using a transformer to step up voltage and then doubling it. B. Using the line voltage and then doubling its peak value.

with other rectifiers of this type, the regulation is poor so that only a relatively light load of a few milliamperes can be supplied. Another doubler commonly used is also shown (see Figure 5-16B). In this circuit $C_1$ must be a larger capacitance than $C_2$. On the negative half cycles, $D_1$ conducts and $C_1$ charges to peak voltage, $V_P$. On the second positive half-cycle, the left-hand plate of $C_1$ goes positive and $D_2$ conducts carrying $C_2$ to almost 2 times the $V_P$. Typically $C_1$ is twice the value of $C_2$ to allow correct charge-sharing between the two capacitors.

# Voltage Triplers

A voltage tripler can be constructed by adding an exra half-wave rectifier as shown in Figure 5-17.

The tripler uses three diodes and three capacitors to achieve producing three times the peak voltage of the secondary of the transformer. The polarity of the diodes is critical in the tripler arrangement as is also the case in doublers. As the number of multipliers is increased, the output impedance increases by the cube of the number of stages. The high values of output impedance limits the available output current. The high impedance also degrades the load regulation.

# Voltage Quadruplers

The voltage quadrupler delivers approximately four times the voltage of the input secondary. The well-known Crockcroft-Walton "ladder" circuit is utilized to produce both triplers

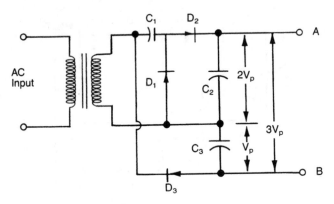

**Figure 5-17.** One type of voltage tripler circuit.

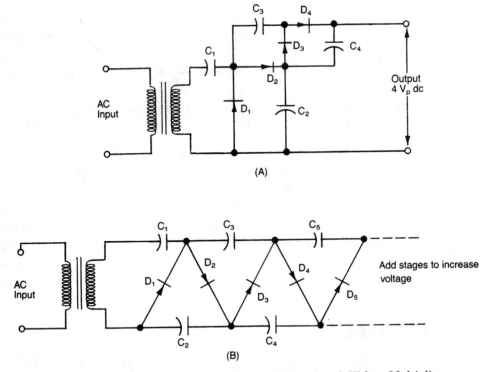

**Figure 5-18.** Voltage quadrupler circuits. A. Quadrupler. B. Crockcroft-Walton Multiplier.

and quadruplers (see Figure 5-18). A quadrupler has an output impedance of about eight times greater than a doubler. As you can see, then as the impedance is increased with the higher voltages, lower amounts of current are available for use from this circuit design.

# Summary

Alternating current is available commercially at low cost. Direct current is more expensive to produce; therefore, a method of changing ac to dc is needed as an inexpensive source of dc for electronic circuits. This is what the rectifier does. The rectifier is composed of either a vacuum tube diode, a semiconductor diode, or a metallic diode. Copper oxide and selenium are used as metal-disk rectifiers.

Silicon and germanium are used to make semiconductor diodes which allow current to flow in one direction only.

Metallic rectifiers may be found in older equipment such as battery chargers, instruments, welders, and electroplating power supplies.

The semiconductor diodes are made in either the point contact or junction configuration. Each has its particular purpose. The point contact is used to rectify small ac signals while the junction diode is used for power supply frequencies.

Three types of rectification are discussed in this chapter. In half-wave rectification the ac waveform has half its hertz or cycle utilized. Full-wave rectification makes use of both halves of the ac hertz or cycle. The two halves are made into pulsating dc. The bridge rectification method produces full-wave also. All three types of output have to be filtered before they are of any value to electronic circuits. However, either full- or half-wave rectification can be used to charge batteries without a filter. The battery produces the filtering action.

The bridge rectifier has an advantage over the regular full-wave rectifier since it allows for twice the voltage output from the same power transformer. The bridge rectifier circuit can withstand twice the full-wave rectifier's peak inverse voltage.

Filters are used to smooth out the pulsating dc current produced by rectifying ac power line frequencies. There are a number of filter circuits which can be used to produce a smoother dc output.

The capacitor input filter produces an average voltage that is greater than that put into it. However, if the resistance of the load is small, a heavy current is drawn by the load. and the average or direct voltage falls. For this reason, the simple capacitor filter is not used with rectifiers that must supply a large load current.

The purpose of the inductor input filter is to filter out the variations in current that are pulsating as a result of the rectification action. The inductor is placed in series with the load to prevent abrupt changes in load current. The capacitor input filter utilized the capacitor directly across or in parallel with the load. A rectifier whose output is filtered by an inductor cannot produce as high a voltage as one whose output is filtered by a capacitor. However, this disadvantage is partly offset because the inductance filter permits a larger current drain without a serious change in output voltage.

The pi filter is used to improve the filtering action of rectified current and voltage. It is a combination of both inductor and capacitor filters. The arrangement looks like or resembles the Greek letter pi ($\pi$), hence its name. It is also referred to as a capacitor input filter since it uses two capacitors and an inductor to separate them. However, the input is considered from the rectifier and output from where the load is connected.

Most pi filters use a resistor instead of an inductor. The inductor is heavy, is more expensive, and is of limited value to circuits which draw low currents.

The $L$ filter is used infrequently. It uses an inductor and a capacitor to filter the pulsating dc output of a rectifier. It is used primarily in high-current applications. Because of the uniform flow of current, the $L$, or inductor input, filter has uses in most high-power circuits. It is used primarily with mercury-vapor rectifier tubes. It also has the advantage of voltage regulation.

Voltage multipliers are used in many electronic devices that use high voltage but need low currents.

# Review Questions

1. When did a need for rectification arise?
2. What were a couple of the early metal-disk rectifiers made of?
3. What is the purpose of a rectifier?
4. Where does pulsating dc come from?
5. What is the difference between a selenium rectifier and a germanium rectifier?
6. Where are, or were, metallic rectifiers used?
7. What are the two materials most often used to make semiconductor diodes?
8. What is the difference between a junction and a point contact diode?
9. What is meant by the term reverse-biased?
10. What is meant by peak inverse voltage (piv)?
11. What is the difference between half-wave and full-wave rectification?
12. What is ripple voltage?
13. How is the ripple smoothed out?
14. What is a bridge rectifier?
15. How does a bridge rectifier differ from a full-wave rectifier?
16. What is the major advantage of the bridge rectifier over the full-wave rectifier?
17. What is the purpose of a filter?
18. Describe the capacitor input filter.
19. What is the disadvantage of using a capacitor input filter?
20. Describe the inductor input filter.
21. What is the disadvantage of using the inductor input filter?
22. Describe the pi filter.
23. What is the advantage of using a pi filter over the inductor-input and capacitor-input filters?
24. What does percentage of ripple mean?
25. Which capacitor has the greatest filtering effect in a pi filter?
26. Why do most pi filters use a resistor instead of an inductor?
27. Where are *L* filters used?
28. What is the disadvantage of using a resistor instead of an inductor in a pi filter?

# Chapter 6

# SEMICONDUCTORS

The term *semiconductor* is applied to both diodes and transistors as well as to certain special types of electronic devices. The word comes from the fact that germanium and silicon perform somewhere between a conductor and an insulator in terms of opposition to current flow. The amount of opposition is programmed into or manufactured into the device by means of controlling the impurities introduced into a pure germanium or pure silicon atom. Germanium and silicon can be purified to better than 99.999 999 percent. Therefore, any introduction of another element is called an *impurity*, or *doping agent*. By controlling the amount of doping agent introduced into each crystalline structure, you can control the amount of opposition to current flow.

## A Bit of History

In 1833 Michael Faraday, an English scientist, made a contribution to the crystalline amplifier by working with silver sulfide. He learned that the resistance of silver sulfide varies *inversely* with the temperature. As the temperature increases, the resistance decreases. This was noted as being different since most conductors will have an increase in resistance along with an increase in temperature.

It took over a hundred years before the Faraday discovery was utilized in any meaningful way. The development of the *crystal amplifier* (the transistor's original name) by three Bell Laboratory scientists utilized the work of Faraday and expanded on it. In June of 1948 John Bardeen, William Shockley, and W. H. Brattain shared an office and rode to work together at Bell Labs in New Jersey. Their work on the development of the transistor led to Nobel Prizes for each. From the announcement day to the present, there has been no letup in the research and development of the semiconductor and solid state physics. Present-day computers and space communications devices are outgrowths of the development of a crystal that would amplify and switch.

## Diodes and How They Work

Semiconductor technology is called *solid state*. This means that the materials are solid as opposed to the vacuum that separates the electrodes in a vacuum tube. Since the vacuum tube was the forerunner of the transistor and diode, it is used as a standard of comparison many times. The uses of the vacuum tube were many, and the purpose of the solid state devices was to replace the vacuum tube and its need for many more watts of energy to heat its filaments and to have the heat dissipate before it causes excess problems for the circuit designer. One of the reasons why Bell Labs worked on the crystal or solid state amplifier was so they could bury the device in a cable, toss it overboard, and let it stay on the bottom of the ocean for overseas communications. Pulling up the cable and repairing the old vacuum tube amplifier had been very expensive and time-consuming. The ideal device was the type of material they developed in the solid state amplifier. Other companies were allowed to manufacture it, after court battles, and develop better-quality devices.

Figure 6-1 shows two types of diodes. The diode comes in many packages and can be bought for almost any purpose. They may be individually packaged or they may be made by the hundreds on a chip. Let us take a look at an individual diode to see how it operates, and then we can look at more details of its packaging and use.

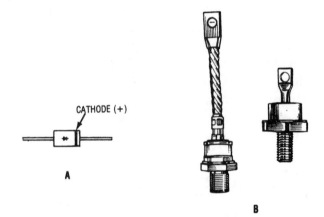

**Figure 6-1.** A. Point contact diode. B. Junction diode.

There are many types of semiconductor materials. We use only two of these for transistors and diodes — silicon and germanium. In their pure forms, both germanium and silicon have properties close to insulators. By nature, both these materials are hard, brittle crystals. They have a type of structure known as a *lattice* (see Figure 6-2).

Outer electrons (called *valence* electrons) of individual atoms of these materials have special traits. They are bonded in pairs to adjacent atoms. Therefore, semiconductor materials are normally poor conductors. To free the outer electrons, it is necessary to apply higher temperatures or strong electron force or pressure in the form of a battery voltage.

## Impurity Donors and Acceptors

In the pure form, germanium and silicon crystals are of no use as a semiconductor device. However, as stated at the beginning, when a certain amount of impurity is added, the crystal can be made to conduct a current. In order to accomplish this, both the quantity and the quality of the impurity must be carefully controlled. Added impurities will create either an excess or a deficiency of electrons, depending on the kind of impurity added.

The impurities that are important in semiconductor materials are those which align themselves in the regular lattice structure whether they have one valence electron too many or one too few. The first type loses its extra electron easily, and in so doing increases the conductivity of the material by contributing a free electron. This type of

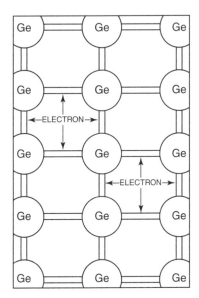

**Figure 6-2.** Germanium crystal lattice showing the bonding of electrons and the sharing with adjacent atoms.

impurity has five valence electrons and is called a *pentavalent* impurity. Arsenic, antimony, bismuth, and phosphorus are pentavalent impurities. Because these materials give up or donate one electron to the material they are also called *donor* impurities.

The second type of impurity tends to compensate for its deficiency of one valence electron by acquiring an electron from its neighbor. Impurities of this type in the lattice structure have only three electrons and are called *trivalent* impurities. Aluminum, indium, gallium, and boron are trivalent impurities. Because these materials accept one electron from the material they are also called *acceptor* impurities.

Semiconductors that have no impurities are called *intrinsic* semiconductors. Semiconductors that have either acceptor or donor impurities are called *extrinsic* semiconductors.

# N Germanium

When a pentavalent impurity like arsenic is added to germanium it will form covalent bonds with the germanium atoms (see Figure 6-3A). The arsenic (As) atom in a germanium lattice structure is shown in this illustration. The arsenic atom has five valence electrons in its outer shell but uses only four of them to form covalent bonds with the germanium atoms, leaving one electron relatively free in the crystal structure. Because this type of material conducts by electron movement, it is called a *negative carrier*, or *N semiconductor*. Pure germanium may be converted into an N semiconductor by doping it with any element containing five electrons in its outer shell. Other pentavalent elements which may be used in place of arsenic as dopants are phosphorus, antimony, and bismuth. The amount of the impurity added is very small. It is in the order of one atom of impurity in 10 million atoms of germanium. The same type of explanation of the manufacture of semiconductors made of silicon is applicable. Just substitute the word germanium for silicon, and the explanation remains the same. Germanium was used here since it was the first of the semiconductors to be developed into diodes and transistors.

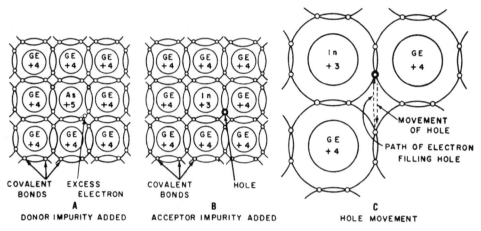

**Figure 6-3.** A. Another way of showing the sharing of germanium electrons in the lattice with a donor impurity added. B. An acceptor impurity added. C. Hole movement.

# P Germanium

A trivalent impurity element can also be added to pure germanium to dope the material. In this case the impurity has one less electron than it needs to establish covalent bonds with four neighboring atoms. Thus in one covalent bond there will be only one electron instead of two. This arrangement leaves a *hole* in that covalent bond.

Figure 6-3B shows the germanium lattice structure with the addition of an indium atom (In). The indium atom has one electron less than it needs to form covalent bonds with the four neighboring atoms and thus creates a hole in the structure. Gallium and boron also exhibit these characteristics. The holes are present only if a trivalent impurity is used. Note that a hole carrier is not created by the removal of an electron from a neutral atom, but is created when a trivalent impurity enters into covalent bonds with a tetravalent (four valence electrons) crystal structure. Because this semiconductor material conducts by the movement of holes which are positive charges, it is called a *positive carrier*, or *P semiconductor*. When an electron fills a hole (see Figure 6-3C), the hole appears to move to the spot previously occupied by the electron.

Both holes and electrons are involved in conduction. The holes are called positive carriers and the electrons are called negative carriers. The one present in the greatest quantity is called the *majority* carrier. The other is called the *minority* carrier. In N material the electrons are the majority carriers, and the holes are the minority carriers. In P material the holes are the majority carriers, and the electrons are the minority carriers.

Pure germanium is an insulator, or a very poor conductor. However, even at room temperature there is enough heat present in the germanium to produce some electron movement. An electron moving out of a covalent bond leaves a hole in the bond. The hole will attract an electron from a nearby atom. This produces a hole in that atom. Thus, both the holes and the electrons appear to move. The holes are positive carriers, and the electrons negative carriers.

Conduction in germanium due to the formation of hole-electron pairs is called *intrinsic conduction*. It occurs even though no voltage is applied across the crystal. It is a random movement (diffusion). Holes and electrons may move in any direction. Intrinsic conduction is kept at a minimum by holding the operating temperature down and shielding the semiconductor from light and from other forms of electromagnetic radiation.

# Current Flow in N Material

Current flow through N material is shown in Figure 6-4. Conduction in this type of semiconductor is similar to conduction in a copper conductor. That is, the application of voltage across the material will cause the loosely bound electron to be released from the impurity atom and move toward the positive potential point.

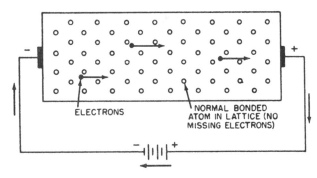

**Figure 6-4.** Current flow in N material.

However, certain differences exist between the N semiconductor and a copper conductor. For example, semiconductor resistance decreases with a temperature increase, because more carriers are made available at higher temperatures. Increasing the temperature releases electrons from more of the impurity atoms in the lattice, causing increased conductivity (decreased resistance). In the copper conductor, increasing the temperature does not increase the number of carriers but increases the thermal agitation or vibration of the structure so as to impede the current flow further (increase the resistance).

# Current Flow in P Material

Current through P material is shown in Figure 6-5. Conduction in this material is by positive carriers, or holes. For the hole to appear to move, an electron in a nearby lattice site must shift to the position where the hole existed originally. Thus the hole moves from the positive terminal to the negative terminal. Electrons from the negative terminal cancel holes in the vicinity of the terminal, while at the positive terminal, electrons are being removed from the covalent bonds, thus creating new holes. The new holes then move

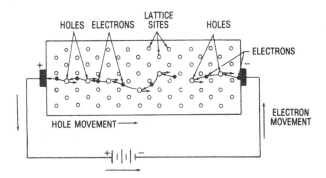

**Figure 6-5.** Current flow in P material.

toward the negative terminal (the electrons shifting to the positive terminal) and are cancelled by more electrons emitted from the negative terminal. This process continues as a steady stream of holes (hole current) moving toward the negative terminal.

In both N and P material, current flow in the external circuit is out the negative terminal of the battery and into the positive terminal.

# Semiconductor Diodes

A semiconductor diode is made by joining a piece of the P material with a piece of N material. The place where the two materials are joined is referred to as the *junction*. This junction is very thin, and each end has a piece of wire attached for connecting the diode thus made into a circuit. Figure 6-6 shows how the two pieces of material form a diode junction.

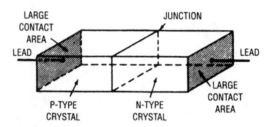

**Figure 6-6.** Pictorial representation of a PN junction diode.

Both holes and electrons are involved in conduction in the PN junction diode. There are minority carriers in both regions: holes in the N material and electrons in the P material. The holes produced in the N material near the junction are attracted by the negative ions on the P side of the junction and pass across the junction. These holes will tend to neutralize the negative ions on the P side of the junction. Similarly, free electrons produced on the P side of the junction will pass across the junction and neutralize positive ions on the N side. This action is an example of intrinsic conduction, which is undesirable.

This flow of minority carriers weakens the potential barrier around the atoms that they neutralize. When this happens, majority carriers are able to cross the junction at the location of the neutral atom. This means that holes from the P material will cross over to the N material, and electrons from the N material will cross over to the P material.

This action results in both holes and electrons crossing the junction in both directions. These motions cancel each other and the net movement contributes nothing toward the net charge or current flow through the junction. Because of intrinsic conduction, the junction is no longer a rectifier when an external voltage is applied across it. It is analogous to an electron tube diode in which not only the cathode emits electrons, but the plate is heated to the point where it also will emit enough electrons to break down the rectifying properties of the diode.

## Operation of the Junction Diode

When an electron leaves the donor atom in the N region and moves over to the P region (see Figure 6-7), the atom has fewer electrons than it needs to neutralize the positive

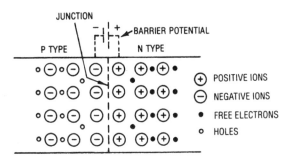

**Figure 6-7.** Location of the ions and carriers in the junction diode.

charge on the nucleus, and it becomes charged (ionized). It has one extra positive charge equal to the negative charge of the electron which it lost.

Similarly, when a hole leaves an acceptor atom in the P region, the atom takes on a negative charge, because the hole has been filled by an electron, and the atom has one more electron than it needs to neutralize the charge on its nucleus.

These charged atoms, or ions as they are called, are fixed in place in the crystal lattice structure, and cannot move. Thus, they make up a layer of fixed charges on both sides of the junction. On the N side of the junction there is a layer of positively charged ions; on the P side there is a layer of negatively charged atoms or ions.

In Figure 6-7 there is a barrier of negative ions on the P side of the junction. This negative barrier will repel electrons from the immediate vicinity of the junction and will prevent the diffusion of any more electrons from the N side over into the P side of the crystal. Similarly on the N side of the junction there is a barrier of the positive ions which will repel holes away from the immediate vicinity of the P side of the junction and prevent diffusion of any additional holes across the junction from the P material into the N material.

The two layers of ionized atoms form a barrier to any further diffusion across the junction. Because the charges at the junction force the majority carriers away from the junction, the barrier is known as the *depletion layer*. It is also known as the barrier layer, or barrier potential.

The charge on the impurity atoms is distributed across the PN junction as shown in Figure 6-7. In the P region the ionized acceptors have a negative charge, and in the N region the ionized donor atoms have a positive charge. At the junction, the charge is zero.

However, in the P region there are holes which have a positive charge, and in the N region there are free electrons which have a negative charge. The distribution of holes and free electrons is shown in the illustration.

The potentials at the junction have driven the holes away from the junction in the P region and the electrons away from the junction in the N region so that the charges in the P region and the N region are moved farther apart. The charge at the junction is zero. The net charges on the crystal in the P region are equal to the difference between the charge on the ionized acceptor atoms and the electrons. These charges cancel except in the immediate region of the junction.

In the area near the junction there is a negative charge in the P region and a positive charge in the N region. As stated previously, they act as a barrier to prevent further diffusion of holes from the P region into the N region and the diffusion of electrons from the N region into the P region. This potential barrier is a potential difference, or voltage, across the junction and is in the order of a few tenths of a volt. It may be represented as a dotted battery with the negative terminal connected to the P material and with the positive terminal connected to the N material (see Figure 6-7).

This barrier potential is like the plate-cathode voltage of a diode. If the plate is made positive and the cathode is made negative, the diode can be made to conduct a current. If the plate is negative with respect to the cathode, the diode will block the flow of current. Thus, the diode tube is a rectifier. The semiconductor diode also is a rectifier.

# Vacuum Tube Diode

The electron tube is considered by many the primary starting point for electronics. It is responsible for the rapid advancement of electronics up to the invention of the transistor.

The electron tube is made up of a highly evacuated glass or metal shell, which encloses several elements. The elements consist of the cathode that emits electrons when hot, the plate, and sometimes one or more grids. The diode does not have a grid. The *di* part of the name means *two*; *ode* is short for electr*ode*. Put them together and you have two electrodes in an envelope. This is not unlike the semiconductor diode that consists of *p* and *n* materials (see Figure 6-8).

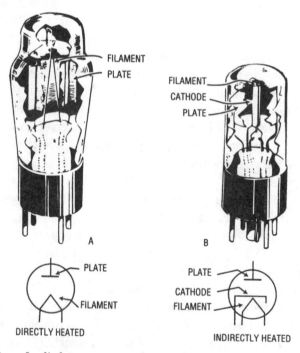

**Figure 6-8.** Cutaway view of a diode or vacuum tube rectifier.

## Construction Details

The original diode was constructed by Thomas Edison, inventor of the incandescent lamp, shortly after his invention of the lamp itself. He added a metal plate inside his evacuated lamp and provided an external terminal from it for use as an electrode. Then he used the heated filament as another electrode and arranged that diode as shown in Figure 6-9.

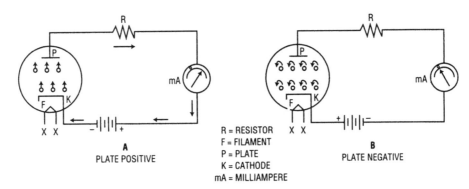

**Figure 6-9.** Operation of the vacuum tube diode circuit.

In an electric circuit the two electrodes of a diode act in the manner of a valve (which is what the vacuum tube is called in England).

## Operation of the Vacuum Tube Diode

The operation of the tube can be observed when it is connected in a circuit. Figure 6-9 is composed of a diode vacuum tube, a battery, a milliammeter, and a resistor. Note what happens to the meter in *A* as compared with *B* in Figure 6-9. As the battery polarity is changed, the action of the vacuum tube changes. When the cathode is negative, the electrons flow through the vacuum from the cathode (K) to the plate (P). The plate has a positive potential applied. When the plate is connected to the negative potential and the cathode to the positive terminal of the battery, the electrons — being of a negative charge —are repelled by the like potential of the plate. Therefore, there is no current flow (or electron flow) through the tube when it is reverse-biased.

Now try to visualize what happens when an alternating current (ac) is applied between the cathode and the plate. As the polarity reverses itself in ac, the tube will conduct and not conduct according to the polarity of the two elements. Thus, the current will flow during one-half of the hertz applied to the tube and not conduct when the other half is applied. That action produces a half-wave rectified output from the circuit (see Figure 6-10).

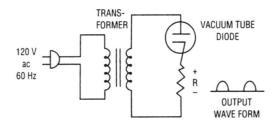

**Figure 6-10.** Output of the rectifier when connected to an ac source.

# Biasing the Junction Diode

*Forward bias* means that the diode is connected to the proper polarity to allow it to conduct current. *Reverse bias* means that the diode puts up a high resistance to the flow of electrons and holes and thereby prevents current flow when the polarity is reversed. This is one of the ways a diode operates. It takes a high resistance in one direction and low resistance in the other to produce a rectification process. By properly biasing the junction, you allow current to flow or not to flow. It is simply supplying the proper voltage polarity to cause it to conduct and the wrong polarity in order to prevent it from conduction.

Probably the best way to explain the biasing arrangement is by using a battery connected across the PN junction. That way you can see the possibilities one step at a time.

If a battery is connected across the PN junction, the battery potential will bias the junction. If the battery is connected so that its voltage opposes the barrier potential across the junction, it will aid current flow through the junction, and the junction is said to be biased in the forward direction or low-resistance direction. If the battery is connected across the junction so that its voltage aids the barrier potential across the junction, it will oppose current flow through the junction, and the junction is said to be reverse-biased. Reverse-bias direction is the high-resistance direction or polarity connection.

# Forward Bias

Forward bias is illustrated in Figure 6-11. The positive terminal of the bias battery is connected to the negative side of the barrier potential or P side of the junction, and the negative terminal of the battery is connected to the positive side of the barrier potential or N side of the junction.

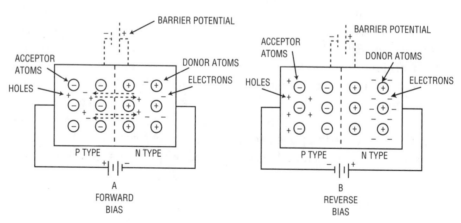

**Figure 6-11.** Forward and reverse biasing of a junction diode.

The positive terminal of the battery is connected to the end of the P side silicon and repels holes toward the junction and attracts electrons from the negative ions near it. The combination of holes moving toward the junction to neutralize charged negative ions on the P side of the junction and electrons taken from the negatively charged acceptor atoms tends to neutralize the negative charge of the barrier potential on the P side of the junction.

On the N side of the crystal the negative terminal of the battery repels electrons toward the junction. These electrons tend to neutralize the positive charge on the donor atoms at the N side of the junction. At the same time, the negative terminal of the battery attached to the N side of the crystal attracts holes away from the charged positive ions (donor atoms) on the N side of the junction. Both of these actions tend to neutralize the positive charge on the donor atoms at the junction, thereby reducing the barrier potential.

The effect of the battery bias voltage in the forward direction is to reduce the barrier potential across the junction and to allow majority carriers to cross the junction. Thus, more electrons flow from the N material across the junction. At the same time, more holes travel from the P material across the junction where they combine with the electrons from the N material.

At the same time that the hole and the electron movement is going on in the crystal, electrons are moving from the negative terminal of the bias battery in the external circuit to the N terminal, and electrons are moving from the P material terminal in the external circuit to the positive terminal of the battery.

It is important to remember that in the forward-biased junction condition, conduction is by the majority carriers (holes in the P material and electrons in the N material). Increasing the battery voltage will increase the number of majority carriers arriving at the junction, and the current flow increases. The only limit to current flow is the resistance of the material on the two sides of the junction. If the battery voltage is increased to the point where the barrier potential across the junction is completely neutralized, heavy current will flow, and the junction may be damaged from the resulting heat. Therefore, the voltage of the bias battery is limited to a relatively small voltage.

## Reverse Bias

Examine Figure 6-11B closely. With reverse bias applied to the junction diode, the negative terminal of the battery is connected to the P material, and the positive terminal connected to the N material. The negative terminal attracts holes away from the junction and depletes the holes in the P material. At the same time, the positive terminal of the battery attracts electrons away from the junction and increases the shortage of electrons on the N side of the junction. This action increases the barrier potential across the junction because there are fewer holes on the P side of the junction to neutralize the negative ions, and fewer electrons on the N side of the junction to neutralize the positive ions formed on this side of the junction. The increase in barrier potential helps to prevent current flow across the junction by majority carriers.

The current flow across the barrier is not zero, however, because of minority carriers crossing the junction. Holes forming in the N side of the depletion layer are attracted by the negative potential applied to the end of the P section. Electrons breaking loose from their outer shells in the atoms of the P material are attracted by the positive voltage applied to the end of the N section of the silicon.

Reverse bias applied across a junction diode increases the barrier potential. This makes it more difficult for the majority carriers to cross the junction. However, some minority carriers will still cross the junction with the result that there will be a small current. This inverse current is what causes heat generation in a diode. When this inverse current is reduced significantly, as is being done with improvements in the manufacture of the materials, the size of the diodes can also be reduced. The heat generated by the inverse current does not have to be dissipated if it is not generated in the first place.

If the inverse or reverse bias is increased beyond a critical value, the reverse current increases rapidly due to avalanche breakdown.

Avalanche breakdown occurs when the applied voltage is sufficiently large enough to cause the covalent bond structure to break down. At this point a sharp rise in reverse current occurs. The acceleration of the few holes and electrons continues to such a point that they have violent collisions with the valence electrons of the silicon crystal atoms releasing more and more carriers. The maximum reverse voltage of the semiconductor diode corresponds to the peak inverse voltage of an electron tube diode. Some diodes are designed to break down. These are called *zener* diodes and are used for voltage regulation circuits. Zener diodes will be discussed later in the chapter.

# The Point Contact Diode

There are a number of diodes. They are designed for special applications in some cases. The point contact diode is a very small, physically speaking, unit that is used for rectifying signals. The junction diode is used for rectifying power line frequencies and higher currents (see Figure 6-12).

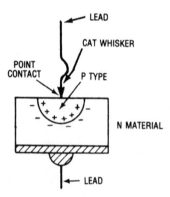

**Figure 6-12.** A point contact diode.

**Figure 6-13.** Symbol for a diode. Note the (+) end is the cathode.

The schematic symbol for the diode is the same (see Figure 6-13). Unlike the junction diode, the point contact type depends on the pressure or contact between a point and a semiconductor crystal for its operation.

One section consists of a small rectangular crystal of N material (either germanium or silicon) and a fine beryllium-copper, phosphor-bronze, or tungsten wire called the *cat whisker* (see Figure 6-12). The cat whisker presses against the semiconductor material and forms the other part of the diode. The reason for using a fine-pointed wire instead of a flat metal plate is to produce a high-intensity electric field at the point of contact without using too large an external voltage source. The opposite end of the cat whisker is used as the diode terminal for connection purposes.

Both contacts with the external circuit are low-resistance connections. During the manufacturing process of the point contact diode, a relatively large current is passed through the cat whisker to the silicon crystal. The result of this large current is the formation of a small region of P material around the crystal in the vicinity of the point contact. Thus, there is a PN junction formed which behaves in the same way as the PN junction described in the junction diode operation.

This very small contact area has a reduced capacitance effect (over the junction type with two pieces of material actually touching along a wide surface) that can be used for rectifying higher frequencies than the junction diode. However, since the size of the cat whisker is limited, the amount of current the diode can handle is also limited.

# Zener Diodes

Reversing voltage can destroy semiconductor diodes under certain conditions. In other situations, the effect of a reverse-biased voltage can make diodes useful. Zener diodes a re devices built around the effects of reverse voltage bias.

There are several important conditions and functions that you should understand in connection with zener diodes. These include

- Avalanche breakdown
- Zener effect
- Voltage regulation
- Switching

## Avalanche Breakdown

This is a condition in which the junction of the diode breaks down. As a result of the breakdown of the barrier put up at the junction, the electrons can then move easily across the junction. This increase in current through the diode is beyond what it can handle, and the diode is damaged beyond repair or recovery in most instances. Figure 6-14 shows the curve for average current voltage of a PN junction of a germanium diode. The curve has been exaggerated to show the biasing. As the electrons streaming across the junction collide with the electrons in the N region of the crystal, the collisions release valence electrons, and the high current through the diode becomes a reality.

Figure 6-15 shows the barrier region of a PN junction of a diode. The polarity is also reversed to make the diode reverse-biased. With a sufficiently high voltage across the diode junction, it breaks down into what is referred to as an *avalanche*. This condition is very dangerous for the circuit since the diode will serve as a short-circuited device and allow excess current to flow in other parts of the circuit causing overheating and damage. The avalanche will occur when there is not enough space area surrounding the semiconductor material to actually dissipate the heat buildup resulting from the breakdown.

When a diode reaches the breakdown point, holes are quickly neutralized. Neutralization results as electrons from the negative bias terminal quickly fill the holes. This means that the opposition represented by the barrier region has broken down. The diode then acts as a path of low impedance. Large currents are allowed to flow in either direction.

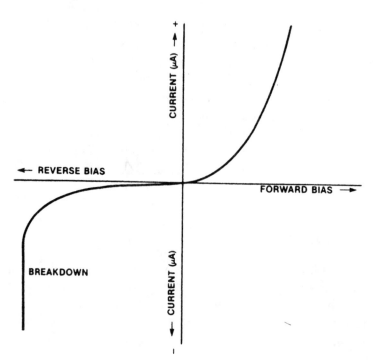

**Figure 6-14.** Curve showing average current voltage of PN junction of a germanium diode.

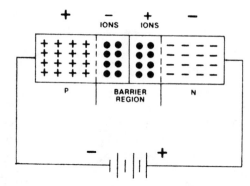

**Figure 6-15.** Schematic representation of a reverse-biased PN junction.

# Zener Effect

The zener effect is produced by a special type of diode. The zener diode is designed pur-
posely to break down. The breakdown takes place when reverse bias exceeds a diode's
rated value. At the point of breakdown, the reverse current increases rather rapidly. This
is demonstrated by Figure 6-16. A zener diode is not damaged during breakdown for it is
designed during manufacture to make sure it can operate at a certain point. Zener diodes
are available with ratings of 2 to 200 volts (V). Wattage ratings are 400 milliwatts (mW) to
50 watts (W). The wattage rating is important, because the physical size determines the
amount of heat dissipated. If the heat is not rapidly dissipated, the junction impedance is
lowered. If this happens, the current increases more rapidly than specified by the manufacturer of the device. The manufacturer will designate the operating temperature of the
zener.

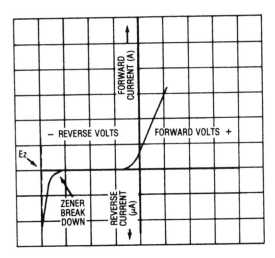

**Figure 6-16.** Curve for a silicon diode zener voltage.

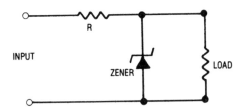

**Figure 6-17.** Zener diode circuit. Note the symbol for the zener diode.

Zeners are used for voltage regulation. A resistor is needed in series with the diode. This prevents the diode from reaching its breakdown point and then exceeding it because it does not have a limiting device in series with it. That is why the resistor is used in series with the diode — to take care of the excess current that would flow if it were not limited by the resistor's opposition in the circuit. Check Figure 6-17 and consider what would happen in a circuit of this type if the current demand is increased by the load. Note that the symbol for the zener diode is slightly different from the standard diode symbol. As the current flows through the diode, it causes a voltage drop across the series resistor equal to the *excess* voltage. Once the voltage has decreased to below the zener's breakdown point, the zener returns to its nonoperating condition. Once another increase of voltage causes the zener effect, the same process is repeated. This way the zener can keep the voltage across it and the load the same or below or at its rated value.

Another use for the zener is in computers. Switching requires rapid changes at high speeds. For instance, some switching operations are in the 100-megahertz (MHz) range. This requires a device that can complete its switching function in 0.01 microseconds ($\mu$s). Within this time, the diode must return to its original condition. That makes it a very valuable device in high-speed switching circuits as well as in voltage regulation circuits.

# Tunnel Diodes

The diodes are somewhat different in their manufacture and their usefulness. The zener diode made use of the breakdown point, and the regular junction diode made use of the fact it did not break down under reverse biasing. However, we now have another entirely different diode. This new type resembles the other diodes in packaging, but performs differently.

Figure 6-18 shows that the bias on the tunnel diode is strictly forward in polarity. Keep in mind that a tunnel diode is made by doping the silicon with an unusually large amount of impurities. The doping leads to high concentrations of charge carriers. This leads to a reduction of the critical voltage for an avalanche effect. Actually, the avalanche breakdown point is reduced to below the zero voltage level. The breakdown point moves into the region of small forward-bias voltages.

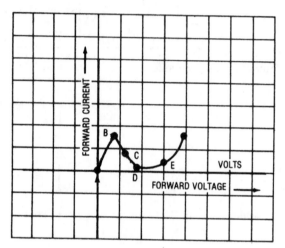

**Figure 6-18.** Characteristic curve for a tunnel diode.

Electrons are said to tunnel through the barrier almost instantly. This is partly because the barrier is extremely small. Its thickness is less than 0.000 001 in. (0.000 025 4 cm). Electrons move through the barrier at rates approaching the speed of light.

Check Figure 6-18. Current is reduced with increases in forward bias over a portion of the diode's operating range. This indicates a negative resistance characteristic. The same behavior is also observed in a tetrode vacuum tube where it has a large dip at certain voltages and currents.

Figure 6-18 shows the negative resistance also in the form of the down curve. This takes in the line from points B to D. A junction diode does not behave this way. This down curve has been produced here intentionally in the tunnel diode. This has an added advantage since the diode can be used as an amplifier, an oscillator, or a very fast switching device. From point E upward the tunnel diode behaves in the same way as a normal junction diode.

Tunnel diodes can be used in extremely small spaces such as part of an integrated circuit (IC) or chip. They can switch at very high rates (2 to 10 GHz). A gigahertz (GHz) is 1000 MHz. A megahertz is *1 million times per second*, and a gigahertz is *1000 times faster than that.*

Tunnel diodes are doped by using gallium arsenide, gallium antimonide, and indium antimonide. Doping changes the reaction of the diodes to produce the operating patterns that were visible in the curve in Figure 6-18.

When forward voltage is increased, the current flow rises to a maximum or peak current value. See point *B*.

A greater increase in forward bias voltage causes the current to decrease. The movement is from points *C* to *D*. This is the negative resistance feature of the tunnel diode.

After the current reaches a minimum at point *D*, it again starts to increase. This is the same behavior as a junction diode. When an increase in the forward-biased voltage is applied, the behavior is shown by point *E* upward.

The voltage range over the negative resistance portion is called the *voltage swing*. The ratio of peak current (point *B*) to the valley current (point *D*) is called the peak-to-valley ratio.

Remember the negative resistance portion of the curve. These changes take place with small voltage variations, in millivolts (mV). The peak-to-valley current ratio is usually around 10.

# The Silicon Controlled Rectifier

Another type of specialized rectifier or diode is the silicon controlled rectifier (SCR). It has another name, but the SCR was coined by General Electric (GE) and has persisted. It was originally the *thyrister*. Inasmuch as GE dominated the marketplace, it soon became known by GE's abbreviated version.

The SCR is a four-layer device. That is, it has either an NPNP or PNPN arrangement for the semiconductor materials. It is a specialized type of device used for the control of current through its cathode-to-anode path. A gate is used to control the resistance between the cathode and anode. By applying a small voltage between the gate and the cathode, it is possible to control that resistance and, as a result, the amount of current flow through the device. An SCR conducts current in the forward direction only. The symbol for the device is shown in Figure 6-19.

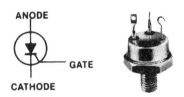

**Figure 6-19.** Symbol for SCR, or thyrister.

Figure 6-20 shows a circuit with an SCR. The function of an SCR is current control. Examples of this are the light dimmer or the speed control for a small hand drill or other handtool that is electrically powered. The resistor is a rheostat. This adjustable resistor is used to control the amount of voltage delivered to the gate of the SCR. The greater the voltage, the less the anode-to-cathode resistance and the more current is allowed to flow through the cathode-anode connection. By adjusting the rheostat, it is possible to control the amount of current flow through the device. As the current increases, the load — if a lamp — will get brighter. As the current increases, the load — if a drill or electric motor — speeds up. Thus, the SCR can be used to control either type of circuit. Other control

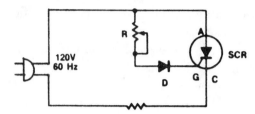

**Figure 6-20.** A circuit with the SCR as a control device.

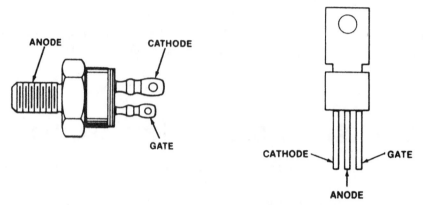

**Figure 6-21.** Two packages used for SCRs.

circuits also use the SCR for their main operating device. Figure 6-21 shows two of the design packages for SCRs.

# Transistors

The word *transistor* comes from two other words: **trans**fer and res**istor**. Thus, it is a transfer resistor or a device that has more impedance (resistance) in the input than in the output, or the other way around depending on its use. By having a difference in impedance between the input and output, it is able to amplify.

There are two ways in which transistors are used. One is switching, and the other is amplifying a signal. The switching ability of a semiconductor has previously been discussed under the diode section of this chapter. However, we will mention it briefly here in the study of transistors. Main emphasis will be on the ability of the transistor to amplify and thereby serve as a replacement for the vacuum tube.

Transistors are made from N and P materials, such as the semiconductor diode. Once they are joined, they resemble two diodes back to back (see Figure 6-22).

Transistors have an emitter, a base, and a collector. These are the connections to the N and P materials that make up the device. We will look at two types of transistors here: the point contact and the junction transistor.

The point contact was developed first. The junction transistor followed later. Transistors are classified as PNP or NPN according to the arrangement of the impurities in the crystal. Symbols used for transistors are shown in Figure 6-23.

The point contact transistor is similar to the point contact diode except that it has two cat whiskers instead of one. The two cat whiskers are placed with their point contacts very

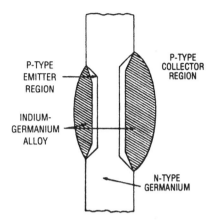

**Figure 6-22.** PNP transistor junction formation.

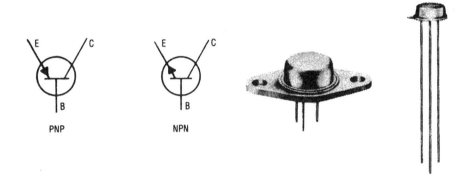

**Figure 6-23.** Transistor symbols.

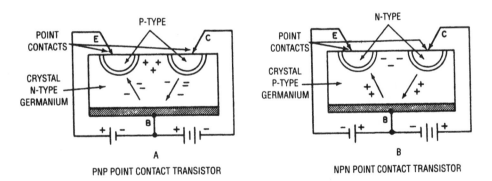

**Figure 6-24.** Point contact transistors.

close together (about 0.002 in.). The diameter of the contacts is about 0.005 in. The contacts are arranged to provide a springlike pressure on the flat surface of the crystal (see Figure 6-24). The crystal may be either N or P germanium. During the manufacturing process a large current is passed from the cat whiskers into the crystal. This action forms a small region around the point contacts of P material when the material is N germanium, or around N material when the crystal is P germanium. The same action occurs in the manufacturing process of the point contact diode previously described. Both point contact transistors consist of two PN junctions indicated as semicircles in the vicinity of the cat whiskers in Figure 6-24. The base (*B*) forms a common connection between the two junctions and the external circuit. One cat whisker is called the emitter (*E*), and the other is called the collector (*C*). Thus one PN junction is between the collector and the base. The point contact transistor is called either a PNP or an NPN according to which material was used for the common or base connection. The other junction in the transistor is between the base and emitter. This emitter-base junction in both transistors is forward-biased. The collector-base junction is reverse-biased.

The emitter-base junctions in both transistors are forward-biased; this means that the barrier potential is lowered, and the flow of majority carriers is increased across the junction. The emitter gets its name from the fact that it emits majority carriers across the junction. In this manner, it is similar to the cathode of the vacuum tube which emits electrons.

Check Figure 6-24 with the PNP transistor. The positive terminal of the left-hand battery is connected to the P material in contact with the emitter. The negative terminal of the battery is connected to the N material in contact with the base of the crystal. In accordance with the previously described action of forward bias across a PN junction, holes are emitted from the P material across the PN junction as electrons are emitted from the N material. Some of these holes will enter the N material as minority carriers. They have an important effect on the reverse-biased collector-base PN junction. The holes will attract electrons assisting them across the collector-base PN junction, increasing the collector current proportionately.

The movement of holes from the emitter region into the collector region is known as *hole injection*. As holes are injected into the collector region of the N material, they exert an attractive force on the electrons, assisting them across the PN barrier around the collector and increasing the collector current. The effect is to neutralize the barrier potential in proportion to the hole injection so that more electrons can flow across the barrier. The action is analogous to that of a space charge or cloud of positive ions in a conducting gas tube. The positive ions neutralize a portion of the negative space charge around the cathode, thereby increasing plate current. In the transistor, the holes injected from the emitter attract additional electrons across the collector PN barrier and increase the flow of collector current.

# Current Gain

In the circuit we have been describing, the total collector current is greater than the emitter current. The emitter is considered the input, and the collector the output. Thus, the action is the same as amplification since it represents a gain caused by the transfer action of the transistor. Current gain is the ratio of change in collector current resulting from a given change in emitter current for a constant collector-base voltage. The symbol for current gain is $\alpha$ce and is read "*alpha*-c-e." The Greek letter $\alpha$ represents the gain. The current gain for point contact transistors is in the order of 2 or 3 $\alpha$ce. Current gain is compared to the vacuum tube amplification factor or the small change in grid-cathode voltage can cause a larger change in cathode-to-plate voltage.

# Junction Transistor

Figure 6-25 shows the current flow in the external circuit of a PNP junction transistor. The junction transistor uses the same semiconductor materials as the point contact transistor

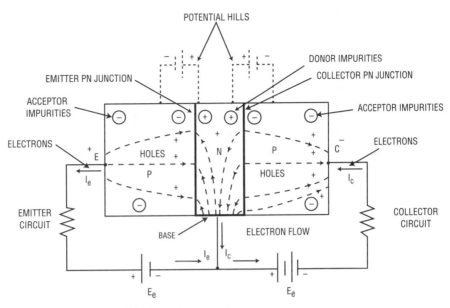

**Figure 6-25.** Current flow in a PNP junction transistor.

but is arranged in the form of a sandwich. The basic PN junction is used. However, such a junction cannot be made satisfactorily by simply bringing two surfaces together because of the difficulty in achieving the necessary smoothness and cleanliness of the surface to be joined. Instead, the junction is formed by a diffusion, or alloy, process.

Current flow in the external circuits consists of electron movement in a counterclockwise direction in both the emitter and collector circuits. Current flow within the P material of the transistor consists of the movement of holes through the P material from the positive terminal to the negative terminal. Current flow in the N material of the transistor consists of the movement of electrons from the collector PN junction to the base and from the base to the emitter PN junction.

Figure 6-26 consists of a transistor with NPN, or the opposite configuration. This means that the polarity of the power source is opposite to that of the PNP transistor. The silicon transistor is usually NPN. A few are made with the PNP, but in most instances the PNP is the germanium.

The paths for current flow in the NPN junction transistor are shown in Figure 6-26. The barrier potentials across the two junction planes are opposite to those of the PNP junction transistor. The dotted battery symbols represent these potential hills. The emitter circuit is biased in the forward direction and the collector circuit is biased in the reverse or high-resistance direction.

Electrons flow in the external circuits of the emitter and the collector in a clockwise direction, or the opposite of the PNP type. The emitter bias neutralizes the barrier potential across the emitter PN junction, allowing electrons to be injected from the emitter through the barriers into the collector P region. The majority carriers in the P material are holes.

As electrons move across the narrow P region of the NPN junction transistor, a certain amount of recombination of electrons with holes will occur. This action somewhat reduces the available current gain of the transistor by causing less increase in collector current for a given increase in emitter current.

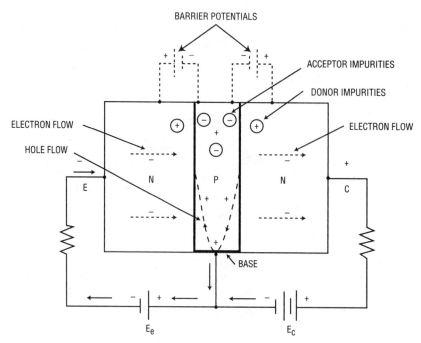

**Figure 6-26.** NPN junction transistor.

The voltage and power gains of the NPN junction transistor are of about the same magnitude as those of the PNP junction transistors. Like those of the grounded base PNP transistors previously described, these gains are principally because of the high resistance of the collector circuit compared to the low resistance in the emitter circuit.

# Other Types of Transistors

A wide variety of types of transistors have evolved during the short period of time that they have been available. In addition to the types just discussed you should be aware of three or four other main types in your study of electronics.

## Alloy Transistors

These transistors are made by alloying metal into opposite sides of a thin piece of semiconductor. The process produces emitter and collector regions (shown in Figure 6-27A).

To achieve uniformity in transistor characteristics, the thickness of the metal pellet must be controlled. Also critical is the quality of the metal. Further, the area of contact between metal and semiconductor and the alloying temperature must be closely controlled. Each of these variables affects the electrical characteristics of the transistor.

## Grown-Junction Transistors

Development of this type of transistor is shown in Figure 6-27B. The grown-junction transistor has one important difference from the alloy type. The junctions are created during the growth of the crystal rather than by alloying after the crystal is grown.

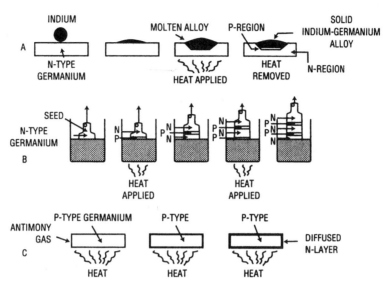

**Figure 6-27.** A. Alloy transistor. B. Rate grown crystal transistor. C. Junction formation by diffusion.

## Microalloy Transistors

These transistors are a variation of the alloy transistor. Figure 6-27C shows the formation of one type of microalloy transistor. The transistor being formed in the illustration is a microalloy-diffused type.

## Germanium Mesa and Silicon Planar Transistors

These transistors use a different construction technique (see Figure 6-28). Use of the diffusion masking materials and photolithographic techniques is shown. This produces a silicon planar structure. All the PN junctions are buried under a protective, passivating layer. The use of a separate collector-contact diffusion reduces the electrical series resistance in the collector.

In the germanium mesa and silicon planar transistors, the original semiconductor wafer serves as the collector. The base region is diffused into the wafer. The emitter *dot*

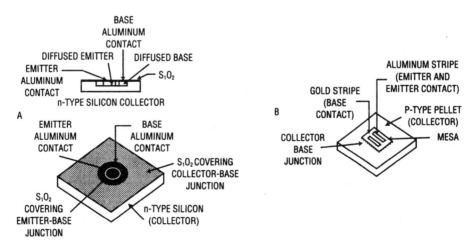

**Figure 6-28.** A. Germanium mesa transistor. B. Silicon planar transistor.

region is alloyed or diffused into the base region. A *mesa*, or flat peak, may then be etched. This reduces the collector areas at the base-collector junction, plus the construction technique is rugged. It has large power-dissipating capacity. It can also operate at very high frequencies. This type of construction is also employed in transistors made for use in integrated circuit configurations.

## Field-Effect Transistors

FETs, as they are called, are small in size and are mechanically rugged. They have low power consumption and high input impedance similar to a vacuum tube.

FETs are unipolar. They operate as a result of one type of charge carrier. These are holes in the P-channel carriers and electrons in the N-channel types. Other types are bipolar and require both holes and electron carriers.

The term MOS means *metal-oxide semiconductor*. The metal control gate is separated from the semiconductor channel (see Figure 6-29). An FET is not affected by the polarity of the bias on the control gate. Changes in temperature affect the FET. They are used in voltage amplifiers, RF amplifiers, and voltage-controlled attenuators. To attenuate means to *reduce*.

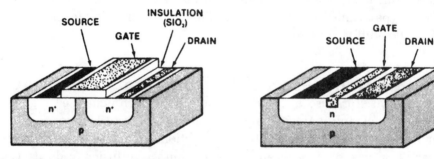

**Figure 6-29.** The PN junctions of FET transistors.

There are four distinct types of MOS transistors. These classifications are based on sources of conduction. The units make use of either electrons (N channel) or holes (P channel) for conduction. Symbols for the four types of MOSFET transistors are shown in Figure 6-30A. Direction of the arrowhead in the symbol indicates the difference between N- and P-channel types. A solid channel line in the symbol indicates *normally on*. A dotted line indicates *normally off*.

MOS field-effect transistors (MOSFETs) are used in broadcast band receivers in low-power circuits. Their use can be seen in RF amplifiers and converters. They are also used in IF stages and the first audio. In other words, they are used in almost all the circuits in a broadcast receiver except the output stage that drives the speaker.

Figure 6-30C consists of a circuit with a MOSFET transistor. Note the parts and how they are connected in the circuit. They can be used in FM receivers as well as AM receivers. The FM receiver uses FETs in the RF amplifier, conversion stages, IF stages, and limiter stages. They are as versatile as the vacuum tube and substitute almost directly in the circuits designed for tubes. However, the voltages are different, and there is no need for a filament in an FET.

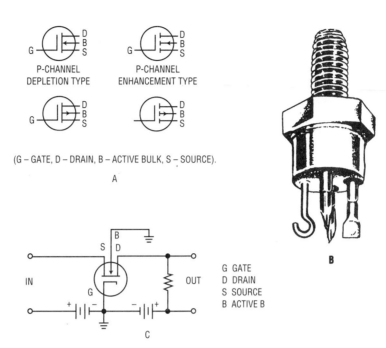

**Figure 6-30.** A. Symbols for FET transistors. B. MOS (metal-oxide semiconductor) transistor. C. An FET transistor in a circuit configuration.

# Integrated Circuits

The integrated circuit (IC) is a single, monolithic chip made of semiconductor material. It was first designed and developed in 1958 a short time after the transistor was introduced and made available for experimentation. J. S. Kilby was responsible for the fabrication of the first IC, or integrated circuit. He was able to make both active and passive circuit elements become part of the chip by using diffusion and deposition techniques. Shortly after his success, Robert Noyce made a complete circuit on a single chip. This led the way to the modern, inexpensive IC.

Transistors, diodes, resistors, and capacitors can now be deposited on a chip. Diodes can be made in many groups to do different things. Photolithography, a combination of photographic and printing techniques, has been used to make the complete circuit. This type of printing has made it possible to produce some very sophisticated ICs with very high reliability.

Figure 6-31 shows how an IC has the resistor, capacitor, diode, and transistor etched on its surface and turned into a complete circuit on a flat surface which can be easily packaged and protected. Figure 6-31A shows how a resistor is placed on the chip. The resistor is made in the form of a thin filament of conductive material. Contacts are made. It is insulated by a reverse-biased junction. It is possible to create a few ohms ($\Omega$) or up to 20,000 $\Omega$ on ICs.

Figure 6-31B shows how a low-loss capacitor is formed. This is done by depositing a thin insulating layer of silicon dioxide on a conducting region. It is then given a metallized layer to form the second plate. It is possible to reverse-bias a diode to form a capacitor with up to 50 picofarads (pF) of capacitance. Note how capacitors require more space on the chip than resistors. The transistors and diodes require less space. They are connected

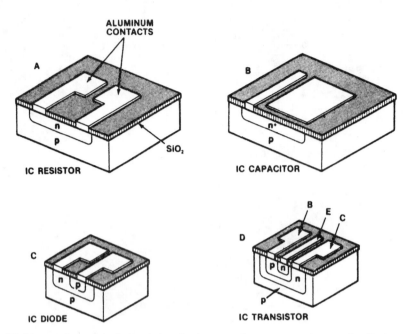

**Figure 6-31.** Making an integrated circuit by placing a resistor, a capacitor, and a diode as well as a transistor on a chip.

to terminals and then brought out to the outside where rigid packaging methods aid in their long life (see Figure 6-32 for packages of ICs).

## IC Configurations

There are three standard packages for integrated circuits (ICs). The multipin circular type is the same as a regular transistor package, but it has more leads than the transistor. A chip can hold more than a thousand transistors in a space that once held only one. The flat pack is a hermetically sealed package, which means it is vacuum-packed and heat-sealed (see Figure 6-32). The DIP, or dual in-line package, is also shown in Figure 6-32. It has many legs like a caterpillar. Each leg is connected to some spot inside that is critical to the functioning of the chip. In most instances, the cost of the packaging can well outstrip the

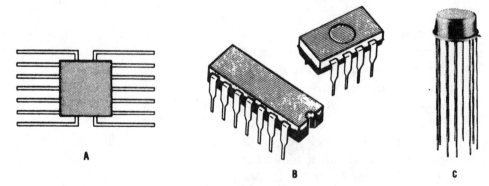

**Figure 6-32.** A. Flat-pack IC with 14-pin connection. B. DIP package with 8-pin and 14-pin configurations. C. TO-5 package with 8 leads.

actual cost of the chip itself. Modern manufacturing technology has improved the quality and reliability of the chip.

# Uses for ICs

ICs are now relatively standardized (see Figures 6-33 and 6-34). This is partly because mass production equipment is available to everyone who wants to buy it. If you want to keep the price low and meet competition, you have to standardize so that your product can be used by a variety of users.

ICs are used in great numbers in computers. They are also used in control circuits, television sets, FM receivers, and in almost every electronic device manufactured today where the latest technology is in demand. Toys and guided missiles use the same types of chips. They have a multitude of applications and the list is growing every day, limited only by the imagination of the engineer and designer.

# Summary

The term semiconductor applies when used to describe diodes and transistors as well as some other devices which use germanium or silicon as the basic material. Semiconductor technology is called solid state. Impurities are inserted into the pure silicon or germanium to control the amount of donors or acceptors each semiconductor will have and thus control its ability to handle current flow.

Impurities used to make semiconductors of silicon and germanium may be arsenic, antimony, bismuth, phosphorus, indium, and aluminum. N semiconductors have an extra electron in each controlled amount of material. The P semiconductors have one electron missing in every four covalent bonded atoms. Pure germanium and silicon are insulators. They have holes or electrons introduced in their four-atom groups in order to control the flow of electrons through the material and thus current flow.

Conduction in a P material is by way of the positive carriers, or holes. Conduction in an N material is by way of the negative carriers, or electrons. A covalent bond is formed when four neighboring atoms share an electron in the outer shell, or orbit. Both holes and electrons are involved in the conduction process in the PN junction diode.

When an electron leaves the donor atom in the N region and moves over to the P region in a junction diode, the atom has fewer electrons than it needs to neutralize the positive charge on the nucleus, and it becomes charged. It has one extra positive charge equal to the negative charge of the electron which it lost. This is how the PN junction diode operates.

Forward bias means that the diode is connected to the proper polarity to allow it to conduct current. Reverse bias means the diode puts up a high resistance to the flow of electrons and holes and thereby prevents current flow when the polarity is reversed. This is one of the ways a diode operates. A diode has a high resistance in one direction and a low resistance in the opposite direction.

The point contact diode has a small piece of N or P material with a cat whisker making contact with it. The junction is formed when a large surge of current is passed through the point where the cat whisker makes contact with the semiconductor material. Point contact

## SYMBOLS, TERMS, AND DEFINITIONS

| NPN BIPOLAR TRANSISTOR | P-CHANNEL ENHANCEMENT MODE MOS FET DUAL GATE | SILICON UNILATERAL SWITCH (SUS) | OPTOISOLATOR WITH PHOTO-DARLINGTON OUTPUT, NO BASE |
| PNP BIPOLAR TRANSISTOR | SILICON N-TYPE UNIJUNCTION TRANSISTOR (UJT) | SILICON ASYMMETRICAL SWITCH (SAS) | OPTOISOLATOR WITH PHOTO-DARLINGTON OUTPUT, AND BASE |
| NPN DARLINGTON TRANSISTOR | SILICON P-TYPE UNIJUNCTION TRANSISTOR (UJT) | PIN DIODE | OPTOISOLATOR TRIAC DRIVER |
| PNP DARLINGTON TRANSISTOR | PROGRAMMABLE UNIJUNCTION TRANSISTOR (PUT) | VARACTOR | OPTOISOLATOR SCR DRIVER |
| N-CHANNEL J FET | RECTIFIER OR DIODE | VARISTOR | AND GATE |
| P-CHANNEL J FET | FULL-WAVE BRIDGE RECTIFIER | METAL OXIDE VARISTOR (MOV) | OR GATE |
| N-CHANNEL ENHANCEMENT MODE MOS FET SINGLE GATE | ZENER DIODE | LIGHT EMITTING DIODE (LED) | NAND GATE |
| P-CHANNEL DEPLETION MODE MOS FET SINGLE GATE | SCR (THYRISTOR) | PHOTODIODE | NOR GATE |
| N-CHANNEL ENHANCEMENT MODE MOS FET SINGLE GATE | TRIAC (THYRISTOR) | NPN PHOTOTRANSISTOR, NO BASE CONNECTION | INVERTING AMP |
| P-CHANNEL ENHANCEMENT MODE MOS FET SINGLE GATE | DIAC (BILATERAL TRIGGER DIODE) | NPN PHOTOTRANSISTOR, WITH BASE CONNECTION | NON-INVERTING AMP |
| N-CHANNEL DEPLETION MODE MOS FET DUAL GATE | SILICON CONTROLLED SWITCH (SCS) (THYRISTOR) | OPTOISOLATOR WITH PHOTOTRANSISTOR OUTPUT | EXCLUSIVE OR GATE |
| P-CHANNEL DEPLETION MODE MOS FET DUAL GATE | SILICON CONTROLLED SWITCH (TRANSISTOR) (SCS) | OPTOISOLATOR WITH PHOTOTRANSISTOR OUTPUT, NO BASE CONNECTION | |
| N-CHANNEL ENHANCEMENT MODE MOS FET DUAL GATE | SILICON BILATERAL SWITCH (SBS) | OPTOISOLATOR WITH PHOTOTRANSISTOR OUTPUT, AND BASE CONNECTION | |

**Figure 6-33.**

## Diodes, Rectifiers, Thyristors

$C_t$ — Total Capacitance - The Total Small-Signal Capacitance Between The Diode Terminals.

$d_i/d_t$ — Rate Of Change Of Current Versus Time.

$d_v/d_t$ — Rate Of Change Of Voltage Versus Time.

$I_F$ — Forward Junction Current - The Value Of DC Current That Flows Through A Semiconductor Diode Or Rectifier Diode In The Forward Direction.

$I_{FRM}$ — Peak Forward Current Repetitive Peak - The Peak Value Of The Forward Current Including All Repetitive Transient Currents.

$I_{FSM}$ — Forward Surge Peak DC Current - Maximum (Peak) Surge Forward Current Having A Specified Waveform And A Short Specified Time Interval.

$I_{GT Min}$ — Gate Trigger Current - Minimum Gate DC Current Required To Trigger The Device Under The Conditions Specified.

$I_{GO Max}$ — Peak Gate Turn-Off Current - Maximum Negative Gate Current Required To Switch Off.

$I_H$ — Holding Current - Anode Current Necessary To Maintain On-State.

$I_O$ — Average Rectifier DC Forward Current - The Value Of The Forward Current Averaged Over A 180° Conduction Angle At 60 Hz.

$I_R$ — DC Reverse Current - Value Of DC Current That Flows Through The Diode In The Reverse Direction. (Leakage Current.)

$I_{RM}$ — Maximum Reverse DC Current - The Respective Value Of Current That Flows Through The Junction In The Reverse Direction.

$I_{trms}$ — Continuous On-State Current.

$I_{tsm}$ — Surge (Non-Repetitive) Peak On-State Current - A Surge Current Of Short-Time Duration.

$L_S$ — Series Inductance - The Inductance Between The Terminals On The Diode.

PRV — Peak Reverse Voltage - Maximum Repetitive Peak Reverse Blocking Voltage That May Be Applied To The Anode-Cathode Of The Device.

$R_S$ — Series Resistance - The Total Small Signal Resistance Between The Diode Terminals.

$r_A$ — Ambient Temperature - The Air Temperature Measured Below A Device, In An Environment Of Substantially Uniform Temperature, Cooled Only By Natural Air Convection And Not Materially Affected By Reflective And Radiant Surfaces.

$T_C$ — Case Temperature - The Temperature Measured At A Specified Location On The Case Of A Device.

$T_j$ — Semiconductor Junction Temperature.

$T_Q$ — Turn Off Time.

$t_{rr}$ — Reverse Recovery Time - The Time Required For The Current Or Voltage To Recover To A Specified Value After Instantaneous Switching From A Stated Forward Current Condition To A Stated Reverse Voltage Or Current Condition.

$V_B$ — DC Breakdown Voltage - Value Of Voltage Measured At The Point Which Breakdown Occurs With The Diode Reverse Biased.

$V_{(BR)R}$ — Static Reverse Breakdown Voltage - The Value Of Negative Anode-To-Cathode Voltage At Which The Differential Resistance Breakdown Between The Anode And Cathode Terminals Changes From A High Value To A Substantially Lower Value.

$V_{DRM}$ — Repetitive Peak Off-State Voltage - Maximum Instantaneous Value Of The Off-State Voltage That Occurs Across The Devices, Including All Repetitive Transient Voltages, But Excluding All Non-Repetitive Transient Voltages.

$V_F$ — Forward Voltage - The Voltage Drop In A Semiconductor Diode Resulting From The Respective Forward Current.

$V_{FM}$ — Maximum Forward Voltage - The Voltage Drop In A Semiconductor Diode Resulting From The Respective Forward Current.

$V_{GFM}$ — Maximum Forward Gate Voltage - Maximum DC Forward Gate Voltage Permitted To Produce A Specified Forward Gate Current.

$V_{GO Max}$ — Peak Gate Turn-Off Voltage - Maximum Reverse Gate Voltage Required To Switch Off.

$V_{GRM}$ — Maximum Reverse Gate Voltage - Maximum Peak Reverse Voltage Allowable Between The Gate Terminal And The Cathode Terminal When The Junction Between The Gate Region And The Adjacent Cathode Region Is Reverse Biased.

$V_Z$ — Zener Regulator Reference Voltage - Value Of DC Voltage Across The Diode When It Is Biased To Operate In Its Breakdown Region.

$\Delta V_Z/\Delta T$ — Change In Zener Voltage To Change In Temperature.

## Transistors

$BV_{CBO}$ — Collector To Base Breakdown Voltage - Voltage Measured Between Collector And Base With Emitter Open.

$BV_{CEO}$ — Collector To Emitter Breakdown Voltage - Voltage Measured Between Collector And Emitter With Base Open.

$BV_{CER}$ — Collector To Emitter Breakdown Voltage - Voltage Measured Between Collector And Emitter When The Base Terminal Is Returned To The Emitter Terminal Through A Specified Resistance.

$BV_{CES}$ — Collector To Emitter Breakdown Voltage - Voltage Measured Between Collector And Emitter With The Base Terminated Through A Short Circuit To The Emitter.

$BV_{CEV}$ — Collector To Emitter Breakdown Voltage - Voltage Measured Between Collector And Emitter When A Specified Voltage (V) Is Applied Between The Base And Emitter.

## Transistors (cont'd)

$BV_{CEX}$ — Collector To Emitter Breakdown Voltage - Voltage Measured Between Collector And Emitter When The Base Is Terminated Through A Specified Load (X) To The Emitter.

$BV_{DSS}$ — Drain To Source Breakdown Voltage - Voltage Measured Between The Drain And Source Terminals With The Gate Short-Circuited To The Source Terminal.

$BV_{EBO}$ — Emitter To Base Breakdown Voltage - Reverse Voltage Measured Between Emitter And Base With The Collector Terminal Open.

$BV_{GSS}$ — Gate To Source Breakdown Voltage - The Breakdown Voltage Between The Gate And Source Terminals With The Drain Terminal Short-Circuited To The Source Terminal.

$C_{ISS}$ — Input Capacitance - The Capacitance Between The Terminals (Gate And Source) With The Drain Short-Circuited To The Source.

$C_{RSS}$ — Reverse Transfer Capacitance - The Capacitance Between The Drain And Gate Terminals.

$f_T$ — Gain Bandwidth Product - Frequency At Which Small-Signal Gain Becomes Unity.

$g_{fs}$ — Forward Transfer Conductance - Common Source Forward Transconductance.

$h_{FE}$ — DC Current Gain - The Ratio Of Collector Current To Base Current At A Specified Collector-Emitter Voltage.

$I_B$ — DC Base Current - Value OF DC Current Into The Base Terminal.

$I_C$ — DC Collector Current - Value OF DC Current Into The Collector Terminal.

$I_{DSS}$ — Zero Bias Drain Current - Amount Of Current Which Flows In The Drain When The Gate Is Connected To The Source.

$P_D$ — Average Power Dissipation.

$P_{IN}$ — Signal Input Power To Device.

$P_{OUT}$ — Signal Output Power.

$r_{DSS}$ — Drain-Source On-State Resistance.

$V_{CC}$ — DC Supply Voltage Applied To The Collector Terminal.

## Special Purpose Devices

$BV_{CER}$ — Breakdown Voltage Between Collector And Emitter With A Specified Resistor Between Base And Emitter.

$BV_{GKF}$ — Gate To Cathode Forward Breakdown Voltage.

$BV_{GKR}$ — Gate To Cathode Reverse Breakdown Voltage.

$h_{FE}$ — DC Current Gain - Ratio Of DC Output Current To The DC Input Current.

$I_{BO + (Max)}$ — Maximum Forward Breakover Current.

$I_{BO - (Max)}$ — Maximum Reverse Breakover Current.

$I_E$ — Value Of The DC Current Into The Emitter.

$I_{EO}$ — Emitter Current With One Base Open.

$I_G$ — DC Gate Current - The DC Current Flowing Through The Gate As A Result Of Applied Gate Voltage.

$I_{T PK}$ — Total Peak Current.

$I_V$ — Valley Current - The Valley Current Is The Emitter Current At The Second Lowest Current Point.

$\eta$ — Intrinsic Stand-Off Ratio.

$R_{BBO}$ — Base 1 To Base 2 Resistance With Open Emitter.

$V_{AK}$ — Anode To Cathode Voltage - The Maximum Value Of Voltage Applied Between Anode And Cathode Without Failure.

$V_{(BO) +}$ — Forward Breakover Voltage.

$V_{(BO) -}$ — Reverse Breakover Voltage.

$\Delta V_F$ — Forward Breakback Voltage.

$\Delta V_R$ — Reverse Breakback Voltage.

$V_{GT}$ — Gate Trigger Voltage - The Gate Voltage Required To Produce The Gate Trigger Current.

## Opto Electronic Devices

$I_D$ — Dark Current - The Current Which Flows In A Photodetector When There Is No Incident Radiation On The Detector.

$I_{FT}$ — Input Trigger Current - Emitter Current Necessary To Trigger The Coupled Device.

$I_L$ — Light Current - The Current That Flows Through A Photo Sensitive Device When It Is Exposed To Illumination.

$P_t$ — Total Device Power Dissipation.

Response Time — The Time It Takes The Device To React To An Incoming Signal.

Rise Time $(t_r)$ — The Time Duration During Which The Leading Edge Of A Pulse Is Increasing From 10% To 90% Of Its Maximum Amplitude.

$V_{ISO}$ — DC Isolation Surge Voltage - The Dielectric Withstanding Voltage Between The Input And Output.

$\lambda_P$ — Wavelength At Peak Emission - The Wavelength At Which The Power Output From A Light-Emitting Diode Is Maximum.

$\theta_{Hi}^-$ — Half-Intensity Beam Angle - The Angle Within Which The Radiant Intensity Is Not Less Than Half Of The Maximum Intensity.

**Figure 6-34.**

diodes handle higher frequencies than junction diodes, but they are limited in the amount of current they can handle as opposed to the higher amounts the junction diode can handle due to its larger-sized junction.

A zener diode is one which breaks down at a given voltage and then recovers when the higher-than-design voltage is removed. It is used for voltage regulation circuits.

Avalanche breakdown is a condition in which the junction of the diode breaks down. As a result of the breakdown of the barrier put up at the junction, electrons can move easily across the junction. Most diodes are destroyed by an avalanche. The zener is the exception. It is designed for operation at the breakdown point repeatedly.

In a tunnel diode the silicon is doped with an unusually large amount of impurities. The doping leads to high concentrations of charge carriers. This leads to a reduction of the critical voltage for an avalanche effect. The breakdown is in the region of small forward-biased voltages. Electrons are said to tunnel through the barrier almost instantly and at the speed of light. This type of diode is used in oscillators and also as very fast switching devices.

The SCR is a silicon controlled rectifier. It has a cathode, an anode, and a gate. The change in voltage between the cathode and gate determines the resistance between the cathode and anode. In a circuit this makes for easy control of the amount of current allowed to flow in a circuit with the anode-cathode as part of the circuit. (Another name for the SCR is thyrister.) SCRs are used in circuits such as light dimmers and speed controls for small-horsepower motors.

The word transistor is a combination of the words transfer and resistor. It is a device which can amplify or switch very fast. Transistors are made from N and P materials such as the semiconductor diode. Once joined, they resemble two diodes back to back. Transistors have three connections — the emitter, the base, and the collector. They are classified as PNP or NPN. NPN or PNP determines the polarity applied to the emitter, base, and collector. The emitter-base junction is usually forward-biased, while the base-collector junction is reverse-biased to obtain a higher impedance in the output part of the circuit.

A number of transistor types have evolved since the first one was introduced in 1948. These types are alloy, grown-junction, microalloy, germanium mesa, and silicon planar transistors as well as the field-effect transistor (FET).

Integrated circuits (ICs) are made of a single chip of semiconductor material. They were first developed in 1958 and have since been the subject of much improvement. Transistors, diodes, capacitors, and resistors can now be deposited on a chip or made into an integrated circuit. They are available in three standardized packages or configurations: the dual in-line package (called the DIP), the multipin circular type resembling the older transistors, and the flat pack.

Integrated circuits are used in almost every type of electronics equipment today. They are used for everything from simple clocks and watches to complicated computers and sophisticated guidance systems for missiles.

# Review Questions

1. How did the discovery with silver sulfide affect the transistor discovery?
2. Why is semiconductor technology called solid state?
3. What is a diode?
4. How are junction diodes made?
5. What are the two types of material used for semiconductors?
6. What is an impurity?
7. What is a donor?
8. What is an acceptor?
9. What is meant by carrier?

10. How does the N and P material differ?
11. What is a trivalent impurity element?
12. What is a pentavalent impurity element?
13. What makes current flow through P material?
14. What makes current flow through N material?
15. How is a semiconductor diode made?
16. What is a barrier potential?
17. What is meant by forward bias?
18. What is meant by reverse bias?
19. How is a point contact diode made?
20. What is a zener diode?
21. What is a tunnel diode?
22. Explain avalanche breakdown.
23. What happens to a diode at avalanche breakdown?
24. What is a valence electron?
25. What does covalence bond mean?
26. What is another name for an SCR?
27. What the function of the gate in an SCR?
28. How did the word transistor evolve?
29. Describe the difference between the point contact and the junction diode.
30. What is the symbol for current gain in a transistor?
31. List at least four types of transistors.
32. What is an FET transistor?
33. Where are FET transistors used?
34. What is an integrated circuit (IC)?
35. How are ICs used?

# Chapter 7

# ELECTRICAL POWER

**E**lectrical power generated in the United States and Canada that is intended for commercial and industrial applications is produced as three-phase (3ϕ) (see Figure 7-1). Power generated for home use is distributed as three-phase, but only, as a rule, one-phase is used to service a given home (or homes). The other phases are used for other homes or groups of homes, so as to balance the total load (see Figure 7-2).

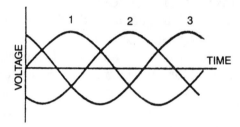

**Figure 7-1.** Three-phase waveform.

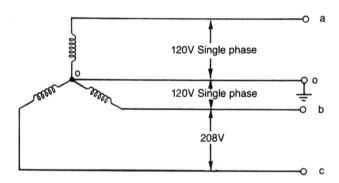

**Figure 7-2.** 120-volts from a three-phase transformer secondary.

# Three-Phase Power

Delta and wye circuits are commonplace in three-phase current sources, such as most commercially generated electrical power. Delta and wye connections are used in connecting transformers and in connecting the loads to these power sources. A three-phase delta resembles the Greek letter delta (Δ or a triangle) in shape, and the three-phase wye is shaped like the letter Y (see Figure 7-3).

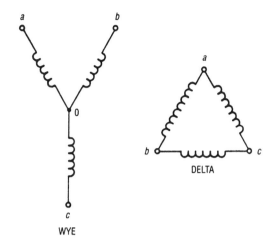

**Figure 7-3.** Delta and wye connections.

# Wye Connections

As can be seen, the wye connection in Figure 7-3 has terminals labeled a, b, and c, and a common point called the neutral, shown at point 0. The terminal pairs, a–b, b–c, and c–a, provide the three-phase supply. In this connection the line voltage is $\sqrt{3}$ (or 1.732 050 808) times the coil voltage, while the line current is the same as the coil current. The neutral point normally is grounded. It can be brought out to the power-consuming device by way of a four-wire power system for a dual-voltage supply. This means it can produce a 120/208-volt system. If three wires are used and connected to points a, b, and c, then three-phase ac is available for use with motors and other loads. If you are using a 208-volt system you know that you have a wye-connected transformer supplying the power.

Single-phase ac is available if you connect from 0 to any one of the points a, b, or c. From 0 to a will produce 120 volts of single-phase ac. From 0 to b will produce 120 volts of single-phase ac. From 0 to c will also produce 120 volts of single-phase ac. Example: $120 \times 1.732\ 050\ 808 = 207.846\ 097$, or the common 208-line voltage.

# Delta Connections

The delta connection has an advantage in current production because two of the coils are in series with each other and are in parallel with the third, and parallel arrangements produce better current availability. The voltage available from any two terminals, a to b or b to c or c to a, is the same. See Figure 7-3 for the delta connection. This voltage is also single-phase.

However, if you wish 240 volts of three-phase ac, all you need to do is bring out three wires — one each from point a, point b, and point c. The current available in a three-phase delta connection is $\sqrt{3}$ times the current capability of any one coil.

There are various methods used to connect three-phase power transformers. They may be delta-to-delta, delta-to-wye, wye-to-wye, or wye-to-delta (see Figure 7-4).

The delta-to-wye connection permits cheaper construction of the transformer in high voltage installations. A wye-to-delta connection, the opposite of the delta-to-wye connection, is often used for step-down voltages.

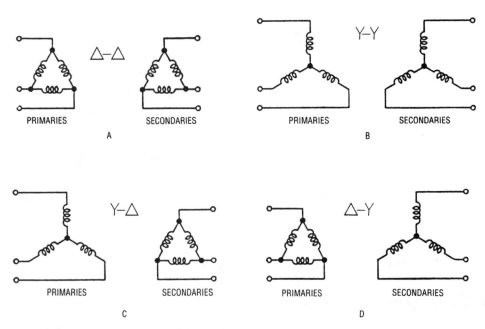

**Figure 7-4.** Different ways to connect delta and wye windings.

# Three-Phase Motors

Three phase motors have good overall characteristics. They are ideal for driving machines in industrial uses. They can be reversed while running. Reversing any two of the three connections to the power line causes the motor to be reversed (see Figure 7-5).

The motor speed under normal load conditions is rarely more than 10 percent below

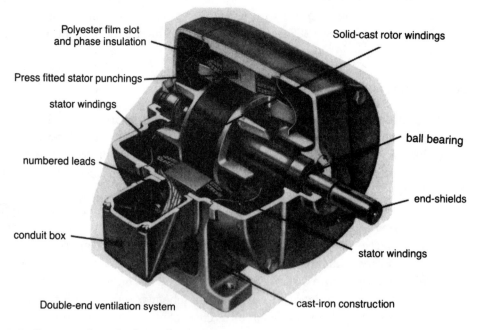

**Figure 7-5.** Cutaway view of a three-phase motor.

synchronous speed. The three-phase motor has high starting and breakdown torque with smooth pull-up torque. It is very efficient to operate. It is available in 208–230/460-volt sizes, with horsepower ratings varying from 1/4 to the hundreds. It can be obtained for operation on 50 or 60 Hz. The three-phase motor is the workhorse of industry.

# Single-Phase Power

Alternating current is constantly changing. These changes are in magnitude and direction. This ac pattern generated by an ac generator is called a waveform. When drawn out or viewed on an oscilloscope it resembles Figure 7-6. This is a single-phase waveform.

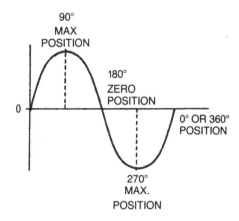

**Figure 7-6.** Sine wave.

In the United States and Canada there are 60 of these waveforms generated per second. In Europe the frequency is 50 Hertz. Motors and transformers have to be designed to operate on either 50 or 60 Hertz according to which country they will be used in.

Electric motors have to be specially designed to operate on single-phase power. There are three basic types of AC motors: universal, induction, and synchronous.

# Universal AC Motor

This is a small dc series motor that can operate on an ac source (see Figure 7-7). There is a difference in function under ac operation. The field coil current and the armature current reverse direction at the same time. Because of this, torque maintains the same direction throughout the cycle. Special field windings and laminated pole pieces are used to reduce heat losses. These special fittings are used when motors are designed for use with both dc and ac. A variable speed control may be provided with small motors. This type of unit utilizes the SCR or silicon controlled rectifier that produces dc. It is discussed in another chapter of this book.

Most hand-held electric drills are variable speed. The speed control unit is nothing more than a variable resistor being controlled as the switch is pulled back toward the handle (see Figure 7-8). The gear operated resistor changes the electronic circuit so that

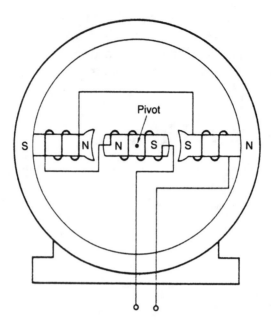

**Figure 7-7.** A universal motor.

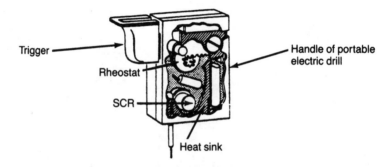

**Figure 7-8.** Variable speed control for a series or universal motor.

the SCR operates properly and increases the speed as you pull harder on the trigger. The SCR controls the amount of current allowed through the coils. A heat sink is used to dissipate the heat for cooler operation of the SCR. Keep in mind that the SCR rectifies the ac and produces dc. This means that the drill now operates on dc. In most instances this also means a greater efficiency since this type of motor is more efficient on dc.

Examples of the universal motor are food mixers, sewing machine motors, and vacuum cleaner motors. Small power hand drills, portable power saws, and other portable power tools use the universal motor.

## Induction Motor

This is the most often used type of ac motor. Induction motors are relatively simple to build. They are rugged, and they operate at constant speed if they are not overloaded. This type of motor is built differently than the universal type. The universal type motor is really a dc series motor adapted to operate on ac.

The field of a single-phase motor, instead of rotating, merely pulsates, and no rotation of the rotor takes place. A single-phase pulsating field may be visualized as two rotating fields revolving at the same speed, but in opposite directions. It follows, therefore, that the rotor will revolve in either direction at nearly synchronous speed, provided it is given an initial impetus in either one direction or the other. The exact value of this initial rotational velocity varies widely with different machines, but a velocity higher than 15% of the synchronous speed is usually sufficient to cause the rotor to accelerate to rated, or running, speed. A single-phase motor can be made self-starting if means can be provided to give the effect of a rotating field.

## Shaded-Pole Motor

This is one type of the induction motor. The shaded-pole motor resulted from one of the first efforts to make a single-phase motor that would start without outside help (see Figure 7-9).

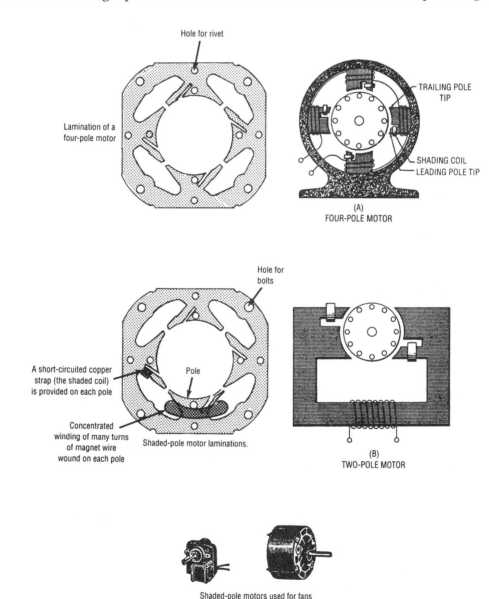

**Figure 7-9.** Shaded-pole motor arrangements.

This motor has salient poles. A portion of each pole is encircled by a heavy copper ring. The presence of the ring causes the magnetic field through the ringed portion of the pole face to lag behind that through the other portion of the pole face. The effect is the production of a slight component of rotation of the field that is sufficient to cause the rotor to revolve (see Figure 7-10).

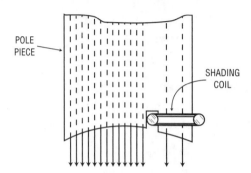

**Figure 7-10.** Action of a shading coil in a shaded-pole motor.

As the rotor accelerates, the torque increases until the rated speed is obtained. Such motors have low starting torque. Their greatest use is in small fans where the initial torque is low. They are also used in clocks, electric typewriters, and electric pencil sharpeners.

## Split-Phase Motors

Many types of split-phase motors have been made. Such motors have a start winding that is displaced 90 electrical degrees from the main or run winding. In some types, the start winding has a fairly high resistance that causes the current in it to be out-of-phase with the current in the run winding. This condition produces, in effect, a rotating field and the rotor revolves. A centrifugal switch is used to disconnect the start winding automatically after the rotor has attained approximately 75% of its rated speed (see Figure 7-11).

Split-phase motors are used where there is no need to start under load. They are used on grinders, buffers, and other similar devices. They are available in fractional-horsepower sizes with various speeds, and are wound to operate on 120-volts ac or 240-volts ac.

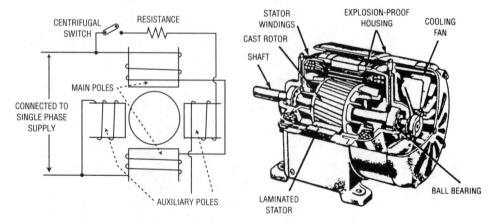

**Figure 7-11.** Cutaway view of a split-phase motor.

## Capacitor-Start Motors

A variation of the split-phase motor is known as the capacitor-start motor. It was made possible by the development of high-quality and high-capacity electrolytic capacitors (see Figure 7-12).

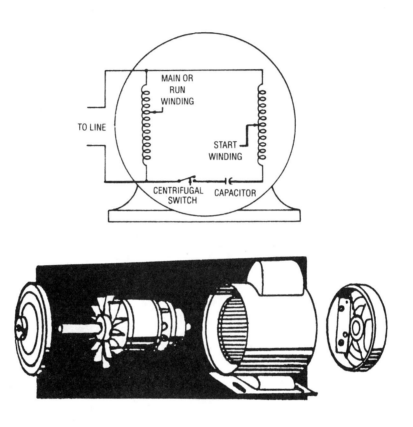

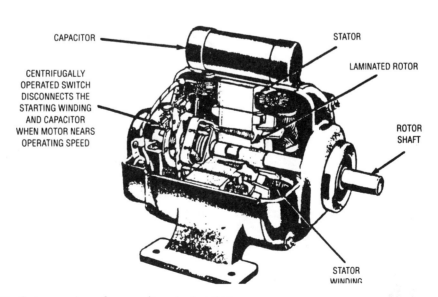

**Figure 7-12.** Cutaway view of a capacitor-start motor.

Almost all fractional-horsepower motors in use today on refrigerators, oil burners, washing machines, table saws, drill presses, and similar devices are capacitor-start. A capacitor motor has a high starting current and the ability to develop about four times its rated horsepower if it is suddenly overloaded. In this adaption of the split-phase motor, the start winding and run winding have the same size and resistance value; the phase shift between currents of the two windings is obtained by means of a capacitor connected in series with the start winding. Capacitor-start motors have a starting torque comparable to their torque at rated speed and can be used in places where the starting load is heavy. A centrifugal switch is required for disconnecting the start winding when the motor speed is up to about 25% of the rated speed.

One of the advantages of the single-value, capacitor-start motor is its ability to be reversed easily and frequently. The motor is quiet and smooth running. If a 5- to 20-horsepower capacitor-start motor is called for, the two-value capacitor is used (see Figure 7-13). It has two sets of field windings in the stator — an auxiliary winding called a phase winding and the main winding. The phase winding is designed for continuous operation. A capacitor remains in series with the winding at all times. A start capacitor is added to the phase circuit to increase starting torque, but is disconnected by a centrifugal switch during acceleration.

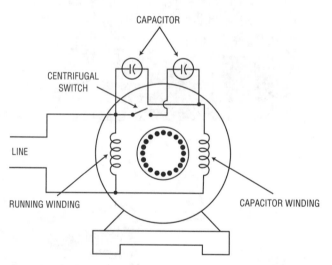

**Figure 7-13.** A two-value capacitor-start motor.

Three-phase motors are less expensive to purchase and maintain than single-phase motors. Single-phase motors are less efficient and their starting currents are relatively high. Both types run at essentially the same speeds. Nevertheless, most machines using electric motors around the home, on the farm, or in small commercial plants are equipped with single-phase motors.

Those who select single-phase motors do so because three-phase power is not available to them. The three-phase motor is rather simple in construction since it has no starting mechanism (see Figure 7-14).

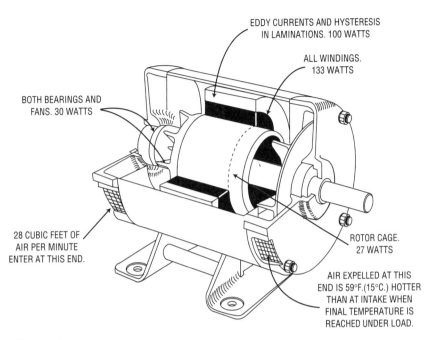

EDDY CURRENTS AND HYSTERESIS
IN LAMINATIONS. 100 WATTS

ALL WINDINGS.
133 WATTS

BOTH BEARINGS AND
FANS. 30 WATTS

28 CUBIC FEET OF
AIR PER MINUTE
ENTER AT THIS END.

ROTOR CAGE.
27 WATTS

AIR EXPELLED AT THIS
END IS 59°F.(15°C.) HOTTER
THAN AT INTAKE WHEN
FINAL TEMPERATURE IS
REACHED UNDER LOAD.

**Figure 7-14.** Three-phase motor.

# Synchronous Motors

The synchronous motor is another type of ac motor that finds special applications for its constant speed characteristic.

A synchronous motor is one of the principle types of ac motors. Like the induction motor, the synchronous motor is designed to take advantage of a rotating magnetic field. Unlike the induction motor, however, the torque developed does not depend upon the induction of current in the rotor.

An understanding of the operation of the synchronous motor may be obtained by considering the simple motor shown in Figure 7-15. Assume that poles A and B are being rotated clockwise by some mechanical means in order to produce a rotating magnetic field. The rotating poles induce poles of opposite polarity as shown in the illustration of the soft iron rotor, and forces of attraction exist between corresponding north and south poles. Consequently, as poles A and B rotate, the rotor is dragged along at the same speed. However, if a load is applied to the rotor shaft, the rotor axis will momentarily fall behind that of the rotating field, but will thereafter continue to rotate with the field at the same speed, as long as the load remains constant. If the load is too large, the rotor will pull out of synchronization with the rotating field and, as a result, will no longer rotate with the field at the same speed. The motor is then said to be overloaded. Figure 7-16 is similar to Figure 7-15, except that it is three-phase rather than single-phase. The magnitude of the induced poles in the rotor in Figure 7-15 is so small that sufficient torque cannot be developed for most practical loads. To avoid such a limitation on motor operation, a winding is placed on the rotor and this winding is energized with dc. A rheostat placed in series with the dc source provides the operator of the machine with a means of varying the strength of the rotor poles, thus placing the motor under control for varying loads.

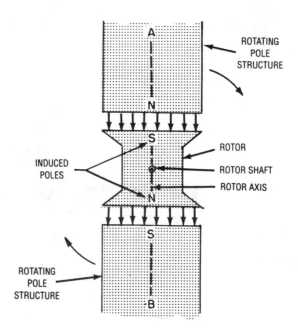

**Figure 7-15.** Synchronous motor magnetic fields.

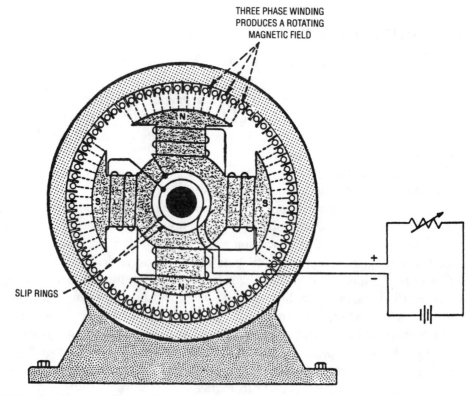

**Figure 7-16.** Synchronous motor with an excited rotor.

## Using the Synchronous Motor as a Capacitor

In large industrial installations where many motors and other inductive load devices are used, it is to the advantage of the user to bring the power factor of the power line servicing the plant back to unity. This lowers the electrical power costs of the plant. One way to do this is to use a large synchronous motor as a rotating capacitor that can be varied to cancel out the effect of the inductance of the many motors and fluorescent lights. Review Chapter 4 where power factor and resonance is discussed.

If a synchronous motor is driven by an external power source and the excitation or voltage applied to the rotor is adjusted to a certain value called 100% excitation, no current will flow from, or to, the stator winding. In this case, the voltage generated in the stator windings by the rotor, or cemf, exactly balances the applied voltage. If the excitation is reduced below the 100% level, however, the difference between the cemf and the applied voltage produces a reactive component of current that lags the applied voltage. The motor then acts as an inductance.

If the excitation is increased above the 100% level, the reactive component leads the applied voltage and the motor acts as a capacitor. This feature of the synchronous motor permits its use as a power factor correction device. When so used it is called a synchronous or rotating capacitor. Large steel mills use this device to correct the power factor and decrease their electrical power costs. Other industrial uses have been found for the synchronous motor since it has a stable speed. It synchronizes on the frequency instead of the voltage.

These are the advantages of the synchronous motor:

1. When used as a synchronous capacitor, the motor is connected on the ac line in parallel with the other motors on the line. It is run either without load or with a very light load. The rotor field is overexcited just enough to produce a leading current which offsets the lagging current of the line with motors operating. A unity power factor (1.00) can be achieved. This means the load on the generator is the same as though only resistance made up the load.
2. The synchronous motor can be made to produce as much as 80% leading power factor. However, because leading power factor on a line is just as detrimental as a lagging power factor, the synchronous motor is regulated to produce just enough leading current to compensate for lagging current in the line.

## Synchronous Motor Properties

The synchronous motor is not a self-starting motor. The rotor is heavy, and from a dead stop it is impossible to bring it into magnetic lock with the rotating magnetic field. For this reason, all synchronous motors have some kind of starting device. Such a simple starter is another motor, either ac or dc, which can bring the rotor up to about 90% of the synchronous speed. The starting motor is then disconnected and the rotor locks in step with the rotating field.

Another starting method is a second winding of the squirrel-cage type on the rotor. This induction winding brings the rotor almost into synchronous speed. When the dc is connected to the rotor windings, the rotor pulls into step with the field. The latter method is more commonly used.

# Summary

Electrical power generated for commercial and industrial use in the United States and Canada is produced as 60 Hz. In Europe the power is generated as 50 Hz. Single-phase power used in homes is produced as three-phase but is obtained by the proper connection of transformers once the power has reached a local distribution center. Phase is designated by a Greek letter, $\phi$.

Delta and wye connections are the two types of common transformer and generator configurations for three-phase power. Wye connections can use three or four wires for transmission to its destination. Delta connections rely upon three wires for transmission. Delta is represented by the Greek letter that resembles a triangle, $\Delta$. Wye is represented by the capital letter Y.

Wye connections produce a voltage advantage of 1.73 times one-phase output. Delta connections produce a current advantage of 1.73 times the current available in one phase of the system.

Single-phase power for home use can be obtained by properly tapping a three-phase transformer. Most power generated in the U.S. is transmitted long distances as three-phase and then broken down to single-phase after it reaches a location near its intended use.

The universal motor can operate on single-phase ac or dc. It is utilized in such devices as hand drills, food mixers, and sewing machines. The field windings and the armature are connected in series and operate at a very high speed if not properly harnessed.

Shaded-pole motors are small, use fractional horsepower, and are used in such things as clocks and some inexpensive record players and for fans on refrigerators. The split-phase motor is the workhorse around the house when the motor can start without a load. If a load is applied from the start, a capacitor-start motor is used. It has high starting torque.

The three-phase electric motor is very simple to construct and operate. It has very little trouble and needs little or no maintenance. It does not have a starting mechanism since the three-phase power does the starting by its very nature.

The synchronous motor can be used as a power factor correction device for industry. It can also be used for a constant speed source since it synchronizes with the frequency of the power line.

# Review Questions

1. What phase is most electrical power generated as in the United States and Canada?
2. How is a delta connection easily recognized?
3. How is the wye connection easily recognized?
4. Which connections have the advantage in current, delta or wye?
5. Which connections have the advantage in voltage, delta or wye?
6. If you have a 208-volt system, it is part of a transformer with a _____ connection.
7. How many wires do you need for a 3-phase system?
8. How do you reverse a three-phase motor?
9. What is the name applied to the ac pattern generated by an alternator?
10. What is the frequency of ac generated in the U.S. and Canada?
11. What is the frequency of ac generated in Europe?

12. What type of electrical power does a universal motor use?
13. What does SCR stand for?
14. How does a shaded-pole motor get its name?
15. Why are three-phase motors less expensive to make and maintain than the single-phase?
16. With what does a synchronous motor synchronize?
17. What is another use industry makes of the synchronous motor?

# Chapter 8

# AUDIO AMPLIFIERS

So far we have been concerned with the basics of electricity and electronics. It is now time for us to put to work some of the information you have acquired. The amplifier is the workhorse of electronics and will serve as a practical application for all that you have learned in the way of theory up to this point.

Amplifiers are usually classified as to range. This means that there are audio frequency amplifiers that operate on the frequencies that can be heard by the human ear. These frequencies usually range from 16 to 16,000 Hz. If the range is higher than the audio range, or in the radio frequency range which cannot be heard, the amplifier is referred to as a radio frequency (RF) amplifier. In this chapter we will deal primarily with the audio (AF) amplifier.

## Distortion

Amplifiers can distort an input signal. Distortion is the type of change the signal undergoes from the time it goes into the stage until it comes out. The amount of distortion depends on the linearity of the dynamic characteristics of the semiconductor device used in the amplifier circuit. Distortion is also dependent on the amplitude of the input signal. This input signal may swing into the nonlinear region of the transistors characteristic family of curves and produce severe distortion. Figure 8-1 shows examples of the six types of distortion.

*Frequency distortion* occurs when some frequency components of a signal are amplified more than others or when frequencies are excluded. *Phase distortion* occurs in most coupling circuits because there is a phase shift of any sine wave, but the complex waveform that makes up the overall waveform may have its phase shifted an amount that depends on its frequency. Thus, the output is not a faithful reproduction of the input waveform. This is not usually a problem in audio amplifiers, but rather in higher frequency, more complex waveforms out of the audio range.

## Hum

Hum is another type of distortion. Hum can be caused by the ac in the amplifier power supply. It may be caused by stray electromagnetic or electrostatic fields. Insufficient filtering of the power supply can also cause hum to be produced and amplified. This is particularly objectionable in audio circuits.

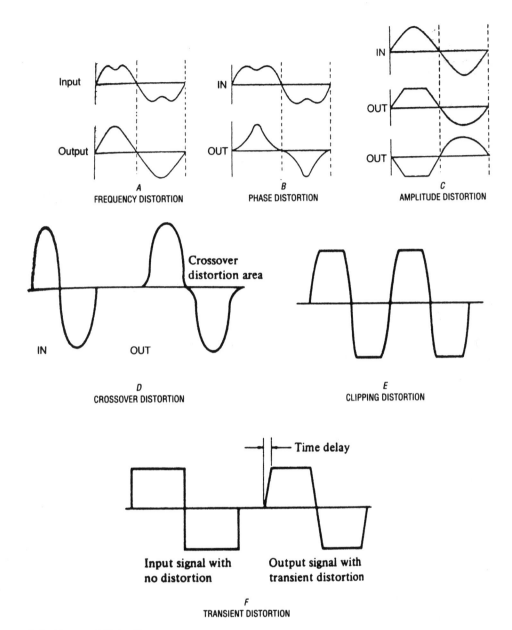

**Figure 8-1.** Types of distortion.

# Intermodulation Distortion

When an amplifier causes the significant distortion of a sine wave, the form of distortion is called intermodulation distortion (IM). It occurs if more than one sine wave is present. The effect is known as heterodyning. With two sine waves of different frequency present in a distorting amplifier, it results in a new sine wave whose frequency is the sum of the two original frequencies. The heterodyning effect will also produce a frequency that is the difference between the two frequencies. Those extra two frequencies plus the two original frequencies can cause some undesirable effects. Intermodulation distortion is found in all amplifiers, but most generally at the lower end of the frequency range. A distortion analyzer is used to measure the amount of intermodulation of an amplifier.

# Classes of Operation

When transistor amplifiers are classified according to their operating point, they are referred to as class A, class B, or class C.

Power amplifiers with the transistor conducting during 100% of the signal cycle is called class A operation. If conduction takes place during 50% of the signal cycle, it is class B operation. If conduction takes place during less than 50% of the signal cycle, then it is class C operation. Conduction that takes place during slightly more than 50% represents class AB operation.

Class A amplifiers usually have linear operation. The bias operating point is set near the center of the active region of the transistor. With a sine wave applied to the input, the output is a complete sine wave (see Figure 8-2). Class A with a resistive load has an efficiency of 25%. Transformer coupled class A has an efficiency of 50%.

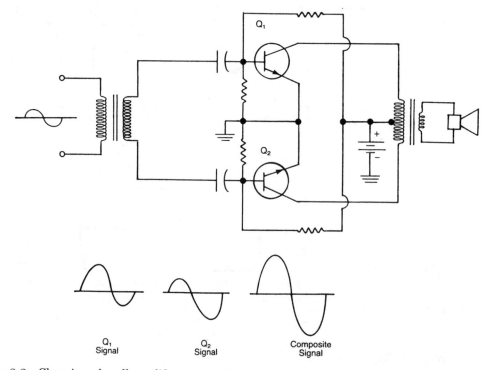

**Figure 8-2.** Class A push-pull amplifier stage with signals.

Class B is not too true in its reproduction of the original signal. It does not have high fidelity. It is more often used as a power amplifier stage. When two transistors are used in push-pull, they are able to put back the signal in its original configuration, and the whole sine wave is available for the next stage. It is similar to class A push-pull. Class B also has the advantage of not requiring a large power supply for a higher output. As you can see, current flows during one-half of the input signal and stops flowing during the other half of the hertz (see Figure 8-3). Class B push-pull has an efficiency of 78.5%.

Class AB is shown in Figure 8-4.

Class C is shown in Figure 8-5. A class C amplifier is so designed that the bias operating point is below cutoff. With a sine wave signal applied to the input, the output is less

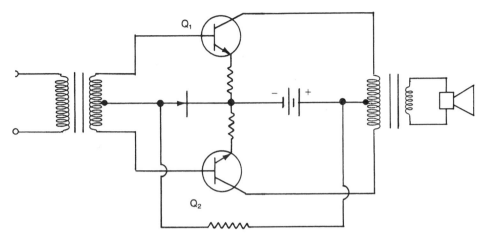

**Figure 8-3.** Class B push-pull amplifier with temperature compensation diode.

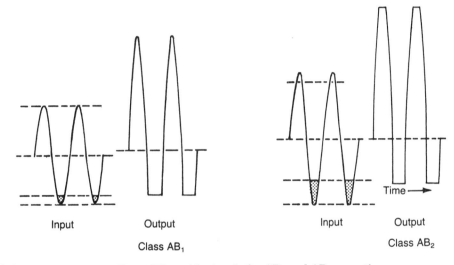

**Figure 8-4.** Class AB push-pull amplifier with signals for $AB_1$ and $AB_2$ operation.

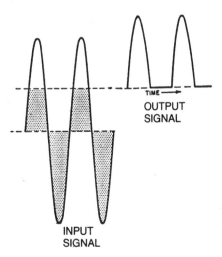

**Figure 8-5.** Class C push-pull amplifier.

than half of one alternation. The class C amplifier is one in which the collector signal flows for less than one-half of the input signal cycle. Today this type of amplifier is used primarily as a radio frequency amplifier and for providing energy to oscillators or switching circuits. The operational efficiency of this amplifier is quite high. It consumes energy only for a small portion of the applied sine wave signal.

Figure 8-6 shows a couple of power amplifiers (PAs) made up of transistor and integrated circuit devices using various classes of operation.

**Figure 8-6.** Power amplifiers.

# Coupling of Transistor Amplifiers

In Figure 8-7 the $RC$ network used for coupling two transistors consists of a collector load resistor, $R_1$, for the first stage, a dc blocking capacitor, $C_1$, and a return resistor, $R_2$, for the input element of the second stage.

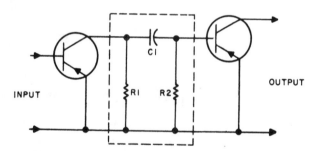

**Figure 8-7.** $RC$-coupled transistors.

There is a difference in the coupling capacitor in the transistor amplifier as compared to the vacuum tube type. The capacitor used in the $RC$ coupling network must have such reactance as to have very low signal loss across it. Usually, speaking from a practical standpoint, the voltage drop across the capacitor should not exceed 10%, while most of the signal voltage — around 90% — should be dropped across the dc return resistor $R_2$. $RC$ coupling is used extensively with junction transistors because of high gain, economy of component parts (the cost of a resistor-capacitor combination is much cheaper than a transformer), and good utilization of board space. The $RC$ coupling network can be used on almost any size amplifier from low-level preamps to high-level amplifiers. The use of $RC$ coupling in battery operated equipment is usually limited to low power operation to limit battery drain.

Transformer coupling is another type of coupling used to connect stages in transistor circuits. They operate basically the same as in vacuum tubes (see Figure 8-8). There are

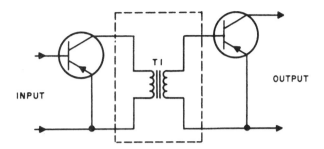

**Figure 8-8.** Transformer-coupled transistors.

certain advantages to this type of coupling. The very low resistance in the base path aids temperature stabilization of the dc operating point. With a swamping resistor in the emitter lead, the current stability factor is ideal. Because there is no collector load resistor to dissipate power, the power efficiency of the transformer-coupled amplifier approaches the theoretical maximum of 50%. For this reason the transformer-coupled amplifier is used extensively in portable equipment where battery power is used.

There are some disadvantages to the transformer being used as a coupling device. It is heavy, bulky, and expensive; it also has frequency response reduction as compared with the *RC* coupling. The transformer is rapidly being replaced in audio amplifier circuits because of expense, frequency response falloff, and its weight and space requirements.

The impedance-coupled amplifier is shown in Figure 8-9. $L_1$ is the load for the input transistor. This inductive load is shunted by the capacitor-resistor combination $C_1$-$R_1$. The main advantage of this arrangement is that it provides high power efficiency since the dc voltage is not dropped across a load resistor.

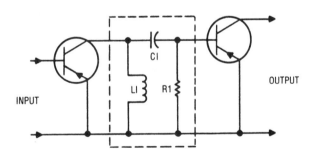

**Figure 8-9.** Impedance-coupled transistors.

The low-frequency response is reduced by the shunt reactance of the inductor. The high-frequency response is reduced by the collector capacitance. Unlike the transformer-coupled amplifier, the impedance-coupled amplifier suffers no loss of high frequencies by leakage reactance. The frequency response of the impedance-coupled amplifier is better than that of the transformer-coupled amplifier, but not as good as that of the *RC*-coupled amplifier.

The direct-coupled amplifier is used to amplify the low frequencies of dc signals — just as was the case with vacuum tube amplifiers. You will no doubt notice by this time that a number of transistor circuits copy those of the vacuum tube. The tube was here first, and the applications were made through the years, and so in order to make sure the transistor had a role to play, it was decided to make it do the same things the tube did, but only better. So, as you study transistor circuits, you will find that some of the later developments do not copy the vacuum tube, but are unique to transistors only.

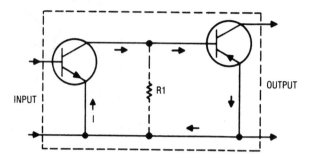

**Figure 8-10.** Direct-coupled transistors.

In the direct-coupled circuit shown in Figure 8-10, you will find two types of transistors. The NPN is connected directly to the PNP type. Current flow is shown by the arrows. If the collector current of the first stage is larger than the base current of the second stage, a collector resistor (shown in dotted lines) must be used.

Since very few parts are used in a direct-coupled amplifier, it makes for a cheaper or less expensive way of getting the amplifier you need. However, there are limitations as to how many stages you can directly couple. Temperature variations of the bias current in one stage is amplified by all the stages. This causes severe temperature instability.

# Transistor Amplifiers

There are three common types of transistor amplifiers: common emitter, common base, and common collector. The transistor is connected in the circuit just as it implies in the common-base, common-emitter, or common-collector designation (see Figures 8-11, 8-12, and 8-13).

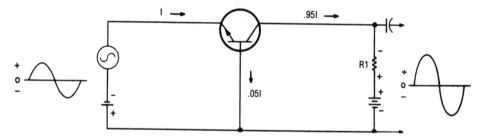

**Figure 8-11.** Common-base transistor amplifier with current flow and voltage waveforms.

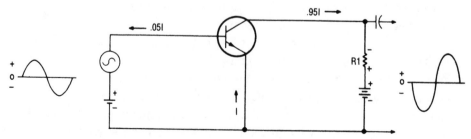

**Figure 8-12.** Common-emitter transistor amplifier with current flow and voltage waveforms.

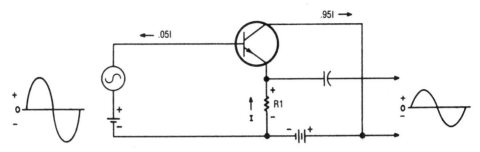

**Figure 8-13.** Common-collector transistor amplifier with current flow and voltage waveforms.

## Common Base

Note the connections made to the base, emitter, and collector of the transistor in Figure 8-11. Current flow through the NPN transistor is indicated by the arrows. About 95 percent of the emitter current reaches the collector. In practical circuits the flow is 92 to 98 percent. Here we use 95 as a compromise; the remainder (5 percent) flows through the base.

Input and output waveforms represent the voltage of the signal. Input signal from the generator is applied in series with the bias (battery) between the base and emitter. The output signal appears across the collector load resistor $R_1$. The output signal has the same shape and phase as the input except that it is amplified.

## Common Emitter

The common-emitter circuit is shown in Figure 8-12. Again, the arrows indicate the current flow and its direction. The input signal of the common-emitter amplifier is 180° out of phase with the output signal that appears across $R_1$.

## Common Collector

The common-collector circuit is shown in Figure 8-13. This is a particularly useful type of circuit when it comes to impedance matching. It does not amplify, but acts as an impedance-matching device. Note that the signal is not reversed but is the same phase as the input. This can be visualized easily if you note that the output signal is taken from the emitter resistor $R_1$.

Table 8-1 will aid you in summarizing the characteristics of the various types of transistor amplifier circuits.

# Integrated Circuits

Chips, ICs, integrated circuits all mean the same thing. The integrated circuit is just what its name says: It is a circuit. This circuit can contain the resistances, capacitances, and inductances as well as the transistors and diodes. All this can be contained in a small chip which is so tiny that it is difficult to see with the unaided eye (see Figure 8-14).

**TABLE 8-1**

| Item | CB Amplifier | CE Amplifier | CC Amplifier |
|------|--------------|--------------|--------------|
| Input resistance | 30–150 Ω | 500–1.5 kΩ | 20–500 kΩ |
| Output resistance | 300–500 kΩ | 30–50 kΩ | 50–1000 kΩ |
| Voltage gain, V | 500–1500 | 300–1000 | Less than 1 |
| Current gain, A | Less than 1 | 25–50 | 25–50 |
| Power gain, dB* | 20–30 | 25–40 | 10–20 |

*dB is a decibel. A decibel is unit of measure of two different amounts of power. Formula: $10 \log (P_2/P_1)$, where $P_1$ = input power in watts and $P_2$ = output power in watts.

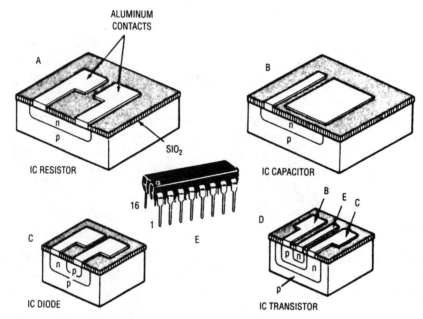

**Figure 8-14.** Construction of an integrated circuit showing the transistor, diode, resistor, and capacitor.

An entire amplifier can be made on one chip. A number of transistors and diodes can be arranged, along with the proper capacitance and resistance, to produce a complete circuit without having to make the circuit on a board. By chemically etching the silicon, it is possible to place all these components on a spot no larger than the head of a pin.

# Operational Amplifiers

The most common linear circuit is the operational amplifier (op-amp). The op-amp is an integrated circuit that is classified as a linear amplifier or digital. Several characteristics that make the op-amp useful in consumer products are high open-loop gain, high input impedance, low output impedance, and an ability to reject unwanted signals.

## Differential Amplifier

See Figure 8-15 for the op-amp used as a differential amplifier that has both inverting and noninverting inputs.

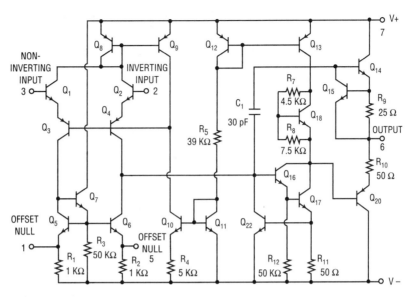

**Figure 8-15.** The 741 op-amp circuitry. It is the same as 20 transistors.

An operational amplifier has an infinite gain, infinite input impedance (open circuit) and zero output impedance. That is, if we consider an ideal op-amp. The op-amp unit is basic to such analog circuits as the summing amplifier, integrator, and inverter. Figure 8-16 shows the symbol for an op-amp with two inputs and a single output. The two inputs are marked with plus and minus signs to indicate the relation between the input at that terminal and the resulting output voltage. A signal applied to the plus input appears in-phase and amplified at the output, while that applied to the minus input is amplified but inverted at the output. While the basic op-amp circuit has very high gain, the useful connection of an op-amp circuit has a gain that is set by an external resistor (see Figure 8-17).

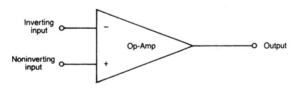

**Figure 8-16.** Operational Amplifier Symbol.

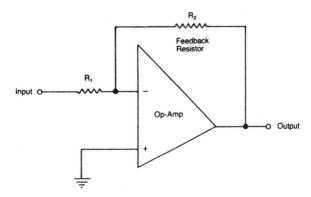

**Figure 8-17.** Constant-gain amplifier.

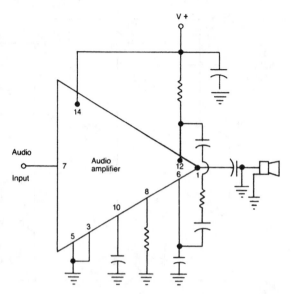

**Figure 8-18.** Linear amplifier used as a final audio output in radio receivers.

With a few added resistors and capacitors the op-amp can be used as a final audio output in receivers (see Figure 8-18). Radios made today utilize one op-amp for the audio output stage. This amplifier section is capable of delivering, in some cases, 5 watts to the speakers. There must be heat sinks on the IC to keep from overheating.

One linear integrated circuit is used in most receivers to replace the entire transistor stages. A linear circuit can be used in the IF section of an AM-FM radio.

Another use for the linear integrated circuit is in voltage regulation.

## Different Forms of Op-amps

A basic form of the op-amp circuit is used in analog computers as an inverting constant-gain multiplier (see Figure 8-19A). The op-amp is also capable of operating as a noninverting constant gain multiplier (see Figure 8-19B). It is also capable of being used as a unity follower that has a gain of 1 or unity. This means it has no polarity reversal and the output is the same polarity and magnitude as the input. The circuit acts the same as the emitter-follower circuit in transistors except that is has the advantage of the gain being set much closer to exactly unity (see Figure 8-19C). This means the circuit can be used for impedance matching. In an integrator circuit the summing unit alone allows addition or subtraction operation. An integrator circuit is needed in solving differential equations. The constant-gain multiplier of Figure 8-17 can be converted to an integrator circuit by using a capacitor as a feedback element rather than using a resistor (see Figure 18-20).

As can be seen from these examples, the op-amp can be used for many purposes. It is a multistage integrated circuit amplifier with open-loop (without negative feedback) volt gains of 50,000 to 1 million.

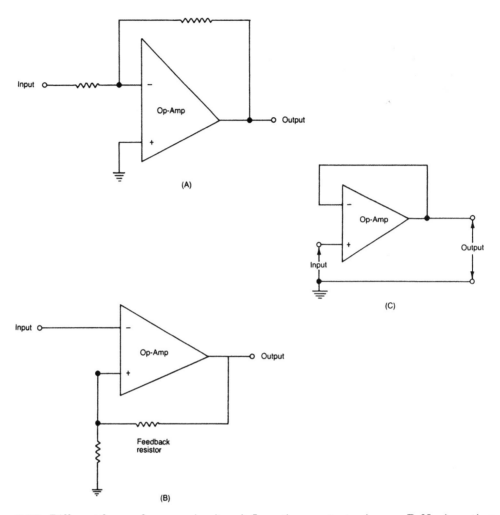

**Figure 8-19.** Different forms of op-amp circuitry. A. Inverting constant gain amp. B. Noninverting constant gain amp. C. Unity follower.

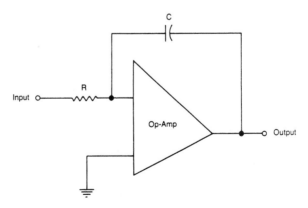

**Figure 8-20.** Op-amp used as an integrator.

# Summary

Each type of amplifier has its own characteristics and can be adjusted to perform a number of operations. Classes of operation of an amplifier determine its method of operation. Class A means that the circuit operation conducts current at all times. Class B conducts only on one-half of the input signal. Class AB has the advantages of both classes A and B with some of its own limitation. Class A has good fidelity. Class B is more of a power amplifier stage. Class AB can be broken down into class $AB_1$ and $AB_2$. Class AB is used for its greater power output than for its fidelity. Clipping is much greater in $AB_2$ than in $AB_1$.

Bias is needed to cause a transistor to operate properly. There are two types of bias— fixed and self. The type most commonly used is self-bias. That is where the bias is developed by using a resistor in the base-emitter circuit of a transistor. The input signal develops bias across the resistor and eliminates the need for an additional bias battery or power source.

Fixed-bias is just that. It is determined by the voltage of the battery, or power source.

Whenever there is need for more than one amplification stage, there is also a need to have the stages coupled. These stages are coupled by a number of methods. One of the methods is *RC* coupling where the series *RC* circuit is connected across the collector load in a transistor circuit. The ac signal voltage is dropped across the capacitive reactance of the capacitor and the resistance of the resistor. The distribution of signal is about 10 percent across the capacitor and 90 percent across the resistor in the circuit of the next stage. This produces a large signal voltage (90 percent) for the input circuit of the next stage.

*RCL* coupling is also called impedance coupling because the first stage uses an inductor for the load. This arrangement is more typical of RF amplifiers than AF amplifiers.

Transformer coupling uses the turns ratio to aid in impedance matching of the two stages and provides a signal boost in most cases. It does have serious limitations such as weight, cost, and bulk.

Direct coupling can be used where a dc voltage or a low frequency ac signal is present and must be coupled. This type of coupling is used when there is a need for high current outputs to drive an instrument or relay.

Transistor circuits can be classified as to their element connection. We have common-base, common-emitter, and common-collector circuits. Each type of circuit has its own input and output impedances and serves a particular purpose in terms of output signal levels.

The input and output signals in the common-emitter circuit are 180° out of phase. The input and output signals for the common-base and common-collector circuits are in phase. The common-collector circuit does not amplify. It is also used for impedance-matching purposes.

The op-amp is a linear circuit amplifier, lt has many commercial uses in consumer products such as radios and computers and calculators. The operational amplifier has an infinite gain, infinite input impedance, and zero output impedance. The op-amp unit is basic to such analog circuits as the summing amplifier, integrator, and inverter. It can also be used in voltage regulation. Many consumer devices make use of its versatility. It can have a voltage gain of 50,000 to 1 million.

# Review Questions

1. What is distortion?
2. How is hum classified?
3. What are the five classes of operation for an amplifier?
4. What are the two types of class AB operation?
5. What is the difference between the two types of AB operation?
6. Where is class C operation used?
7. Name the two types of bias.
8. Describe grid-leak bias.
9. What are the components needed for *RC* coupling?
10. What are the components needed for impedance coupling?
11. Name an advantage of using transformer coupling.
12. Why do you need direct coupling?
13. What are the classes of operation for transistor amplifiers?
14. Name the types of coupling used for transistor amplifiers.
15. What are the three types of common transistor circuits?
16. What type of transistor amplifier circuit is used for impedance matching only?
17. What kind of transistor amplifier circuit is used for current gain that is less than 1?
18. What is an op-amp?
19. What is a differential amplifier?
20. What are the different forms of op-amps?

# Chapter 9

# OTHER TYPES OF AMPLIFIERS

Radio frequency amplifiers are used in both receivers and transmitters. Transmitters are used to generate radio frequencies and then to broadcast them. Receivers pick up the transmitted signals and convert them to intelligence that can be used in a number of ways. In order to obtain the power needed for long-distance transmission at frequencies above human hearing, it is necessary to obtain amplifiers to boost weak signals. Voltage amplifiers and power amplifiers are used for this amplification. The circuits can be made using vacuum tubes or transistors.

If a circuit amplifies signals above human hearing, it is usually referred to as a radio frequency (RF) amplifier. Since this type of amplifier is usually tunable, it can operate over a wide range of frequencies. Very high frequencies (VHF) are between 30 and 300 megahertz (MHz). Ultrahigh frequencies (UHF) are from 300 to 3000 MHz. These are above the 16-kilohertz (kHz) range of human hearing, and so they qualify as radio frequencies. Intermediate frequency (IF) is also in the radio frequency grouping. IF is used in receivers and may be 455 kHz for AM broadcast radio and 10.7 MHz for FM. Television receivers use even higher frequencies for IF. Video amplifiers are very high frequency amplifiers. They are used for amplifying the picture information frequencies of television.

## Tuned Circuit Coupling

Radio frequency stages are coupled by tuned stages in the IF amplifiers and similar circuits. Figure 9-1 shows that type of circuit. The primary is $L_1$, and the secondary is $L_2$. The capacitor $C_1$ is variable and can be used to tune the coil capacitor combination to a given frequency as needed. $C_1$ tunes $L_1$ to resonance at the signal frequency. The coil and capacitor tuned to resonance produce a large signal voltage across the tank circuit. The large circulating current in the primary of the tank circuit forms a magnetic field which induces voltage into the secondary winding.

The induced voltage in the secondary circuit made up of $C_2$ and $L_2$ is considered in series since it is generated by the coil reacting to the magnetic field.

The secondary circuit is not tuned to resonance. The current that flows is in phase with the induced voltage. The voltage and current induced is 180° out of phase with that which was in the primary. Transformer coupling has some advantages. The resonant conditions of the tank circuit result in a gain in signal voltage that is very selective. This type of

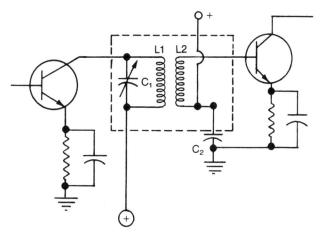

**Figure 9-1.** Transformer coupling.

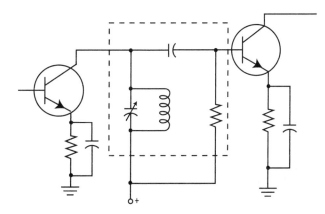

**Figure 9-2.** Tuned impedance coupling.

coupling is frequently used in IF amplifiers in receivers and in output stages of transmitters. Figure 9-2 shows how a single-tuned stage with *RC* coupling reduces the number of tunable capacitors and an inductor. This is known as impedance coupling.

# Voltage Amplifiers

As you already know, the transistor is not a voltage-amplifying device; however, we shall look at it here as an example of a small-signal amplifier. Just keep in mind that the transistorized RF amplifier will handle small signals.

The RF amplifier deals primarily with tuned circuits. This means that either one or two tuned circuits exist in each of the RF amplifiers. In a transistor circuit, such as that shown in Figure 9-3, you can see the tuned circuits in both the input and output of the stage. This type of circuit must develop maximum signal at the tuned frequency and must also match the transistor input and output impedances. Note how the tuned circuits are tapped for connection to the base and collector. These are not always center taps, but they are tapped where they will cause an impedance match for the particular transistor being used.

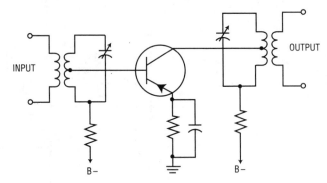

**Figure 9-3.** RF amplifier with tuned circuit.

Tuned circuits used as inputs to grounded-emitter and grounded-base amplifiers have a very low input impedance. To overcome some of the problems of making the impedance match the transistor, the coils are tapped so that the transistor will see the proper input or output impedance. The lower portions of the tank circuit coils are used for impedance matching (see Figure 9-3).

# Power Amplifiers

Power amplifiers that boost radio frequencies are needed in transmitters. They are used to amplify the carrier frequency so that it can reach the proper power level to cover the distance allocated for that frequency or station.

In almost all cases power amplifiers require larger input signals. They usually operate class B or class C, which means that current flows during only part of the input signal. The output is sinusoidal, though, since the tank circuit has a coil that will have a collapsing magnetic field after the first portion of the signal flows through it. The magnetic field collapses and puts back the energy in almost the same amount it took to make the magnetic field. Therefore the output is the same shape as the input signal.

Figure 9-4 shows a class D amplifier used with pulse width modulation. It can run as high as 90% efficient. If a transistor is used as an RF amplifier, it is usually operated class C. This means that for most of the input cycle only a small amount of cutoff current flows in the collector circuit. In Figure 9-5 the circuit is a grounded-emitter. This means that it only conducts during the positive peaks of the input signal, and then only in bursts. The bursts of input signal are converted in the tank circuit to sinusoidal output by the nature of that tank. If you noticed closely, you saw that the transistor was biased in reverse of what it usually would be. This is done to achieve a bias point considerably beyond cutoff to operate as a class C amplifier. The emitter is biased positive with respect to the base.

# Feedback in Amplifiers

There are two types of feedback to look at closely in regard to the operation of amplifiers — regenerative and degenerative. In most instances they are called positive and negative feedback. In special amplifier applications, it is desirable to feed back a signal from the

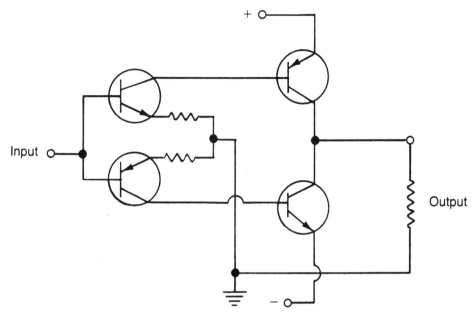

**Figure 9-4.** Class D power amplifier.

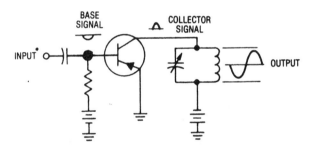

**Figure 9-5.** Reverse-biased base-emitter circuit.

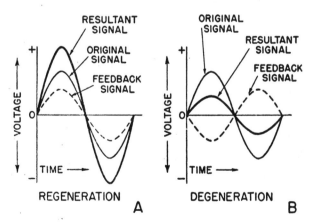

**Figure 9-6.** Regeneration and degeneration signals.

output stage to the input of the same stage or to a preceding stage. This fed-back signal can take either of two forms, shown in Figure 9-6, where the voltage amplitudes are plotted against time. The principle of the feedback is shown in the block diagram of Figure 9-7. The phase of the signal that is fed back (with reference to the input signal)

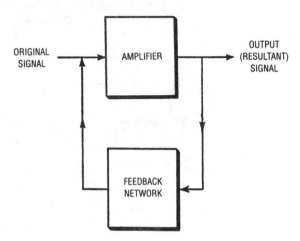

**Figure 9-7.** Block diagram of feedback amplifier.

determines the type of feedback that results. If the feedback aids the original input signal by increasing the amplitude, it is called *regenerative*, or *positive*, *feedback*. If it opposes the original input signal and decreases in amplitude, it is called *degenerative, negative*, or *inverse feedback*.

To produce regenerative feedback, the original feedback signals must be in phase with each other, as in A of Figure 9-6. Adding these two waveshapes, or waveforms, produces the regenerative signal, which is larger in amplitude than the original signal. In B, the original and feedback signals are opposite in phase, and degenerative feedback is produced. The waveform is smaller in amplitude than the original signal.

In positive feedback, the voltage output of an amplifier is increased because the effective input voltage is increased. The greater amplification also usually increases the amount of distortion and noise in the amplifier. Sometimes the amount of regenerative feedback is so great it produces sustained oscillations. In negative feedback, the voltage gain of an amplifier is decreased because the effective input voltage is decreased. In practical applications, this type of feedback is used to reduce the effects of distortion. Degenerative feedback also improves the frequency response and stability of amplifiers.

# Integrated Circuits

Integrated circuits (ICs) are used in all types of modern electronic devices. They are *integrated* circuits, meaning that they are made as a total circuit and housed in one enclosure. The enclosure may take a number of shapes. It may be similar to a TO-5 transistor package with 8 leads instead of 3, or it may be what is referred to as a dual in-line package (DIP) with as many as 24 leads (see Figure 9-8). All components are manufactured as a common unit.

## Packaging ICs

The size of the package to be used for the IC, the function to be performed, and the type of application determines the construction of the chip.

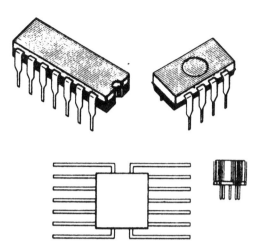

**Figure 9-8.** IC packages.

There are three categories of IC packages: small-scale integration (SSI), medium-scale integration (MSI), and large-scale integration (LSI). The SSI package generally has fewer than 200 components in it, MSI packages have between 200 and 1000 components, and LSI packages may have anywhere from 1000 to 256,000 or more components.

Keep in mind that transistors, diodes, resistors, and capacitors are referred to as discrete components. An IC may have thousands of these discrete components located on one chip. The concept of having thousands of parts on one chip so small that the human eye cannot see is hard to believe when you have been working with vacuum tubes and transistors for some time. Technological advances continue to increase this number of components on a chip.

Integrated circuits can be used to do any number of things electronically. They are classified further according to their function. The two broad categories of classification here are digital and linear.

*Linear* means there is a direct relationship between input and output signals for an integrated circuit. Linear devices are classified as *analog*. This type of device is used for operational amplifiers, sensing amplifiers, signal drivers, voltage regulators, and signal comparators and as linear amplifiers. They can also be adapted to other uses by adding nonlinear devices or components.

There is a definite difference in the application of digital and linear ICs. The digital circuits are concerned primarily with computers and devices which use calculations to operate properly. The linear devices will be examined in this chapter with the digital left for another chapter.

# Operational Amplifiers

Of the linear integrated circuits, the operational amplifier is the most popular type. The *op amp*, as it is called, produces very high gains in a frequency range from 0 to 1 MHz (see Figure 9-9 for its symbol). The triangle shape is used to indicate an amplifier. You will see more of this symbol as you progress in the electronics field. Inside is a number of

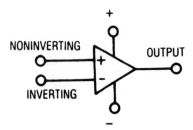

Figure 9-9. Symbol for operational amplifier.

transistors, diodes, capacitors, and whatever else is needed to make the circuit function as a high-gain broadband amplifier.

Op amps are further classified into two broad circuit configurations. They are either open-loop or closed-loop.

# Open Loop

Figure 9-10 shows the typical open-loop op amp at work. This circuit does not have feedback. Gain comes directly from the op amp. The gain of the op amp in an open-loop configuration can be obtained by using the formula

$$A_{\text{vol}} = \frac{E_o}{E_{id}}$$

The $E_o$ is used to indicate the output signal voltage while $E_{id}$ indicates the differential input to the IC.

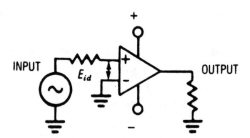

Figure 9-10. An op amp operated as an open-loop amplifier.

# Closed Loop

The closed-loop op amp has feedback (see Figure 9-11). Feedback can be either positive or negative. Positive feedback reinforces the input signal. Properly adjusted for phase and amplitude, positive feedback can cause the amplifier to oscillate. A positive feedback loop is connected to the noninverting side of the input signal. This positive feedback loop increases the output of the circuit. The increase is caused because the input and output are in phase and the feedback is aiding the input and increasing its level of input to result in a greater output.

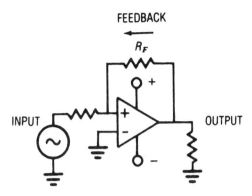

**Figure 9-11.** An op amp operated as a closed-loop amplifier.

Closed-loop gain can be expressed in the formula

$$A_{\text{vol}} = \frac{E_o}{E_s}$$

The $E_o$ is the signal voltage output while $E_s$ represents the input signal voltage.

There is positive and negative feedback. Negative feedback can be used for a number of purposes. In this closed-loop op-amp circuit, it is used to increase the circuit stability and to reduce distortion. The negative feedback is 180° out of phase with the input signal. This cuts or drops the input signal. This reduces the output signal, and it has a better quality since some of the distortion was present in the upper peaks of the input signal.

Op amps are used in equipment that has many uses in commercial, medical, industrial, and educational fields. One of the advantages of the op amp over other devices is its compactness and ability to have a high gain. The sensitivity to weak signals and even to dc makes it a desirable device for many applications. Op amps can be used for sensors in many types of instruments and as signal generators.

One of the most popular op amps today is model 741 made by Fairchild Semiconductors (see Figure 9-12 for its internal circuitry). It has eight pins in a DIP. The first op amp was produced by Fairchild in 1965.

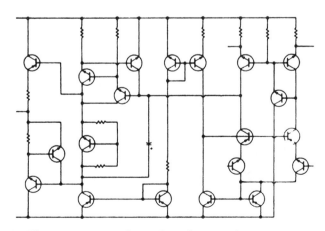

**Figure 9-12.** The 741 op-amp schematic and connections.

# Differential Amplifiers

The differential amplifier does not include capacitors in its composition. It is a linear amplifier and is abbreviated as diff amp. Figure 9-13 shows the symbol used for the diff amp. Note the differences between this symbol and that in Figure 9-9 for the op amp.

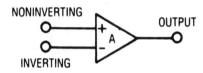

**Figure 9-13.** Differential amplifier (diff amp) symbol.

The two inputs of the differential amplifier make it capable of distinguishing between the two sources. A difference output is generated from the two inputs. Components inside the amplifier respond to the differences between the two signals. The like signals are suppressed. A variety of differential amplifiers are available from manufacturers.

Differential amplifiers require two transistors as nearly identical as possible. Thus, a commercially available method whereby they are made together is preferable to one that makes the two transistors separately and then tries to match them. IC manufacturers have the ability to make two transistors nearly identical or at least close enough to operate properly in circuits that demand close tolerances. Figure 9-14 shows how two transistors are connected in order to produce a differentiating circuit. Many of this type circuit can be reproduced in one chip and enclosed in a single package. Figure 9-15 shows how three diff amps are connected in sequence and enclosed in a single package. The circuits are connected in what is referred to as cascade. *Cascade* means in series or one after the other. The output of one circuit is fed directly to the input of the other with maximum utilization of the coupling process.

Integrated circuits have millions of applications. More of these applications will be discussed as we progress further in your exposure to the study of circuits.

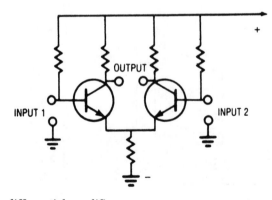

**Figure 9-14.** Circuit for a differential amplifier.

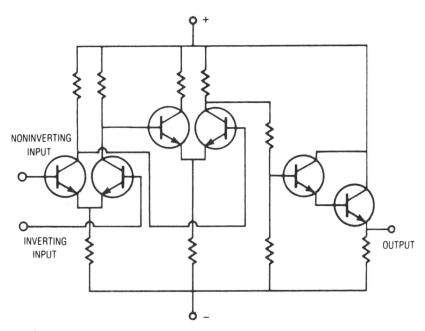

**Figure 9-15.** Three differential amplifiers connected in cascade.

# Field Effect Transistors

The field effect transistor (FET) has high input impedance compared with the low input impedance of the bipolar transistor. The FET operates on low dc supply voltages. This type of transistor is used in many consumer devices inasmuch as it is lightweight, rugged, and very small in size.

The FET has three terminals with slightly different names from those encountered with transistors. It has a gate (G), drain (D), and source (S). The drain could be called the anode and the source called the cathode, with the gate being the current controlling connection (see Figure 9-16).

FETs can be classified as to two major types: junction FET (JFET) and metal oxide semiconductor (MOSFET). Sometimes the MOSFET is also referred to as the insulated gate FET (IGFET). The JFET is further classified according to the type of material that makes up its *channel*. The channel is either P-type or N-type material. However, the MOSFET is classified according to its *mode* of operation — that could be either depletion mode or the enhancement mode. Figure 9-17 shows the schematic for MOSFETs.

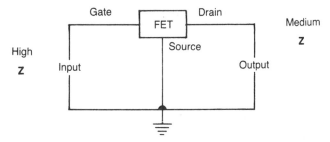

**Figure 9-16.** Single gate FET circuit configuration.

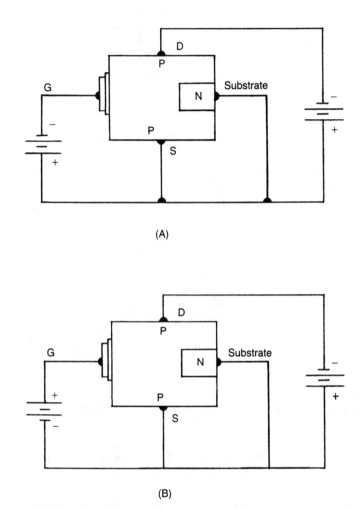

(A)

(B)

**Figure 9-17.** A. Depletion Mode MOSFET. B. Enhancement Mode MOSFET.

The most commonly used of the field effect transistors is the MOSFET. It was designed to be used in circuits that were originally vacuum tube types. The MOSFET matches the characteristics of the tubes pretty well in regard to input and output impedances. The MOSFET also has a drain, source, and gate and *operates much the same as the JFET*. The drain is connected to the positive voltage power source and the source is connected to the negative voltage source. The substrate that makes up the transistor is connected to the source voltage supply. The device is biased to set up an electron flow between the source and the drain. Current flows through the narrow channel created by the substrate (see Figure 9-18 for both JFET and MOSFET).

# Depletion Mode

In order to understand the functioning of the FET it is best to recall your knowledge of capacitors. The gate and N-type material are likened to the two plates of a capacitor. The metal oxide then would be the dielectric material between the plates. By applying a negative voltage to the gate, a negative charge is developed on the gate. Using the capacitor

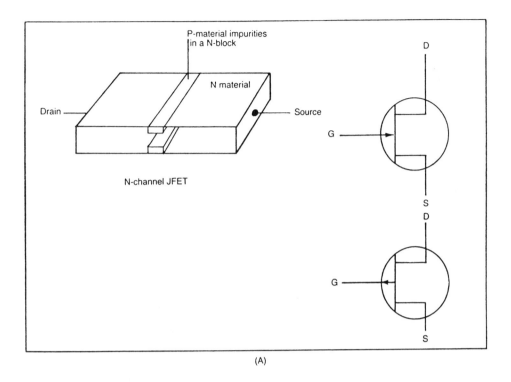

(A)

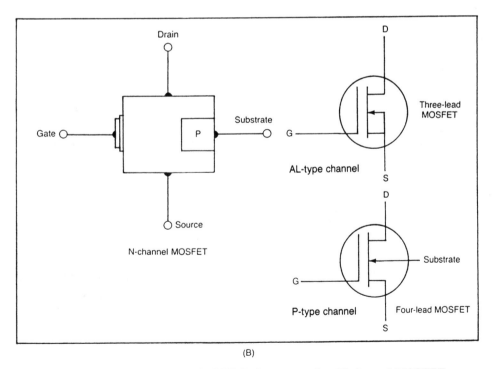

(B)

**Figure 9-18.** A. Structure of an N-channel JFET. B. Structure of an N-channel MOSFET.

similarity again, we see that this assembly acts as a capacitor and the other plate develops a positive charge. That means the positive plate creates a depletion area. That in turn restricts current through the narrow channel. The more negative the gate voltage, the wider the depletion region between the N channel. When enough negative voltage is applied to the gate, current between the source and the drain can be cut off. The N channel is depleted of electrons by this action of the positive charge. This is the type of operation that gives the MOSFET its *depletion mode* of operation.

## Enhancement Mode

Note that the gate of the MOSFET is insulated from the channel. That means a negative or positive voltage can be applied to the gate. A positive voltage is applied in the enhancement mode of operation, and a positive voltage is also applied to the gate. In the depletion mode, the gate, insulator, and channel acted as a capacitor. Note though that in this case the gate has developed a positive charge. This means that the channel has a negative charge. Negative charges that develop in the channel are current carriers in the N material. They improve the conditions so more electrons reach the drain. That means the current increases and the current flow in the channel is *enhanced*. As the gate voltage becomes more positive, it increases the current flow through the drain.

# Common FET Configurations

Transistors with common emitter, common base, and common collector circuits are used for different purposes and so are FETs. Circuits are designed to be used for various purposes with a common source, common drain, and common gate.

A *common source* configuration is shown in Figure 9-19. This is the most common type of circuit used in consumer products. This configuration provides a high input impedance, a medium to high output impedance, and a voltage gain greater than one. The input signal is between the gate and the source. The output of the amplifier is between the drain and the source. The fact that the source is common to the input and the output signal gives this configuration its name.

A *common gate* circuit is also shown in Figure 9-19. This is a configuration with an input signal applied between the source and gate. The amplifier made with this arrangement develops a medium input impedance. The output signal is taken from between the gate and the drain. This means a high output impedance for this particular arrangement. One important characteristic of this type of stage is that it can operate at high frequencies. Because of its high frequency operation and medium input impedances, this amplifier offers low gain to the signal. That is one reason why the amplifier does not need large amounts of neutralization that would be provided by feedback capacitors.

A *common drain* circuit is shown in Figure 9-19. It is also known as a source follower. The name comes from the vacuum-tube-type circuit that was a cathode follower and was used to match impedances and not amplify. The input impedance is high in this configuration. It is higher than in the common source circuit. The output impedance is very low. There is no signal shift between the input and the output. The voltage gain

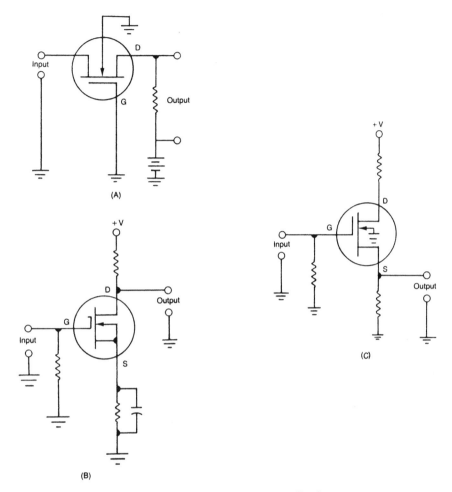

**Figure 9-19.** A. Common Gate. B. Common Source. C. Common Drain.

for this type of amplifier is less than one, and it is used primarily for impedance-matching purposes.

# Gallium-Arsenide FET

There are a number of devices that use a gallium-arsenide field effect transistor (GASFET). They are constructed for special purposes such as the *triac driver output coupler*. It is optically coupled to a silicon bilateral switch designed for applications requiring isolated triac triggering such as interfacing from logic circuits to 110/120 V rms line voltage. These devices offer low current, isolated ac switching, and high output blocking voltage and are small in size and low in cost (see Figure 9-20).

The *digital logic coupler* is a gallium-arsenide IRED (infrared emitting diode). It is optically coupled to a high-speed integrated detector. And, it is designed for applications that require electrical isolation, fast response time, and digital logic compatibility such as interfacing computer terminals to peripheral equipment, digital control of power supplies, motors, and other servo machine applications. It is intended for use as a digital inverter;

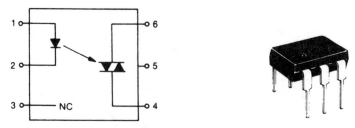

**Figure 9-20.** Triac driver output coupler. Copyright of Motorola, Inc. Used by permission.

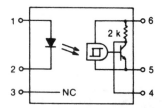

**Figure 9-21.** Digital logic coupler. Copyright of Motorola, Inc. Used by permission.

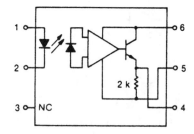

**Figure 9-22.** Optically isolated ac linear coupler. Copyright of Motorola, Inc. Used by permission.

the application of a current to the IRED input results in a low voltage. With the IRED off the output voltage is high (see Figure 9-21).

The *optically isolated ac linear coupler* is a gallium-arsenide IRED optically coupled to a bipolar monolithic amplifier. It converts an input current variation to an output voltage variation while providing a high degree of electrical isolation between the input and output. It can be used for telephone line coupling, peripheral equipment isolation, and audio applications (see Figure 9-22).

# Infrared Emitting Diodes

Infrared (wavelength is 900 nm) gallium-arsenide emitters are available for use in light modulators, shaft or position encoders, punched card and tape readers, optical switching and logic circuits. They are spectrally matched for use with silicon detectors (see Figure 9-23).

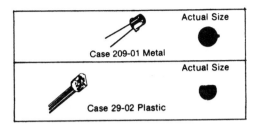

Figure 9-23. Infrared emitting diodes. Copyright of Motorola, Inc. Used by permission.

# Silicon Photo Detectors

A variety of silicon photo detectors are available for a wide range of light-detecting applications. Devices are available in packages offering choices of viewing angle and size in either low-cost, economical plastic cases, or rugged, hermetic metal cans. Their advantage over phototubes are high sensitivity, good temperature stability, and proven silicon reliability. Applications include card and tape readers, pattern and character recognition, shaft encoders, position sensors, counters, and others. Maximum sensitivity occurs at approximately 800 nm. Photodiodes are used where high speed is required [1.0 nanosecond (ns)] (see Figure 9-24).

Phototransistors are used where moderate sensitivity and medium speed (2 ns) are required. Figure 9-25 shows what they look like packaged.

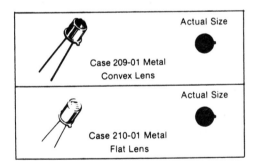

Figure 9-24. Photodiodes. Copyright of Motorola, Inc. Used by permission.

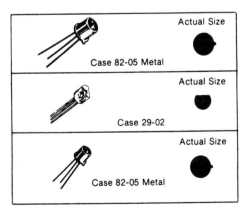

Figure 9-25. Phototransistors. Copyright of Motorola. Inc. Used by permission.

Photodarlingtons are used where maximum sensitivity is required with typical rise and fall times of 50 microseconds (μs). They are also packaged in metal and plastic cases such as that shown in Figure 9-25.

Photo triac drivers contain a light-sensitive IC acting as a trigger device for direct interface with a triac (see Figure 9-26). As can be observed, this is the same case used for the phototransistor and the photodarlington. It is impossible to tell which is which unless you know which device is used in a particular type of circuit. Physically they look the same.

**Figure 9-26** Photo triac driver. Copyright of Motorola, Inc. Used by permission.

# Summary

Radio frequency amplifiers are used in receivers and transmitters. Transmitters are used to generate a signal that can be broadcast, and receivers are designed to receive transmitted signals and process them for any intelligence on the received waves.

A circuit that amplifies frequencies that are above human hearing is referred to as a radio frequency (RF) amplifier. VHF frequencies are located between 30 and 300 MHz. UHF frequencies are located between 300 and 3000 MHz. RF stages are coupled by tuned stages in IF amplifiers and similar circuits. Double-tuned transformer coupling has some advantages. Resonant conditions of the tank circuit result in a gain in signal voltage that is very selective. This type of coupling is frequently used in IF amplifiers in receivers and in output stages of transmitters. The arrangement is also available in a stage with only one tuned circuit.

The RF amplifier deals primarily with tuned circuits. Tuned circuits used as inputs to grounded-emitter and grounded-base amplifiers have a very low input impedance. Power amplifiers that boost radio frequencies are needed for use in transmitters. In almost all cases they require larger input signals.

If a transistor is used as an RF amplifier, it is usually operated at class C. This means for most of the input hertz only a small amount of cutoff current flows in the collector circuit. It conducts only during the positive peaks of the input signal and only then in bursts. The emitter is positive biased with respect to the base.

There are two types of feedback to look for in amplifiers — regenerative and degenerative. These are also called positive and negative, or inverse, feedback. Regenerative means positive, and degenerative means negative, or inverse. Regenerative produces a higher output signal. Degenerative, or inverse, feedback produces better fidelity and less distortion.

Integrated circuits (ICs) are used in all types of modern electronic equipment. Integrated circuits are just what the name implies: they are made up of a number of transistors, diodes, resistors, and capacitors — all located on one chip and housed in a complete package. There are three categories of IC packages — small-scale, medium-scale, and large-scale integration. Each type designates the number of components that are housed in the enclosure or package. Integrated circuits can be used to do any number of things electronically. They are classified further according to their function. The two broad categories of classification are digital and linear. The op amp is the most popular of the integrated circuits. The op amp produces very high gains in a frequency range from 0 to 1 MHz. The symbol for an op amp is the triangle. It can be connected in an open- or closed-loop arrangement. The closed loop has a feedback circuit externally connected. The differential amplifier (diff amp) is a linear amplifier with two inputs. They use two identical transistors. They can be produced commercially identical on an IC chip.

The field effect transistor has high input impedance compared with the low input impedance of the bipolar transistor. The FET operates on low dc supply voltages just like any other transistor. It has three terminals labeled gate, drain, and source.

The FET can be classified as to two major types: JFET or junction FET and MOSFET, which is a metal oxide semiconductor. The JFET is further classified according to the type of material that makes up its channel. It is either a P-type or N-type. The MOSFET is classified according to its mode of operation.

FETs can be connected into three configurations. These are the same, basically, as any other transistor: common gate, common source, and common drain. The configuration varies according to the job the FET is expected to perform.

Gallium-arsenide FETs are called GASFETs. They are made for use in a number of current high-technology circuits and perform various functions according to their construction.

# Review Questions

1. Where are radio frequency amplifiers used in electronics?
2. What is the name of the frequency range above human hearing?
3. What is the range of frequencies designated VHF?
4. What is the range of frequencies designated UHF?
5. Decribe tuned circuit coupling.
6. What is a double-tuned amplifier?
7. What type of circuits does the RF amplifier contain?
8. How do you obtain low-impedance inputs for RF amplifiers?
9. Where are power amplifiers used in RF circuits?
10. What class of operation do RF power amplifier stages have?
11. Why is the emitter reverse-biased in a transistor power amplifier stage?
12. What are the two types of feedback employed in RF amplifier circuits?
13. What is the difference between positive and negative feedback?
14. What is a DIP?
15. How many components can you put into an MSI package?
16. What is a linear amplifier?
17. What is an op amp?
18. Why are op amps so popular?
19. What does closed loop mean?
20. What is a differential amplifier used for?

# Chapter 10

# STEREO EQUIPMENT

For a "stereo" effect, we need at least two amplifiers and two speakers to reproduce sound which has good fidelity to the original, giving you a feeling of *being there* when the music was recorded.

Recording and reproducing high-quality sound dates back to the 1930s when Edwin Armstrong transmitted high-fidelity musical programs over his FM station. Some record companies, such as Victor, produced long-playing $33^1/_3$ records during the thirties. They were available to those who wanted to pay the high prices and had the equipment to play them. World War II stopped the development of high-fidelity equipment until about 1946 when the FM band was shifted, and people were able to hear the better-fidelity sound produced by FM equipment. In 1947 the variable reluctance pickup for the phonograph gave the high-fidelity movement a boost forward. Diamond- and sapphire-tipped styli replaced the old-fashioned steel needle for record players in 1948 when Columbia produced the long-playing (LP) microgroove record, and plastic records replaced the shellac. The development of the Williamson amplifier and the use of two-speaker systems launched the movement toward today's high-quality sound.

The word stereo is derived from the Greek meaning "solid," or "three-dimensional," space. When properly reproduced, stereophonic sound creates an aural perspective and produces a feeling of presence and an illusion of depth. It causes the ear to reject distortion and helps it to hear a wider range of frequencies. A primitive attempt to demonstrate stereo sound reproduction was made at the Paris Exposition of 1881 when engineers used two telephone circuits for the transmission of programs from the stage of the Paris Opera.

For a better understanding of stereo equipment and the task it is called upon to perform, we must take a closer look at the basics of sound.

## Nature of Sound

The origin of sound is always a vibrating body. A good example of a vibrating body which produces a sound is a tuning fork. When a tuning fork is struck, the tines begin to vibrate (see Figure 10-1). When the tine marked A moves to the right, it pushes the molecules in that region to the right, thereby producing a *compression* of air molecules. When the same tine moves to the left, it produces an area of reduced pressure called a *rarefaction*. Every sound wave is composed of these compressions and rarefactions of air. If the tines are permitted to vibrate continuously, a continuous sound would be generated. The frequency of the sound produced by the tuning fork is dependent upon the length and mass of the tines. The longer they are, the lower will be the frequency of the sound produced.

When the compressions and rarefactions produced by the tuning fork reach the

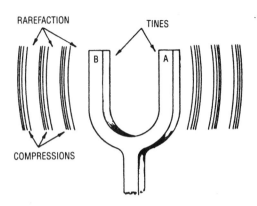

**Figure 10-1.** Tuning fork and its tines.

eardrum of the listener, they produce small inward and outward motions of the eardrum. The process of hearing the sound is thus begun. The velocity at which the sound waves travel is dependent upon the medium through which they travel. If the medium is air, the velocity of sound is approximately 1080 feet per second (ft/s). The response of the human ear to sound frequencies is given as from 16 Hz to 16 kilohertz (kHz). It is an especially good ear that can hear above 16 kHz.

The sound produced by the tuning fork is composed of many frequencies. There is the normal vibrating frequency, which is called the *fundamental frequency*, and other frequencies, which are called *overtones*. In musical instruments, the overtones are usually multiples of the fundamental. These multiples are called *harmonics* (see Figure 10-2).

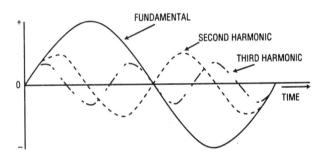

**Figure 10-2.** Note how the second and third harmonies relate to the fundamental frequency.

A sound may be described as having three characteristics: pitch, loudness, and quality. The *pitch* of a sound generally depends on the frequency of the fundamental. As the frequency increases, the pitch also increases. *Loudness* describes the magnitude of the auditory sensation produced by the sound. The sound may be so loud as to produce physical pain and permanently damage the ear.

*Quality* is a comparison of different notes. A *note* is defined as a tone of a definite pitch. The pitch, or note, can be played on a guitar and a trumpet, and the listener will be able to distinguish between them. The reason for this is that both the guitar and the trumpet produce a note that is not only composed of the fundamental frequency but also includes the harmonics. When the two notes differ in quality they also differ in the harmonic frequencies produced and the relative intensity of their various overtones.

In order to make these sounds useful to the field of electronics you must be able to convert them to electric impulses. A microphone is a device that converts sound waves to electric impulses. Various types are available and used for different types of recording or as inputs to audio amplifiers for public address systems.

# Microphones

Sound energy can be converted to electric energy by the use of a microphone. There are a number of types of microphones available today, but we will limit our discussion to three basic types so that you can see the principles on which they work (see Figure 10-3).

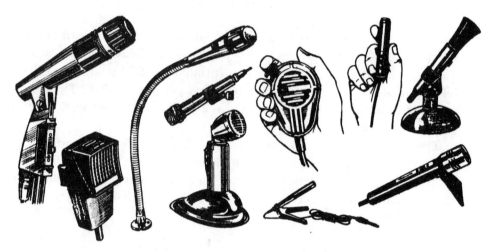

**Figure 10-3.** General-purpose microphones.

When you speak into a microphone, the audio pressure causes a diaphragm to move in accordance with the pressure applied to it. The diaphragm is attached to a device that causes current to flow in proportion to the instantaneous pressure applied to the diaphragm. Figure 10-4 shows the construction of a carbon microphone where the diaphragm moves the carbon granules back and forth to compress them and change the internal resistance of the carbon material. This in turn causes the current through the resistance (carbon granules) to vary with the pressure generated by the audio striking the diaphragm.

Microphones are rated according to their frequency response, impedance, and sensitivity. For good quality, the electric waves from the microphone must correspond closely to the magnitude and frequency of the sound waves that cause them, so that no new frequencies are introduced. The frequency range of the microphone (range of frequencies over which the microphone is capable of responding) need be no wider than the desired overall response limits of the system with which it is to be used. The microphone response should be uniform, or flat, within its frequency range, and free from any sharp peaks or dips.

The actual impedance of a microphone is of importance chiefly because it is related to the load impedance into which the microphone will operate. If the load has a high impedance, the microphone should also have a high impedance, and vice versa. Of course, impedance-matching devices may be used between the microphone and its load (see Figure 10-5).

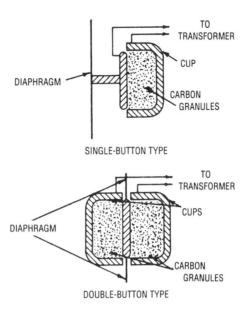

**Figure 10-4.** Carbon microphone construction.

**Figure 10-5.** Microphone line transformer matches the low-impedance microphone to high-impedance input of amps.

The sensitivity or efficiency of a microphone is usually expressed in terms of the power level which the microphone delivers to a terminating load. It is important to have the microphone sensitivity as high as possible. High sensitivity means a high-power output level for a given input sound level. High microphone output levels require less gain in amplifiers used with them and thus provide a greater margin over thermal noise, amplifier hum, and noise pickup in the line between the microphone and the amplifier.

Microphones presently available make use of the properties of resistance, inductance, and capacitance. Microphones that make use of the piezoelectric effect are also available. The types of microphones that will specifically be discussed in this chapter are the carbon microphone; the moving coil, or dynamic, microphone; the velocity microphone; and the crystal microphone. The function of any of these microphones may be expressed graphically as shown in Figure 10-6.

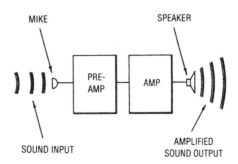

**Figure 10-6.** Block diagram of an audio system.

# Carbon Microphone

The carbon microphone is made of a diaphragm that vibrates with the impulses of sound and turns the vibrations into electric energy. It turns the electric current through the circuit into electric impulses by compressing the carbon granules. As the density of the granules changes, so does its electrical resistance. As the resistance changes, so does the current flow through the carbon deposit. Figure 10-4 shows the construction of the carbon microphone. If the source of the pressure variations is the human voice, the compressions and rarefactions of air applied to the carbon pile will cause its resistance to vary. In fact under these conditions, the resistance of the pile will vary at an audio rate.

In the single-button type of carbon microphone, the carbon granules are placed in a cup or button and are permitted to make contact with the suspended perpendicular element, which is the diaphragm. If a stress is placed on the diaphragm, the pressure exerted on the carbon granules is increased, and the resistance of the carbon piles decreases.

In the double-button microphone, there is more of a push-pull action. Any movement of the diaphragm increases the pressure in one cup while decreasing the pressure in the other cup by approximately the same amount. The word *approximately* is used here because it is highly unlikely that the two cups would be filled with exactly the same amount of carbon granules.

Figure 10-7 shows how the carbon microphone is connected in a circuit. Any current flowing in the microphone circuit will also flow through the primary of the transformer. With no movement of the diaphragm, there will be no change in the resistance of the cup. Under these conditions, the dc flowing through the cup will be constant, and the resulting magnetic field about the primary of the transformer will not fluctuate. Thus, no voltage will be induced in the secondary of the transformer.

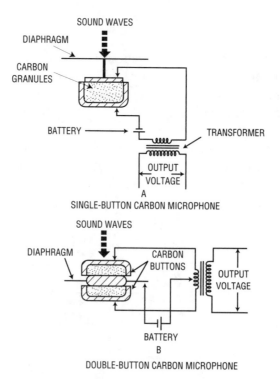

Figure 10-7. Circuit for a carbon microphone.

When the diaphragm is compressed, the resistance of the cup is decreased, and the current flow through the circuit is increased. The field that was stationary about the primary will expand and cut the secondary windings. A voltage is induced in the secondary of the transformer. If the pressure on the diaphragm is reduced, the resistance of the carbon pile increases, and circuit current decreases. The voltage induced in the secondary of the transformer is then of a reverse polarity. If the diaphragm is moved at an audio rate, the voltage induced in the secondary of the transformer will also vary at an audio rate. Therefore, the amount of voltage induced in the secondary is dependent on the pressure applied to the microphone diaphragm. The frequency of the output voltage is dependent on the frequency of the input. The frequency limitations of the microphone are governed by the ability of the carbon granules to change their density.

When the double-button microphone is used, the amount of possible distortion realized through the constant shuffling of the carbon granules is reduced. The push-pull effect realized by the use of the double-button microphone and the center-tapped transformer tends to cancel the even-order harmonics.

The operating range of the carbon microphone is between 100 and 5000 hertz (Hz).

The carbon microphone is used in the mouthpiece or transmitter of the telephone (see Figure 10-8). The felt washers in the microphone furnish damping. The low frequencies applied to the microphone are usually more intense. The tendency of the diaphragm to maintain the low-frequency vibrations after the force is removed is very great. By using the felt washers, the vibrations after the force has been removed are minimized or damped.

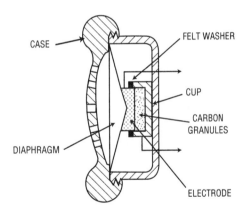

**Figure 10-8.** Carbon microphone: cross-sectional view.

It will be seen that the carbon microphone possesses certain advantages that make it a valuable device. In comparison to other microphones that will be discussed, the carbon microphone produces a relatively large output. It also has the advantage of being lightweight, inexpensive, portable, and rugged. It is used in places where wide frequency response may be sacrificed for good sensitivity. In voice communications, the quality of the speech is not critical. Therefore, the carbon microphone enjoys wide popularity in telephones.

Because of the low impedance of the carbon microphone, an impedance-matching transformer is required when it is desired to send the output of the microphone to the high-impedance input of a speech amplifier (see Figure 10-5). However, since the output is in the order of about –50 decibels (dB), the amplification required for the carbon microphone is less than that required by other types of microphones. One disadvantage apparent with the use of the carbon microphone is the noise generated by the loosely packed carbon granules. One other disadvantage of the carbon microphone is that it does require an external dc source.

# Dynamic Microphone

The dynamic microphone is really a moving-coil type. The name *dynamic* comes from the fact the microphone has a moving part (see Figure 10-9). The coil winding that is wound around the pole piece is able to move up and down the pole. It is attached to the flexible diaphragm and is caused to move by sound waves striking the diaphragm. As it moves, it passes through the magnetic field set up between the poles of the magnet. Since the diaphragm will move at an audio rate, the voltage induced into the coil when it moves will also vary at an audio rate. The diaphragm in Figure 10-9 shows a cross section of the moving-coil microphone.

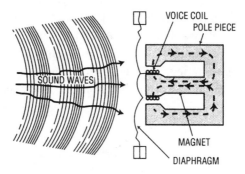

**Figure 10-9.** Moving-coil microphone.

In Figure 10-10 the diaphragm is located close to the pole piece. The diaphragm is held in place by the ring and washer. The area under the diaphragm is completely enclosed except for the narrow slit designated $S_1$. This slit serves to control the response of the microphone by imposing a load on the diaphragm as a result of the increased air resistance of the enclosed air spaces designated as $O_1$ and $O_2$. The cavity marked $C$ improves the damping action and the frequency response. It also serves to increase the faithfulness of signal conversion. At the low frequencies, the stiffness of the diaphragm governs its motion

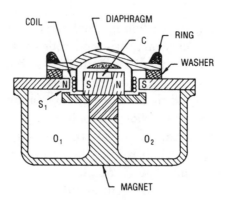

**Figure 10-10.** Moving-coil microphone: cross-sectional view.

and results in a response that normally decreases as frequency applied decreases. The decreased frequency response begins at about 200 Hz, and if left uncorrected, the sensitivity of the microphone falls off rapidly below 200 Hz. However, if the effective force on the diaphragm is increased at a rate corresponding to the decreased motion, caused by stiffness, a uniform response is obtained. In this moving-coil microphone, the uniform response is obtained by having an air passage provided by the tube which connects the diaphragm to the air chamber within the magnet.

The tube length and diameter can be proportioned to make it possible to reduce the internal pressure at the desired frequency. Therefore, at low frequencies, the effective force on the diaphragm increases to offset the increased stiffness.

This type of microphone has low impedance between 50 and 100 ohms ($\Omega$). With this type of microphone, an impedance-matching transformer is used. The microphone can be designed so that its frequency response is from 30 Hz to 18 kHz. It has an output in the order of about –85 dB. The sensitivity of the moving-coil microphone is high. It is light, rugged, moisture-proof, and small, and is not subject to the effects of temperature and humidity. It does not require an external source of dc voltage. Keep in mind that a permanent-magnet speaker can also be used as a dynamic-type microphone. As we examine speakers you will see the similarity. This is why the speaker can be used as a microphone in an intercom system.

## Velocity Microphone

Microphones do pick up noise. The two previously discussed microphones are subject to picking up noise. That makes them limited for use in some cases. This disadvantage of noise pickup can be eliminated by using a velocity microphone (see Figure 10-11 to see how it operates). The operation of the velocity microphone is fundamentally the same as that of the moving-coil type. Instead of the moving coil, a strip of metal is caused to vibrate in the magnetic field (see Figure 10-11). This is because the metal strip, called the *ribbon*,

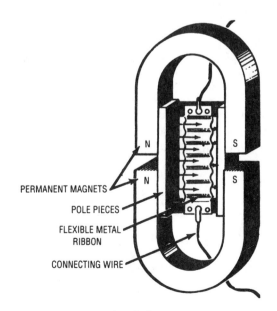

**Figure 10-11.** Velocity microphone: cross-sectional view.

performs the function of the diaphragm. The metal strip is arranged in such a way that its length is perpendicular to, and its width is in the plane of, the magnetic lines of force. Notice that the pole pieces are constructed in such a way that air may pass freely through the microphone.

The ribbon passes through the magnetic field as it is caused to vibrate by the sound waves striking it. An emf is induced in the ribbon in proportion to the strength of the magnetic field that strikes the ribbon, the velocity that the ribbon cuts through the field, and the length of the ribbon in the field. Since the ribbon is caused to move through the field at an audio rate, the voltage induced in the ribbon will also vary at an audio rate.

This type of microphone is subject to damage by sudden gusts of wind. That is why it is covered by a fine mesh of silk and protected physically by having a screen wire covering its outer surface. It also has to be protected from materials that may be attracted by the strong permanent magnets.

The name *velocity* microphone comes from the ability of the very thin ribbon to move rapidly with any small amount of wind pressure or with any pressure created by the human voice. This is a bidirectional microphone since it can respond to air pressure changes from either the front or back. The ability of the ribbon to respond to various pressure changes causes it to have a very good frequency range or response of between 20 Hz and 15 kHz. This is a very sensitive microphone and can pick up very weak sounds. A person talking into the microphone can stand about 18 in. from it easily without having to shout or talk in a loud voice. The impedance is very low but can be raised with a transformer to between 20 and 600 Ω. Output is about –90 dB.

# Crystal Microphone

The piezoelectric effect is used to advantage when the crystal microphone is used. The piezo effect refers to certain crystals that give off a small current when pressured. By taking the characteristics of the crystal (Rochelle salts or quartz) and having a diaphragm attached so that it touches the crystal, it is possible to utilize this electric generator as a microphone (see Figure 10-12).

There are two types of crystal microphones: the directly actuated and the diaphragm-actuated. In the directly actuated type, the sound acts directly on the crystal (see Figure 10-12). Figure 10-13 shows the diaphragm-actuated crystal microphone. In the directly actuated type, thin pieces of metal make contact with the crystal and the foil on each side, causing the crystal to be activated directly by the pressure waves of the voice. This is not a very efficient arrangement. When the sound waves strike the surface of the crystal, the force is dissipated over its entire surface. To obtain higher efficiency, a diaphragm is used.

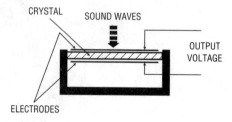

**Figure 10-12.** Directly actuated crystal microphone.

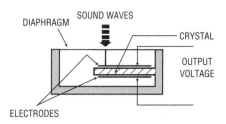

**Figure 10-13.** Diaphragm-actuated crystal microphone.

In the diaphragm-actuated microphone, the full force of the diaphragm is exerted on a small area of the crystal. Therefore, as the diaphragm moves back and forth, the voltage produced by the crystal varies at the same rate. Since the diaphragm is mechanically connected to the crystal, considerable stress will be placed on the crystal. The output of the diaphragm-actuated microphone is higher than that of the directly actuated type.

Frequency response of crystal microphones is not very uniform. This is due to the inertia of the crystal. However, for applications such as amateur radio or any noncritical application, they are widely used. Because of its high impedance, the crystal microphone may be directly connected to the input of the speech amplifier. The output of the crystal microphone is in the order of –55 dB. The diaphragm type has a frequency response of 80 to 6000 Hz. The type of crystal most widely used is Rochelle salt because of its sensitivity. The disadvantages of the crystal microphone are its sensitivity to temperature and humidity change and its susceptibility to damage from rough handling.

# Turntables and Record Changers

Every stereo system has more than one type of input. The ability to play records was once paramount to most people purchasing a stereo system. That meant the person had to choose between a record changer and a turntable. Certain distinctions made a real difference in the quality of sound.

## Record Changers

Record players came in many sizes and shapes. It all depended upon what you wanted for quality of sound and how well you treated your records. A record changer not only played the records, but it also had the ability to play more than one record. This is an outgrowth of the days when records ran for only 2 min., and it took a stack of them to last 24 or 25 min. With the advent of the long-playing (24 to 30 min.) $33^{1}/_{3}$ rpm record and its acceptance, the turntable became the choice of people who really wanted to take care of their records and who were concerned about the quality of sound reproduction.

Figure 10-14 is a typical record changer. The pickup arm is rather short in comparison with a turntable. This causes some distortion due to the way the stylus is held in reference to the grooves in the record. This record changer has four speeds: 16, 33, 45, and 78. The needle has to be changed for the 78 rpm records since the grooves are wider than those in the other records with lower speeds. The stacking of records on top of one another can cause damage to the surface of the records. Dropping the records also can scratch or

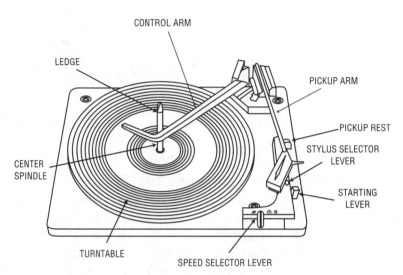

**Figure 10-14.** Record changer with four speeds.

damage the surfaces of both records at the time of impact. Then, there is the possibility of slippage between the two surfaces. Some models had a cueing and pause control. The device allowed you to manually raise and lower the stylus under positive control at any desired point on the record. The pressure exerted by the stylus on the record should be adjusted with a pressure gauge.

# Turntables

Turntables and record players have been included for their historical significance and are seldom used today. In an argument as to whether a turntable is better than a record changer, it is the turntable that wins in most instances (see Figure 10-15). Here are some of the advantages of the turntable over the record changer: The record changer has a problem with its shorter pickup arm inasmuch as tracking error and distortion are produced by it. The stacking of records can cause damage to them. Loading the records on the changer can get to be time consuming, and damage can result in record surfaces. The stylus is designed to contact the record groove at an angle of 90°. As records pile up on the changer, the stylus contacts each succeeding record with a different angle of incidence and with a different amount of stylus pressure. This increases the amount of wear on both the record and the stylus, and produces distortion.

The speed of the turntable can be adjusted in most of the more expensive devices. Figure 10-15, for instance, has an adjustment for obtaining the proper speed so that the turntable will rotate precisely at the right speed and reduce wow and noise. Figure 10-15B shows the strobe that is used to adjust the speed. An adjustment is made to obtain the correct speed when the marks seem to be stationary. Placing only one record at a time aids in eliminating some of the problems associated with record changers. Handling the records by the edges also keeps fingerprints and dust from accumulating and causing damage to the recording. In order to obtain a constant speed turntable, a frequency generator servo dc motor is used. It can be adjusted slightly as shown in Figure 10-15. Most turntables have only two speeds: $33^{1}/_{3}$ and 45 rpm.

45rpm
33-1/3rpm | 50Hz
45rpm
33-1/3rpm | 60Hz

The strobe-line
for 60 Hz, 33-1/3 rpm
seems to be stationary.

In the U.S.A. and CANADA use
60 Hz lines.
The 50 Hz lines are for European
countries.

A

B

speed
33    45

C

**Figure 10-15.** A. Turntable. B. Speed adjustment. C. Speed selection.

# Pickups

There are at least six types of record pickups or styli (see Figure 10-16). Note how each one of the methods uses the movement generated by the needle following the grooves in the record to produce a small emf which can be amplified by the stereo amplifier. Construction details on the moving-coil dynamic pickup (Figure 10-17) shows how this popular device uses the moving coil in a magnetic field to generate an ac voltage which can be amplified. Figure 10-18 shows the simple magnetic principle that uses an air gap reluctance as a means of generating an output for the amplifier. More details on this type of pickup are shown in Figure 10-19. Details on how the crystal pickup is made and how it is able to produce a higher output than most other types are shown in Figure 10-20. Newer ceramic pickups use barium titanate in place of Rochelle salts to produce a flat response from 20 to 20,000 Hz. Zirconium dioxide has also been used to make some ceramic pickups. In good tone arms a tracking force of only 2 to 4 grams (g) is needed for full output.

All these record players and pickups are designed to be used in conjunction with an amplifier. They may be monaural, or one-channel, output, or they may be stereo with two channels. They all serve as another source of input to an amplifier. However, another type of input is the tape recorder. It may be of the reel-to-reel type, cassettes, or cartridges.

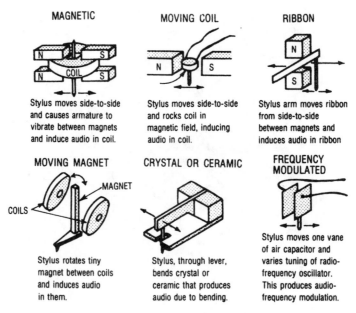

**Figure 10-16.** Different types of pickups.

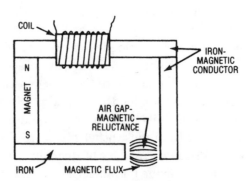

**Figure 10-17.** The moving-coil dynamic pickup.

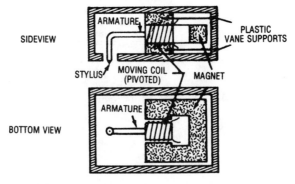

**Figure 10-18.** Variable reluctance pickup.

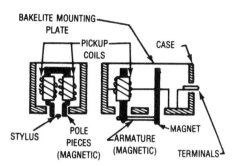

**Figure 10-19.** Variable reluctance pickup with construction details.

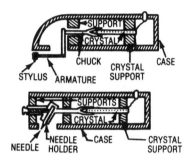

**Figure 10-20.** Two types of crystal pickups with construction details.

# Tape Recorders

It was in Denmark in 1898 that Valdemar Poulsen patented the first workable magnetic recorder. The quality was poor, but the heavy piano wire used for the recording was later improved and served during World War II as a means for pilots to keep track of their victories while dog fighting. Poulsen's machine used a magnet that was moved along a length of wire. It was set up so that one pole of the electromagnet surrounded the wire. When the microphone produced the electric impulses from a voice or loud noise, they were fed to the electromagnet and recorded on the wire in the form of a continuous series of transverse magnetizations varying in polarity and strength. Playback of the recording was accomplished by connecting the electromagnet to a telephone receiver, and the electromagnet was moved along the wire.

During World War II the Germans developed the magnetic recorder to such a degree that it allowed Hitler to record his messages and have them played back at various places over the air when he was actually somewhere else in hiding. The German recorders used a plastic tape coated with a thin layer of fine particles of iron oxide. Today's recorders are a direct result of those early models.

The tape used in tape recorders may have an oxide coating that is about 0.0006 in. thick. Most of the plastic is a Mylar made by DuPont Corporation. Acetate is also used as the film for tape. Take a look at Figure 10-21 to see how the tape is recorded. The recording head is an electromagnet that has the sound energy converted to electric impulses fed to it. This varying magnetic field causes the iron filings or particles on the tape to become small magnets. The output of the tape player can be reproduced by pulling the magnetic

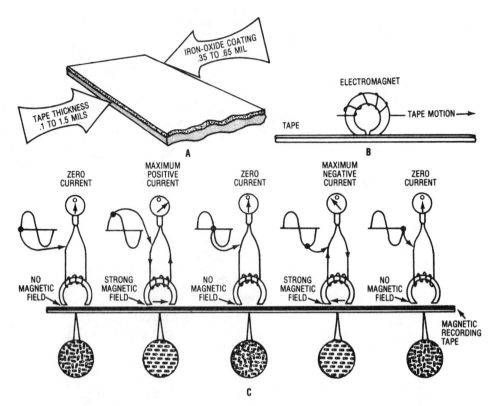

**Figure 10-21.** A. Magnetic type with construction details. B. Magnetic head recording. C. Producing a sine wave voltage onto a magnetic tape.

tape over a playback head. An erase head with ac fed into it causes the tape to be erased before it is rerecorded (see Figure 10-22B).

Uses for tape recorders are unlimited. Medicine, industry, business, and computers all use magnetic tape recorders in one form or another. They are very much present today in video recorders where programs from the air can be taped, and played back later on the television set at the viewer's leisure.

# Tapes

Tapes are available in cassettes and reels. Cassettes are very much in evidence when portable tape decks are used by young people to take their music with them wherever they go. Some very small cassettes are becoming popular and may replace the larger ones (see Figure 10-23). Figure 10-24 shows how information can be recorded on tape that is easily read by computers. Storing computer information on tape was common in the 1980s; however, with today's improvements in technology, microcomputers now use disk storage. However, some commercial use is made of tape storage.

Tapes come in $1/4$-, $1/2$-, and $3/4$-in. sizes, while broadcasting stations use 1- and 2-in. widths. The $1/4$-in. tapes can record stereo or two tracks in one direction and two tracks in the other. Figure 10-25 shows how they are recorded so that they are not subject to adjacent channel cross talk. There has to be some spacing between channels to protect the output from stray signals. Keep in mind that speed and tape quality make a difference in the

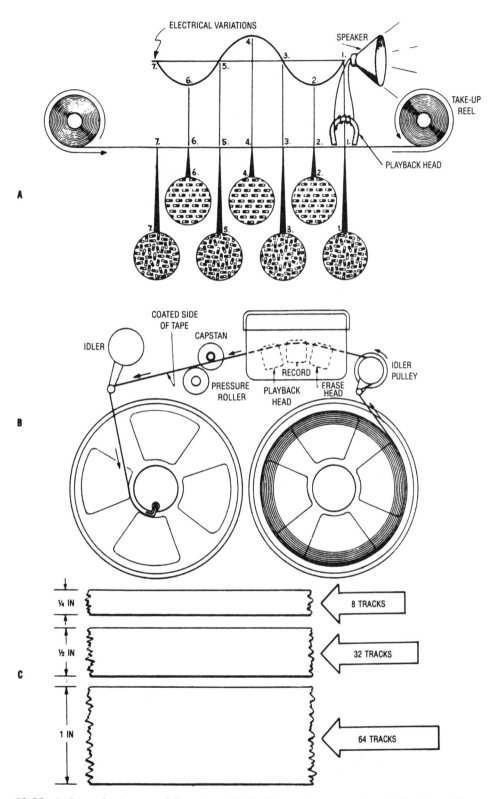

**Figure 10-22.** A. Reproducing sound from tape. B. Reel-to-reel tape recorder-playback machine. C. Tape widths and more space to record various tracks. Up to 128 tracks can now be recorded on 1-in. tape.

**Figure 10-23.** Four-track cassette ready to drop into the recorder.

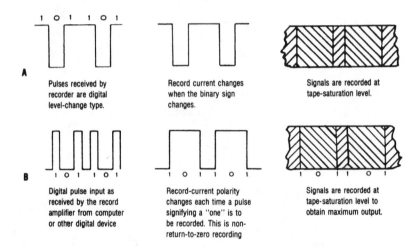

**A**

Pulses received by recorder are digital level-change type.

Record current changes when the binary sign changes.

Signals are recorded at tape-saturation level.

**B**

Digital pulse input as received by the record amplifier from computer or other digital device

Record-current polarity changes each time a pulse signifying a "one" is to be recorded. This is non-return-to-zero recording

Signals are recorded at tape-saturation level to obtain maximum output.

**Figure 10-24.** Early methods of recording computer information. A. The PCM method. B. The NRZ (nonreturn-to-zero) method.

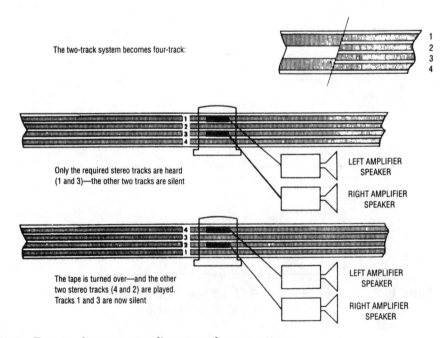

The two-track system becomes four-track:

1
2
3
4

Only the required stereo tracks are heard (1 and 3)—the other two tracks are silent

LEFT AMPLIFIER
SPEAKER

RIGHT AMPLIFIER
SPEAKER

The tape is turned over—and the other two stereo tracks (4 and 2) are played. Tracks 1 and 3 are now silent

LEFT AMPLIFIER
SPEAKER

RIGHT AMPLIFIER
SPEAKER

**Figure 10-25.** Four-track stereo recording as used on cassettes.

fidelity of the output. Wider tapes used by video recorders, both home and commercial types, allow for more signal to be recorded and less interference between the various information sources. No matter what is recorded, it has to be played back through an amplifier — and, in most instances, a preamplifier.

# Compact Disc

## How It Works

The compact disc has all but displaced the record as the primary source of recorded music. Record companies are now in the business of producing discs instead of records. All this rapid displacement of the record as a means of entertainment came about with the development of a compact disc player that could be transported with little or no effort and adapted to use in automobiles and home stereo systems with little or no additional expense. As in almost all electronics equipment revolutions the price goes down as the volume or demand increases.

Advantages of the compact disc over the record are many. The compact disc is not as susceptible to the accumulation of dust on its surface (as it is sealed between two layers of transparent plastic) as are the tiny grooves in the record. The disc is smaller, not subject to breaking or cracking, and has a smooth surface. The CD player has the ability to reproduce sound free of hiss, record pops, and noises of various kinds associated with needles riding over plastic. Now that multiple CD playing equipment has been developed with the easy disc-handling capability, it is possible to play hours of music without interruption.

### The Disc

Today's compact disc looks something like the old 45 rpm records. The CD hole (15 mm) is in the middle and has a 120 mm outside diameter (less than 5 inches). The disc is 1.2 mm thick. It is possible to place a little more than an hour of music on one disc (see Figure 10-26). There is also a 3-in. CD, but due to the lack of space here, we can only describe one. Therefore, we will concentrate on the 5-in. size. Some disc players will play both the 3- and 5-in. discs.

The silver CD has a pickup stylus rather than grooves. A laser beam is used in the stylus to aid in the pickup of the music on the disc. The CD inside diameter speed is around 500 rpm but slows down to approximately 200 rpm at the outside rim. This compares to the old 45 rpm record's speed being constant. The stylus starts reading the information on

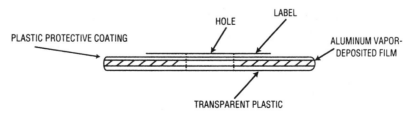

**Figure 10-26.** Construction details of the compact disc.

the disc near the center hole and moves outward toward the edge. The phonograph records did the opposite by starting at the outside edge and coming inward toward the center hole.

## Comparison of Phonograph Records and CDs

Phonograph records use a needle against the plastic disc to pick up the recorded information that has been placed in grooves. Instead, the CD stylus's laser beam rides through a groove that has been *coded* with varying amplitudes that correspond to the sound signal. The CD information groove is sealed between two layers of transparent plastic. The information is embedded in an aluminum vapor deposited film that is then coated with two layers of plastic, one on top and one on the bottom.

Compact discs have a track of microscopic indentations called *pits*, instead of grooves. These pits and the space between the pits are the *encoded digital representation* of the original audio information that was in analog form. The high-density information on the tracks is read by a laser pickup device. The pickup has no physical contact with the surface of the disc (see Figure 10-27).

The phonograph record has the music recorded or embedded in the plastic in grooves that have been distorted in an analog manner. The CD utilizes digitized coding to record the music. By digitizing the music signals it is possible to eliminate both deterioration of the signals through the recording process and playback process as well as the mechanical restrictions or physical wear (see Figure 10-28). The CD incorporates a high density and high fidelity that could not be achieved with conventional systems (see Figure 10-29).

Phonograph records contain two channels of information. One channel of information is placed on each side of the groove. One stylus is used to pick up both channels of information. There is a great amount of cross talk between channels. With the compact disc method of recording, right and left channel information are in *serial* sequence. That means

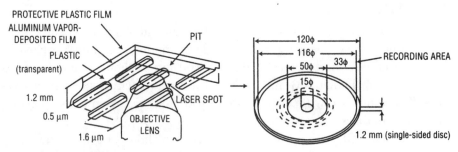

**Figure 10-27.** Details of the CD.

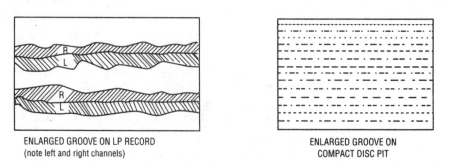

ENLARGED GROOVE ON LP RECORD
(note left and right channels)

ENLARGED GROOVE ON
COMPACT DISC PIT

**Figure 10-28.** Comparison of the LP record groove and the compact disc groove with pits.

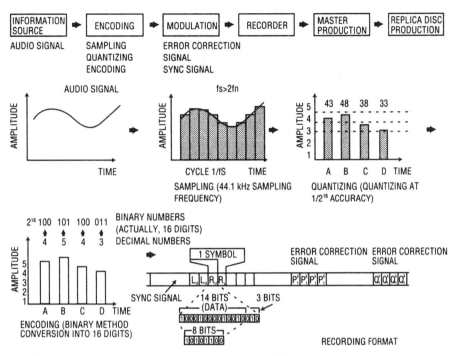

**Figure 10-29.** Step-by-step procedure for the recording of a CD.

one channel is recorded and then the other, in series or one after the other. This produces channel separation that is extremely good. Good channel separation is important for accurate stereo reproduction.

# Digitizing the Audio

Digital electronics is a field of study all its own. It has developed rather rapidly with the advent of computer technology being applied to the audio recording and playback field. The following is an attempt to explain in detail in simple terms some of the aspects of the process. By studying Figure 10-29 you should be able to decipher the process.

The audio signal to be recorded is a smoothly changing or analog waveform that is sampled at a 44.1 kHz rate (see Figure 10-29). The analog audio signal is converted to a digital pulse train without any loss in information. This happens as long as the sampling rate is at least two times the highest frequency to be reproduced. Sampled audio is converted to 1's and 0's using 16 bits of resolution. This conversion provides 65,536 (or $2^{16}$) possible voltage level representations. The dynamic range of the system using 16-bit resolution is greater than 90 dB. It is possible that some sampled voltage levels might fall between two steps of the 65,536 combinations. Then the voltage level is rounded off to the closest 16-bit level. Keep in mind that 65,536 is also $2^{16}$.

## Encoding

In order to make sure the recording and playback of the music is exactly the same, a very complex *encoding scheme* is used to transform the digital data to a form that can be placed on the disc. Each 16-bit word is divided into 8-bit symbols (see Figure 10-29). These symbols are arranged in a predetermined sequence with error correction, sync, and sub-code information added. The subcode is used to store index and time information. This information is then modulated by a process known as *eight-to-fourteen-bit modulation* (EFM). The 8-bit data is changed to 14-bit data through the use of a ROM-based (*read-only memory*) IC (integrated circuit or chip). The EFM reduces the disc system's sensitivity to optical system tolerances in the disc player. The three merging bits are added to each word to produce a 17-bit unit that contains the sync and subcode data. The encoded data is then recorded onto the disc as a series of small pits of varying lengths. During the playback process, the laser pickup reads the transition between the pit and the mirror — called the *island* — not the pit itself. A disc can hold up to 2.5 billion pits. The pits are about 0.5 microns wide. A micron is one-millionth of a meter.

## The Laser Pickup

Needless to say, the laser pickup is rather delicate. It has to be, in order to track correctly the small bits of embedded information in the disc. There are bascially two types of laser pickups. One type is the arc, or swinging-out arm mechanism, and the other is the slide or sled mechanism that glides along metal rods straight out from the center of the disc. The pickup assembly is made up of the objective lens, focus-tracking coils, collimating lens, beam splitter, semitransparent mirror, photodetectors, monitor, and laser diode (see Figure 10-30).

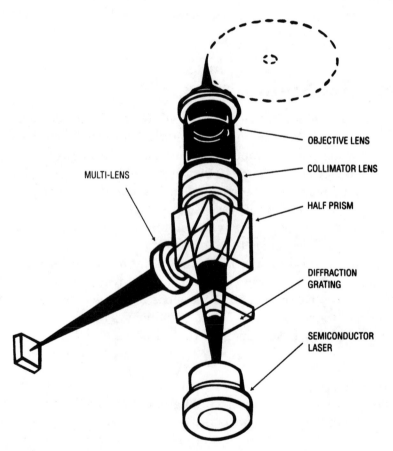

**Figure 10-30.** Identification of the optical path and the parts of a CD stylus or pickup.

## Optical Selection

There are two types of optical pickups. They can have either one or three beams. There is very little difference in the sound output of the player, but most players use the three-beam system that is slightly more complicated in design. The semiconductor laser light source has a wavelength from 750 to 850 nm. A *nanometer, nm*, is one-thousandth of a millionth of a meter, or 0.000 000 001 meter, also read as $1 \times 10^{-9}$.

The path of the laser beam and the arrangement of the optical elements in the optical system shown in Figure 10-30 uses the semiconductor laser to emit a beam of light with a wavelength of 780 nm. It is barely within the range of visibility. The beam is produced from an extremely small point and has an elliptical distribution. It is dispersed in a conical shape.

The beam used to detect tracking error is produced by having the beam pass through a diffraction grating that splits it into three separate beams. The beams are the primary beam or *zero order* and two side beams (*plus or minus order*). Some higher *order* elements are also produced, but they are lost and not used. Then the beams are passed through a half prism where 50 percent of the energy is lost (see Figure 10-30).

The *collimator lens* produces a completely collimated or *parallel* beam (Figure 10-31). The collimated beam diameter is large enough to cover the movement of the *objective lens*. The beam is then condensed to a spot with an extremely small diameter by the objective lens before it is radiated to the disc. Part of the beam is then reflected back from the disc, diffracted, and then routed back through the objective lens to be re-collimated and condensed.

When the reflected beam reaches the half prism, 50 percent passes through the grating and returns to the laser diode. The other 50 percent is reflected by the prism to the

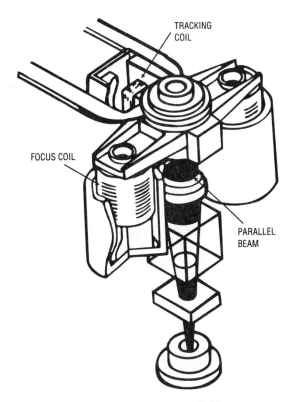

**Figure 10-31.** Focus and tracking coils control the laser parallel beam.

multiple lens that has the functions of both a concave and cylindrical lens. This beam then goes to the photodiode alley where an electrical signal with a strength proportional to the intensity of the beam is produced.

## Video and Audio CDs Compared

The optical path of the compact disc can be compared to that of the video disc player. The first feature is that the outgoing path is a straight line, so no auxiliary parts are needed to alter the light path. This way, overall tolerances can be minimized. The development of the double shaft activator for use in the parallel drive method allows the objective lens unit to be reduced in size. This makes it possible to maintain very satisfactory performance while using compact optical parts.

Another feature is the half prism. In the video disc player optical system, the outgoing and incoming light paths are separated by a $^1/_4$ wavelength panel and polarizing beam splitter. The primary reasons the half prism can be used in a CD player, but not in a video disc player, are:

- The semiconductor laser diode is much smaller than the HeNe laser. However, it has a fairly high optical power output. That means the energy loss caused by the half mirror is not a problem.
- The video and compact discs have a tendency to polarize light because they are made of a resin-based material that is not perfectly flat. In video discs, the amount of polarization is carefully checked against an established standard. In compact discs, the limitation is not very strict. Because of the lack of a strict standard, CD players normally use an extremely accurate $^1/_4$ wavelength plate. In actual use, however, this plate cannot function properly due to polarization of the laser beam caused by the disc. Since a half prism is not affected by polarization of the laser beam, a very stable optical path can be made.

Variations in compact discs make it necessary to pay close attention to the optical section of a CD player. The most important concern is accommodating any differences between various compact discs. This can be done by using a very short wavelength. The 780-nm wavelength is used because this is the shortest wavelength possible today with mass-produced pickups.

## Laser Beams

A very small beam is needed to reproduce signals encoded as a series of tiny pits. CD players use a laser beam spot of about 1.6 micrometer (1.6 millionths of a meter) in diameter. By rotating the disc and shining the laser beam on the series of pits, an *optosensor* or photodetector can be used to detect the presence or absence of the pits within a fixed period of time. Changes in the reflected light correspond to the recorded signals.

The laser beam is produced by a *laser diode* with a 780-nm wavelength and a 3-nW optical output (see Figure 10-32). The output of 3-nW is *three-thousandth of a millionth of a watt*. The beam from the laser diode is divided into three beams. All three beams pass through a half mirror and become a parallel beam. The collimator lens makes it parallel, then it gets refracted by a prism, passes through the objective lens, and focuses on the disc.

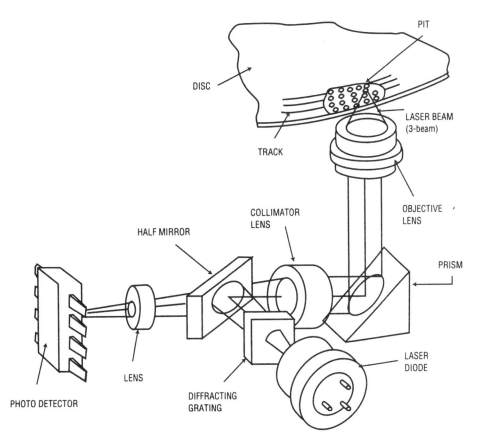

**Figure 10-32.** A three-beam pickup assembly.

Light focused on the disc reads the disc data and is reflected. It then passes back through the objective and collimator lens, through the half mirror and the flat concave cylindrical lens, then the beam strikes the photodetector. In order to be in line with the vertical *fluctuations* and *aberrations* of the disc, the objective lens moves up, down, left, and right so that the series of pits on the disc are always in focus.

## One-Beam and Three-Beams

The one- and three-beam lasers both use the objective lens, collimator, laser, and photodiodes. The one-beam system can have a semitransparent mirror and optical wedge besides those components already mentioned. The three-beam system can also have a subbeam, quarter-wave plate, polarization beam, splitter, diffraction grafting, and cylindrical lens besides those elements common to both systems. All of these components are located in the optical pickup section. Nevertheless, very little difference in sound is noticeable between the two systems. Most commercially available units utilize the three-beam system.

## Handling the Disc

The compact disc is a little more delicate than most people realize. It should be handled with care. Small pinholes in the aluminum coating can cause *dropout*, or errors in playing.

These small pinholes are hard to see. Hold the disc up to a strong light and take a look. Of course, the label might obstruct some holes, but return the disc if you see several pinholes.

Always hold the compact disc by the edges. Do not touch or scratch the rainbow side (opposite the label). Some players can play through a smear of fingerprints, but don't take any chances. Keep in mind that the side of the disc with the rainbow reflection is the side where the audio information is stored. Keep it clean. Do not stick paper or adhesive tape on the label side. Don't write on it.

Keep the compact disc free of dirt and dust. To clean a record, you wipe it with a circular motion. **Do not try to clean off the CD with this method.** If fingerprints and dust stick to the disc, wipe it with a soft cloth. Start from the center out. Excessive dust in the player can gum up the disc drive and its delicate mechanism.

Prevent scratches from covering a large area where the data bits are located. **Do not go around the disc in a circular wiping motion.** If it is hard to remove the smudges, wipe them off with a moist cloth dipped in clear water. Discs can become scratched when they are wiped clean with a dry cloth. Excessive cleaning of dust can even grind particles into the soft plastic. If it's not visibly dirty, leave it alone. You can blow off dust with a can of photo dust spray.

Do not clean the disc with benzene, alcohol, thinner, record cleaner, or antistatic agents. There are many different types of CD cleaners on the market. Some might do more damage than good, so you have to judge for yourself. Some of these units suggest that you clean in a circular motion, but this should be avoided. Make sure the commercial CD cleaner wipes outward on the disc surface. However, simply cleaning the disc with water and a soft cloth does an acceptable job.

Do not bend the disc. Do not store the disc where there is excessive heat or cold. Cold can cause the disc to become brittle and too much heat can cause it to warp. Keep the disc in its plastic case. Protect it from direct sunlight and humidity. The disc can be stored vertically or horizontally. However, keep it in its case.

# Digital Audio Tape

Digital audio tape (DAT) recording has taken some direction from video tape recording. By using digital electronics so readily available from ICs, the world of recorded music has become much better. The R-DAT system that Sharp has developed has a tape transport system, Figure 10-33, that works just like that of a video cassette recorder. Once inserted in the deck, the cassette's protective lid opens and the tape is extracted and wrapped 90° around the head bearing drum. As the tape moves past the drum from left to right at 0.815 cm ($\frac{1}{3}$-inch) per second, the drum rotates counterclockwise at 2000 rpm. This combination yields a recording speed of 3.133 meters (123 inches) per second, which is 65 times faster than analog cassette decks. Because the tape is held at an angle to the drum in a helical pattern, the drum's two magnetic heads "write" and "read" information in diagonal tracks across the width of the tape instead of longitudinally along its length, as in analog recording (see Figure 10-34). This space-saving arrangement can produce two hours of information on a matchbox size cassette. Because each of the two heads is mounted at a different azimuth, the information bearing tracks are laid down in an alternating pattern (see Figure 10-35).

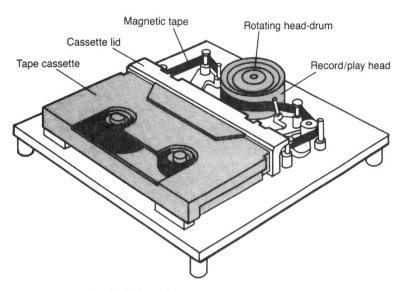

**Figure 10-33.** Tape transport for digital audio tape.

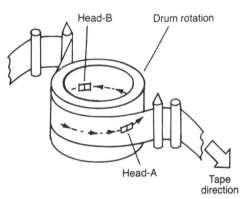

**Figure 10-34.** Rotating head of a DAT system.

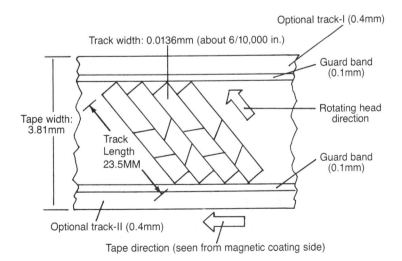

**Figure 10-35.** How the digitized information is recorded on the tape.

## How It Works

DAT decks sample the music at a rate of 48 kHz or 48,000 times per second and then assign each sample a 16-digit code comprised of ones and zeros. The results are just as good as anything available today. It has a frequency response from 4 Hz to 22 kHz.

Recording studios use digital recording for many special effects. Some boards have the ability to take as many as 56 inputs from microphones and similar sound sources. By using multitrack mixing systems, new avenues of creativity for recording stars and technicians have been created.

# Preamplifiers

Most signals from record players, tape heads, and microphones need to be amplified before they can be heard or used. This is the job of the preamplifier. The preamplifier amplifies the weak signals so that they are strong enough to drive the regular power amplifier. The power amplifier is used to drive the speaker or headsets, or whatever output is desired. In some cases that would be the computer.

The schematic shown in Figure 10-36 includes a good example of a preamplifier. It has $Q_1$ and $Q_2$ as the two transistors used to amplify the input from the tape head or from the microphone or auxiliary input. $Q_3$ is the phase splitter or driver stage since it drives the transformer which causes that part of the input signal to $Q_4$ to be different from that presented to the base of $Q_5$. The first two transistors are needed to amplify the signal sufficiently to cause $Q_3$ to operate properly. The push-pull arrangement of $Q_4$ and $Q_5$ are sufficient to drive the speaker connected to the transformer $T_2$. This complete circuit is typical of a portable tape recorder. Note how the switches, $S1$-1 and $S1$-2 are attached to the same shaft and are moved at the same time. In the position shown, they are in tape position. When they are *up*, they put the microphone jack into the circuit. When no outside microphone is attached, the capacitor microphone is in the circuit.

When $S1$-1 and $S1$-2 are in the *up* position, the output of the bottom half of the output transformer $T2$ is fed to the record head of the recorder. The microphone jack is connected to the input of $Q_1$ by the action of $S1$-1. If the microphone jack is not used, the capacitor microphone built into the recorder unit is activated as a source of input.

This is a *single* preamplifier and one tape head, one microphone jack, and one auxiliary jack. If you use a stereo tape recorder, you will find that the schematic has twice the equipment this one does. It will need two tape heads, two microphone jacks, and two auxiliary jacks for other inputs such as record players, and a duplication of the five transistors and their associated component parts. There will, of course, be the need for two speakers to produce the two separate signals to give the stereo effect.

# Power Amplifiers

Power amplifiers take many variations. They can be used to drive speakers, relays, and other equipment. In this chapter we are primarily concerned with driving speakers to produce the stereo effect. In order to produce the stereo effect, we must have two amplifiers

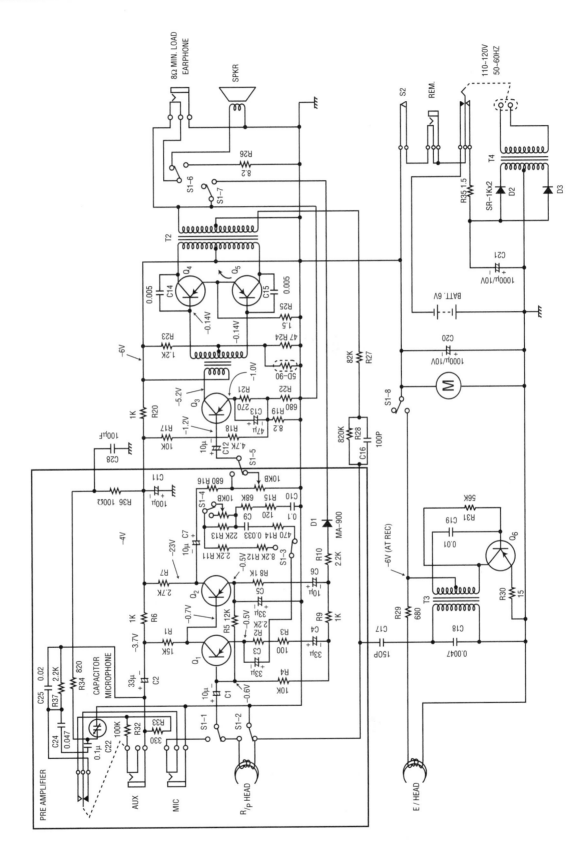

**Figure 10-36.** Schematic of a tape recorder. Note the preamplifier section.

which are completely separated from one another, because the stereo signal has at least two signals which originate from two different sound tracks on a record or from two different records. Of course, some stereo systems use up to seven, and sometimes eight, amplifiers with a corresponding number of signals and speakers.

Today's integrated circuits (ICs) have been employed to make the power amplifiers more compact and less power hungry. They are able to produce large amounts of output power with a higher efficiency than ever before. However some circuits still call for heat dissipation at such levels that heat sinks (Figure 10-37) have to be added to get rid of the heat before it damages the IC or individual transistors.

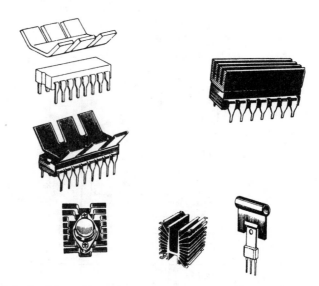

**Figure 10-37.** Heat sinks for ICs and transistors.

# Comparison of Power Amplifiers

In order to obtain a working knowledge of power amplifiers, it is necessary to take a look at three types. One type, and the oldest, is the vacuum tube circuit; another is the transistorized version of the tube circuit; and still another is the up-to-date version of the IC. All three of these circuits operate with the basic theory of the vacuum tube with some modifications for the IC and its many transistors in one package. These amplifiers are used in the making of stereo amplifiers with their two or more outputs.

## Transistor Push-Pull Circuit

The circuit for a push-pull transistorized amplifier is similar to that shown in Figure 10-38. Figure 10-39 shows the similarities. It is instantly recognizable as being push-pull, since you are already familiar with the vacuum tube configuration. Much of the distortion introduced in large-signal amplifiers can be eliminated by using two transistors in push-pull similar to Figure 10-39. The input transformer $T_1$ receives a sinusoidal voltage from a low-level source. The signals applied to the two transistors are 180° out of phase, and the resulting collector currents are 180° out of phase. The output transformer $T_2$ delivers to the load a current that is proportional to the difference of the two collector currents. Any

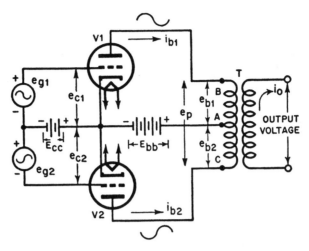

Figure 10-38. Triode push-pull circuit.

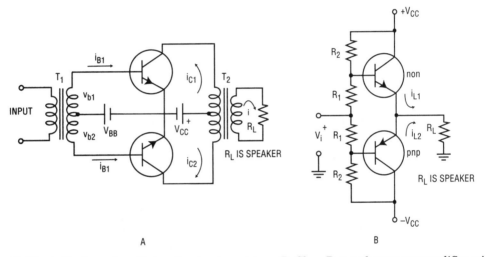

A

B

Figure 10-39. A. Basic push-pull circuit using transistors. B. Class B complementary amplifier with no transformer needed.

even-harmonic distortion components tend to cancel, and the only distortion is that due to odd harmonics. Also the performance of transformer $T_2$ is improved because the dc components of $i_{c1}$ and $i_{c2}$ cancel, and magnetic core saturation and the accompanying nonlinearity are avoided. The push-pull circuit is particularly useful for class B operation. Since the heat losses in a transistor cannot exceed the rated heat dissipation, two transistors in class B can supply nearly six times the power output of one similar transistor in class A. The expensive and heavy transformers can be eliminated by using a PNP and an NPN transistor in a complementary circuit such as Figure 10-39B. In IC circuits of this type the resistors $R_1$ are replaced by diodes whose forward voltages "track" the base-emitter voltages of the transistors. This means the high efficiency of class B can be achieved at low cost.

# Integrated Circuit Amplifiers

Figure 10-40 shows an operational amplifier IC which has 20 transistors and 10 resistors all in one package.

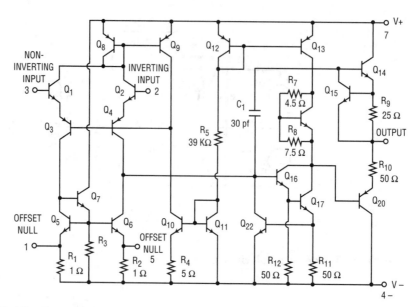

**Figure 10-40.** The 741 operational amplifier.

Figure 10-41 is a schematic of a stereo tape recorder that takes an eight-track cartridge plus the input from a record player and an AM-FM stereo receiver. The eight-track recorder is no longer made. It had a rather short life before being replaced by more advanced electronics. The main thing to notice here is the triangular-shaped enclosures. Those in the center of the schematic are the output amplifiers which drive the speakers or headsets. See if you can trace the output of the amplifiers, the pointed end of the triangle to the speaker and headphone connections. This schematic emphasizes the success achieved with the design of such a piece of equipment. The size of the unit and the cost of the unit have been reduced tremendously since the days of the vacuum tube amplifier. Keep in mind that ICs contain many transistors and diodes as well as the required resistors to make the circuits operate and drive the speakers. The large triangular-shaped unit at the top right of the schematic is the amplifier for the output of the AM and FM receiver sections. It amplifies this weak AF signal and passes it on to the output ICs so that they can process the signals and drive two speakers, each with a different signal. Other examples of IC that serve as audio amplifiers are shown in Figure 10-42.

# Speakers

Much has been written about speakers. There are improvements being made on enclosures for speakers to enhance their ability to respond to the whole audio range of frequencies — 16 to 16,000 Hz.

Two types of speakers have been made: permanent magnet and electromagnet. The names imply the types of magnetism used to aid in the conversion of AF signals to actual sound waves that can be heard by the human ear. The electromagnetic speaker is no longer in use. It was used in early model radios and electric organs. Today, the permanent magnet speaker is in common use. The construction of this type of speaker can be seen in

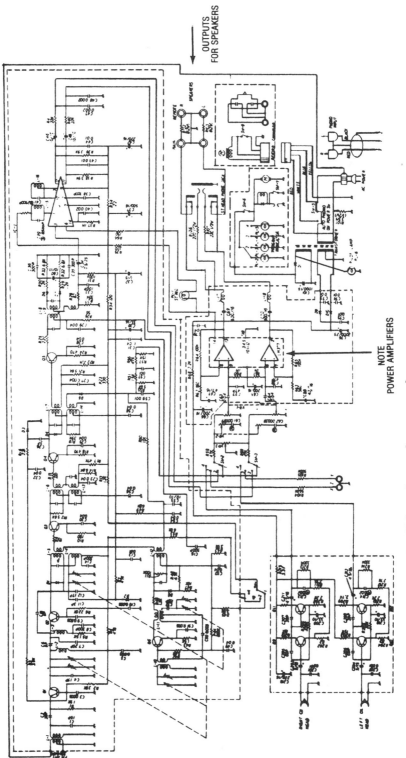

Figure 10-41. Schematic of a stereo AM-FM receiver and tape recorder.

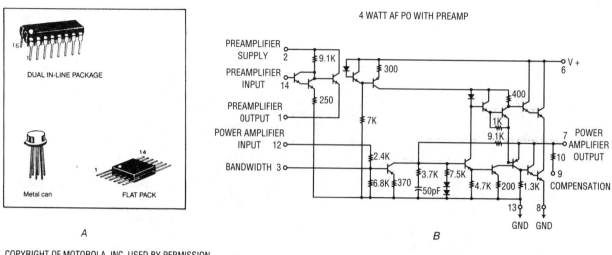

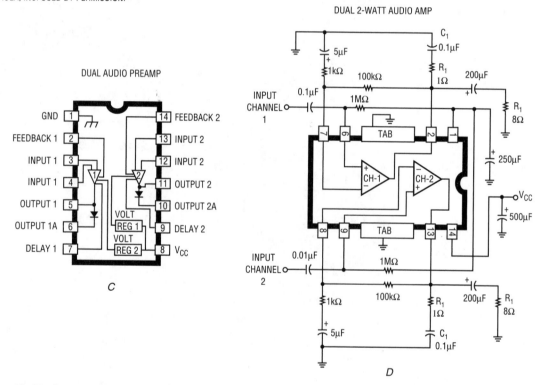

**Figure 10-42.** Integrated circuits. A. Three types of IC cases or packages. B. A 4-watt audio power amplifier with its own pre-amplifier in a 14-pin DIP. C. A dual audio pre-amp in a 14-pin DIP. D. A dual 2-watt audio amplifier in a 14-pin DIP.

Figure 10-43. Note how the coil is shown wrapped around the base of the cone of the speaker; this coil is called the *voice coil*. The size of wire used in this voice coil determines the wattage rating of the speaker since the size of the wire is directly related to the amount of current it can handle. Watts equal the current squared times impedance. Since the impedance is small (4, 8, or 16 Ω), the current has to be rather large in order to have any large amounts of power. Thus, higher wattage speaker ratings call for voice coils of some

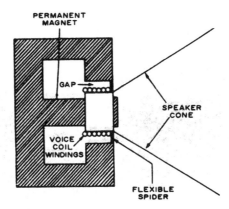

**Figure 10-43.** Construction of a PM (permanent magnet) speaker.

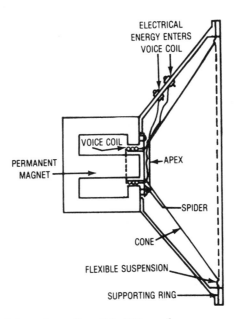

**Figure 10-44.** Note location of the voice coil on this PM speaker.

fairly large size wire. Usually turning up the volume on an amplifier results in overdriving the speaker and burning out the voice coil (see Figure 10-44).

The AF current from the amplifier is fed to the voice coil which sets up a varying magnetic field in step with the current variations. Since the voice coil is located between the poles of the permanent magnet, the magnetic field alternations about the voice coil will interact with the stationary field established by the permanent magnet. The voice coil is wound around a bakelite core, and has the ability to move back and forth. Since the voice coil is physically connected to the paper cone, any movement of the voice coil will cause a corresponding movement of the cone. Because the paper cone has a large diameter, it will possess a large surface area. Any movement by it will displace a large volume of air, causing sound. Any audio current variations will cause a displacement of the voice coil due to the interaction of the magnetic fields. The moving voice coil will cause the paper cone to move resulting in an audible sound. The higher the value of current, the louder the sound. The paper cone and voice coil are suspended from a metal frame by a flexible device called a *spider*. The spider ensures that the voice coil remains centered around the permanent magnet but allows back and forth movement of the voice coil and cone. It also returns the coil to its original position when no current is flowing through the coil.

## Speaker Voice Coil

A voice coil is a low-impedance device, as previously mentioned. It is usually connected to the output of a power amplifier that has a high output impedance. An impedance-matching transformer is used to match the output impedance of the amplifier to the input impedance of the speaker. The connection of the speaker to the amplifier is shown in Figure 10-45. This shows the push-pull vacuum tube amplifier with $T_2$ used as the output transformer. Movement of the voice coil and the paper cone depends on the strength of the magnetic field about the voice coil and the strength of the fixed magnetic field. The strength of the magnetic field produced by the voice coil is a function of the magnitude of the current flow through it. The strength of the field around the permanent magnet is determined by the size of the magnet and its composition. The permanent magnet may be quite large, and is a rough measure of the quality of the speaker. The heavier the magnet, the higher the quality of reproduction of the low frequencies.

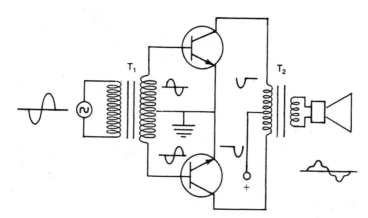

**Figure 10-45.**  Speaker connected to the push-pull amplifier output transformer.

## Permanent Magnets

The permanent magnets used for speakers are made of various combinations of ferromagnetic materials. Alloys are used in the construction so that the magnet will retain its magnetization long after the magnetizing force has been removed. The common metals used in the construction are aluminum, nickel, cobalt, steel, and lead. The proportions in which these metals are used governs the magnetizing ability of the magnet. One of the more popular compositions is *Alnico* which is made of aluminum, nickel, and cobalt.

# Speaker Enclosures

There are a number of different types of speaker enclosures. There is the flat baffle, the infinite baffle, and the ducted-port, or bass-reflex, types which we will look at here. Since others are constantly being developed, you will have to visit a local stereo store to keep up with the latest. Magazine articles are released periodically when new advances or new sounds are possible with speaker enclosures. It is the enclosure that makes the speaker behave the way you want it to. In other words, if you want to accentuate the highs or lows or the in-between frequencies for a particular application, you must use the proper enclosure and crossover network to make sure the right range of frequencies reaches the right speaker in the enclosure.

Figure 10-46 shows the *flat*, or *plane surface*, *baffle*. Its advantage is that it emphasizes low frequencies if you are more than 11 ft. from the front of the speaker. It has some disadvantages: Low-frequency dropout is noticeable if you are closer than 11 ft. or so. This is the simplest of the speakers since only a moderate amount of design and mount effort is involved.

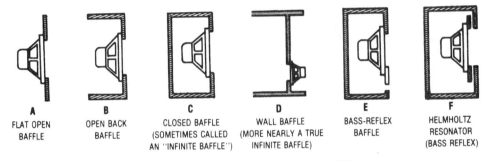

**Figure 10-46.** Various enclosures that fit the classification of flat baffle.

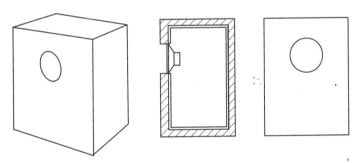

**Figure 10-47.** Construction of the infinite baffle speaker enclosure.

In Figure 10-47 the *infinite baffle* is shown. Here the enclosure can be a box or a closet. The cabinet encloses the speaker, and the back waves from the speaker are not ducted or released. This loads the speaker and tends to produce a boomy sound for the low frequencies. A volume of 10 to 15 ft$^3$ is desired for this type of enclosure. It is a low-efficiency arrangement since some of its power is absorbed by the lining of the baffle.

The *bass-reflex baffle* has a ducted port in front which allows the back pressure, or back waves, created by the speaker to exit through the port. This back wave is 180° out of phase with the front existing waves. This means they are in phase and aid in producing a more efficient speaker enclosure. Higher frequencies have a tendency to be absorbed by the lining in the cabinet. Thus, the low-frequency response is enhanced. It is an inexpensive type of enclosure and easy to make. The location of the port is not critical, but it should not be closer than 4 in. to the speaker (see Figure 10-48 for a bass-reflex enclosure).

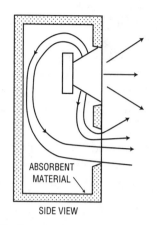 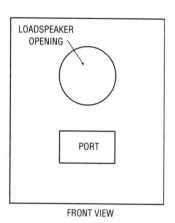

**Figure 10-48.** The bass-reflex baffle enclosure.

# Crossover Networks

Crossovers are designed for the proper routing of frequencies to speakers that can efficiently handle them. The crossover network is made up of a capacitor alone, or it may be any combination of capacitance, resistance, and inductance. The filters made by these combinations are low-pass or high-pass and can direct the proper frequencies to the proper load. For instance, the 12-in. speaker is called a *woofer*. The woofer is accustomed to and designed for handling low frequencies (about 16 to 300 Hz). The midrange is usually 8 in. but can be anywhere from a 5- to an 8-in. size. However, it handles the midrange of frequencies best. This is the 300- to 3000-Hz range. In order to make sure that they get to the 8-in. speaker, a network has to be specially designed. Figure 10-49 is an example of this arrangement. Now, in most speaker enclosures there will also be a high-frequency (3000- to 20,000-Hz) horn or 4-in. speaker which needs to get only those frequencies so it is not damaged. This can be done by a properly designed network. These networks are

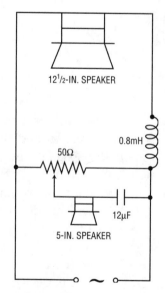

**Figure 10-49.** Using a crossover network to make sure the right frequencies reach the correct speaker.

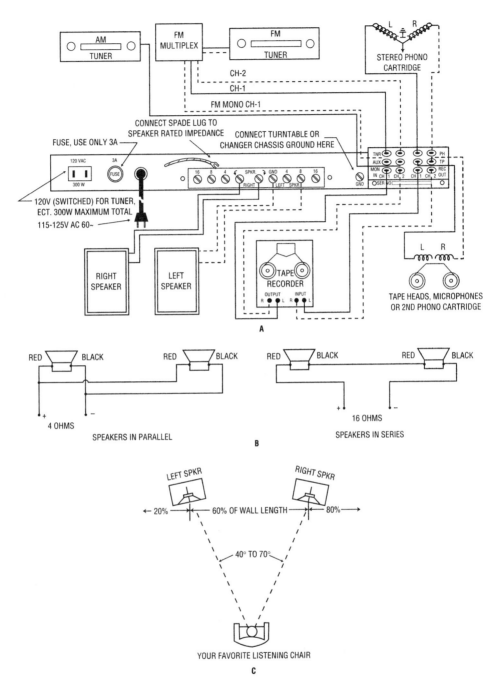

**Figure 10-50.** A. Connections to a typical stereo high-power amplifier. B. Connecting speakers in series or parallel so that two may be operated from a single tap. C. Arrangement of speakers for the stereo effect.

available commercially at stores that sell electronics parts, or they can be designed and made.

There are any number of ways to hook up a stereo amplifier to its outputs and inputs. However, each has its own labeled instructions. Take a look at Figure 10-50A and B to see how this particular unit is hooked up to provide for the stereo effect using two speakers and inputs from a record player, AM tuner, FM tuner, and microphones. This is typical of

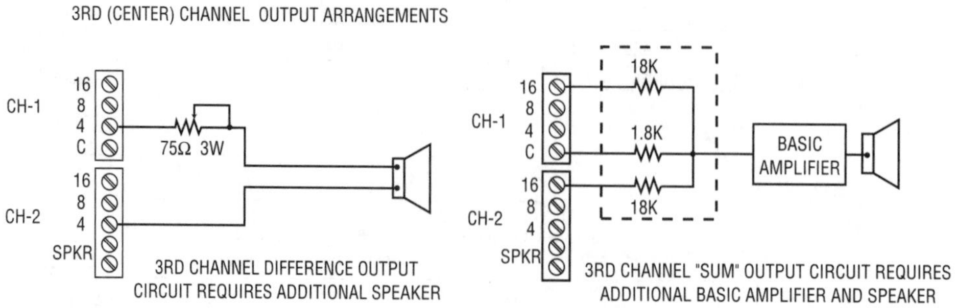

3RD (CENTER) CHANNEL OUTPUT ARRANGEMENTS

**Figure 10-51.** Connections for a "phantom" third channel.

the amplifiers available today. Just keep in mind the necessity to match the impedance of the speaker or speakers to that of the amplifier for maximum transfer of energy and better frequency response and reproduction.

Figure 10-50C shows how the stereo effect is obtained by using two speakers. Of course, the sound can be enhanced by adding another speaker to fill in between the two, creating a *"phantom" third channel* (see Figure 10-51). This is done using only the two outputs of an amplifier. Much can be learned from studying the current magazines on the topic of stereo and the latest devices and improvements.

# Summary

Stereo means solid, or three-dimensional. In terms of electronics it means that there are at least two amplifiers and two speakers to the system that produces a sound which appears to surround you. The origin of sound is always a vibrating body. Sound produces compression of air molecules and rarefaction of air. This compression and rarefaction causes a microphone to produce electric impulses. Electric impulses are amplified until they are sufficient in power to drive a speaker. A speaker is used to produce sound from electric energy.

There are a number of types of microphones — carbon, dynamic, velocity, ribbon, and crystal. Turntables and record changers are used to produce music from recorded disks. Turntables can produce a higher quality sound than a record changer. There are at least six types of record pickups or styli. The first magnetic recorder was produced in Denmark in 1898 by Valdemar Poulsen. Recording tapes today are available in $1/4$-, $1/2$-, $3/4$-, 1-, and 2-in. widths. Everything is recorded on tape from music to television pictures and sound to computer data. Preamplifiers are usually needed to amplify the inputs to stereo power amplifiers. The vacuum tube push-pull amplifier was one of the standbys of olden days for power amplification. Today the transistor and integrated circuit (IC) have replaced the vacuum tube in most applications.

Just as the vacuum tube has been replaced by the transistor, the phonograph record has been replaced by the compact disc, CD. The ability to digitize the analog waveform has improved the quality of recorded music available to anyone willing to purchase an inexpensive CD player. They are available in portable units with headphones and in automobiles as well as in large expensive home entertainment systems. The compact disc is now being utilized in the computer and for various video programs. Whole dictionaries and encyclopedias are available with pictures on video CDs. This is another example of how fast the electronics field grows once a product is accepted by the public.

However, recent events have caused a new surge of interest in the old vinyl records such as 33 rpm and 45 rpm. Some people prefer the sound of the plastic record to that of the compact disc. Some say it has a more mellow sound than the CD. Therefore, it may be some time before the plastic disc, used for so many years to record music, is entirely replaced.

The permanent magnet (PM) speaker is the most popular today and comes in a wide variety of diameters and shapes. The size of voice coil wire limits the amount of current a speaker can handle. Crossover networks control which speaker reproduces the various frequencies. There are a number of different speaker enclosures. One of them is the bass-reflex type which has a ducted port or hole in the front. Other types include the infinite baffle and the plane baffle.

# Review Questions

1. What does the word stereo mean?
2. What is the minimum number of speakers and amplifier channels needed for the stereo effect?
3. What is meant by compression and rarefaction?
4. What is the frequency range the human ear can hear?
5. How is the sensitivity of a microphone expressed?
6. What is the main difference between a carbon microphone and a dynamic microphone?
7. What is the frequency range of the dynamic microphone?
8. Describe the velocity microphone.
9. How does the crystal microphone work?
10. What is the crystal used in a crystal microphone?
11. What is the difference between a turntable and a record changer?
12. Name at least six types of styli.
13. Who invented the magnetic recorder?
14. What is the difference between a cassette and a cartridge?
15. What is a CD?
16. How big is the CD?
17. How fast does the CD travel?
18. What is digitizing?
19. What is a pit?
20. What is meant by encoding?
21. How does the CD pickup work?
22. What is a nanometer?
23. What does a collimator lens do?
24. What is an optosensor?
25. How much power does the laser diode produce in the 780-nm wavelength?
26. What is the best way to handle a CD?
27. What is the difference between a preamplifier and a power amplifier?
28. What advantages does the push-pull circuit offer over other types of power amplifier circuits?

29. Name two types of speakers.
30. Describe the speaker voice coil.
31. What is a speaker enclosure?
32. Name three types of speaker enclosures.
33. What does a crossover network do?
34. What is the frequency range of a 12-in. speaker?
35. What is another name for a 12-in., or bass, speaker?
36. What is a midrange speaker?
37. What is a horn speaker?

# Chapter 11

# OSCILLATORS

Oscillators are producers of oscillations, or usable frequencies. They are used in communications applications and in other fields where certain types of energy are needed. There are mechanical oscillators and electronic oscillators. The electronic types are those to be studied in this chapter. Their outputs will be examined in terms of putting them to work to both communicate and detect moving objects and to cook foods by radar frequencies.

Mechanical oscillations are produced by such things as the alternator, or ac generator. The frequency of the oscillations depends upon the speed of the generator. Special generators can produce as high as several hundred kilohertz. However, this frequency is still not high enough for most types of communication. Because of the high frequencies required, a nonmechanical system is needed to produce the oscillations which form the carrier wave. Transistor circuits have been designed to fulfill this function. These circuits are called *oscillators*. There are several basic arrangements which produce oscillations.

# Conditions Needed for Oscillations

The *LC* circuit is the best basic source of oscillations. Any combination of inductor and capacitor will produce a frequency if properly energized and a source of feedback is arranged. Oscillations produced by a parallel *LC* tank circuit are described here as a basis for simple oscillation. Other methods will be shown as we progress in the chapter.

Refer to Figure 11-1. If *C*, a charged capacitor, is placed across *L*, an inductor, the capacitor discharges through the inductor as seen in A. The capacitor acts as a source of emf and forces current through the circuit. The flow of current through *L* creates a magnetic field about the coil. This magnetic field opposes the rise in current through the circuit. As *C* discharges, the current reaches a maximum and then declines. Between the time the current is zero and the time the current is maximum, the magnetic field builds up. In other words, energy is stored in the magnetic field.

When the capacitor is discharged, the current attempts to fall to zero. This attempted change in current flow is opposed by the inductance. The magnetic field now acts as a source of emf and continues to force current through the circuit in the same direction as in *B*. This current serves to charge the capacitor to a voltage whose polarity is opposite to the polarity it had when it was first placed in the circuit. As the magnetic field collapses, the current slowly falls to zero.

The capacitor, in Figure 11-1C, now discharges in the opposite direction, reversing the original path of current flow. This current flow again is opposed by the inductor, which builds up a magnetic field opposite in direction to the original magnetic field. The current reaches a maximum and then declines.

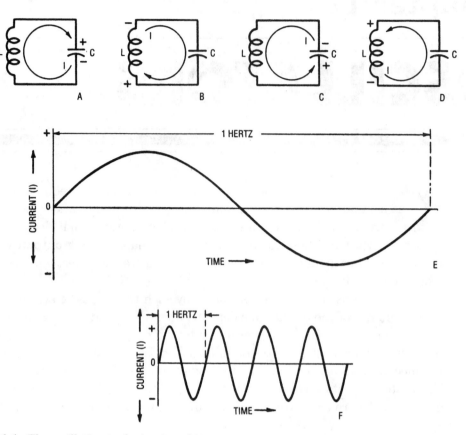

**Figure 11-1.** The oscillating tank circuit and its output.

When the capacitor has discharged, the current attempts to fall to zero. This change is opposed by the inductor, which uses the energy stored in the magnetic field to prevent the current from dropping to zero at the instant the capacitor is discharged. Therefore, the current is sustained, and it serves to charge the capacitor to a voltage whose polarity is the same polarity as it was originally (Figure 11-1D).

The entire process now is repeated for as long as energy remains in the circuit. The current in the *LC* circuit (Figure 11-1E) has the form of a sine wave. It has changed from zero to a maximum in one direction, through zero to a maximum in the opposite direction, and back to zero. Several hertz of output are shown in Figure 11-1F.

The output described here is possible only in an *LC* circuit which loses no energy. If any resistance is present in the circuit, energy is lost in the form of heat. Because no actual *LC* circuit is free of resistance, the circuit can be represented as shown in Figure 11-2A. The resistance of *R* is the dc resistance of the wire forming the coil, plus the leakage resistance of the capacitor, plus the resistance of the connecting leads.

In an ideal situation, the energy is transferred back and forth between the inductor and the capacitor without loss. In the actual *RLC* circuit, each transfer of energy involves a loss. The current which accomplishes the transfer of energy must pass through R, and each time the current flows, a part of the energy is dissipated. Consequently, less and less energy is transferred on each succeeding alternation of current flow. This means that the oscillations die out as the energy is dissipated by the resistance. The current in the *RLC* circuit is shown in *B*. Note how the amplitude of each succeeding alternation is decreased. This is known as a *damped* oscillatory current.

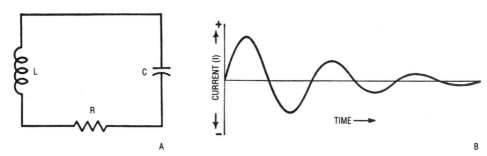

**Figure 11-2.** Damped oscillatory current in an *RLC* circuit.

## Circuit Conditions

If a fully charged capacitor is substituted for *C* on each hertz or cycle, the oscillations can be sustained. In other words, if the circuit can be supplied with sufficient energy to make up for the resistive loss, the current continues to oscillate so long as the extra energy is supplied.

This extra energy, however, must be injected into the circuit so that it aids the flow of current. The current flows in one direction for a time, and then in the other direction. This means that the source of energy must cause the current to flow in the right direction and at the right time. The source, therefore, must be an emf producing a current of the same frequency and time relations as the oscillatory current. This current must also be large enough to replenish the energy loss. Energy of this nature can be obtained from a similar *RLC* circuit. How this is done is the topic of the next part of the chapter.

# Feedback

One of the most important parts of an oscillator circuit is the feedback path. An oscillator will not continue to oscillate unless it has feedback. Remember that the feedback signal must be there at the right time and in the proper direction. In order to ensure that this feedback occurs at the right time, the resonant frequency of the *LC* tank circuit in the output of this stage is made approximately the same as the *LC* tank circuit of the input *LC* circuit.

The feedback must occur at the right time. If the current in the input *LC* circuit is flowing in one direction and the feedback induces a current in the opposite direction, the oscillatory current is damped more quickly than if no feedback were present. This type of feedback is *degenerative* (see Figure 11-3). Note that the oscillatory current is more highly damped than the natural damped oscillation shown in Figure 11-2.

If the feedback induces a current in the same direction as the oscillatory current in the input *LC* tank circuit, the feedback is *regenerative*. Its effect is to sustain current oscillations. If the feedback is just sufficient to replace the losses, each oscillation is of the same size (see Figure 11-3B).

If the regenerative feedback is too small, the oscillations in the input tank circuit die out, although the oscillations are not damped as highly as when there is no regenerative feedback. If the feedback is too large, the oscillations in the input tank circuit build up until the output *LC* tank current is swung alternately from zero to saturation. This

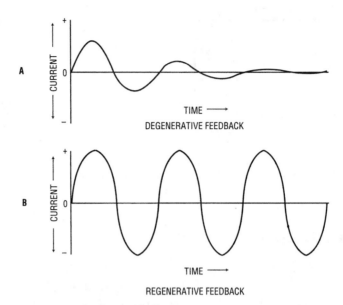

**Figure 11-3.** Waveforms showing the effect of feedback.

results in a highly distorted oscillation. The regenerative feedback, therefore, must have the proper amplitude.

# Oscillator Output

An oscillator must produce a usable output. The output can be small, provided it is sufficient to drive the input circuit of the following stage. The following stage, or stages, can be used to build up the oscillations to the desired amplitudes. This means that the oscillations in the output circuit must be large enough to provide both the proper amount of feedback and a useful output.

A certain amount of energy is needed to drive the transistor. This energy must be replaced by the energy taken from the collector circuit in an oscillator. Since the output also is taken from the collector circuit, the energy available in the output circuit must be greater than the energy needed in the input circuit. In other words, the total output must be greater than the input. This means that the transistor must be capable of amplifying. Consequently, any transistor with an amplification factor *greater than one* can be used in an oscillator.

These are the basic fundamentals of oscillators. From here on we will deal with the specific types of oscillators. Some of them, you will notice, have names attached. The men who invented them attached their names, which stuck through the years. In other cases the oscillator name may be such that it describes its type of circuit and/or the type of feedback.

# Crystals

The crystal oscillator has certain advantages that other types do not have. A crystal, for instance, can be cut to an exact thickness. It will resonate within a very small percentage

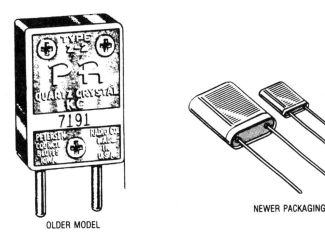

**Figure 11-4.** Crystals used in transmitters.

of a given frequency. It also possesses a far higher $Q$ than any $LC$ network. $Q$, $Q = X_L/R$, is a figure of merit and is found by dividing inductive reactance of a coil by its resistance. This means that the circuit can have both frequency precision and frequency stability when using a crystal (see Figure 11-4).

Broadcast transmitters (AM, FM, and TV) make use of the crystal to stay on frequency. They have a heated box that contains the crystal, and it is kept at a constant temperature within ±0.2°C. Since the humidity is constant, it is very accurate. CB operators also used crystals to produce the frequency stability they needed for staying on their assigned channels.

## Types of Oscillators

Many types of oscillators are used. These include the tri-tet, the dynatron, and the transitron. The transitron can be grouped under the heading of negative-resistance oscillator. In ultrahigh frequency work, such oscillators as the klystron and the magnetron generally are used. (They will be discussed later in this chapter.)

# Transistor Oscillators

The input and output impedances of an electron tube circuit are high; the feedback signal suffers little loss, because of the mismatch in feedback network. In the common-base configuration of the transistor, the input impedance is low and the output impedance is high. Coupling the feedback signal from the output to the input requires a feedback network to match the unequal impedances. Sometimes the loss due to mismatch may be compensated for by providing more feedback energy. The use of the other transistor configurations involve similar problems.

## Tuned Base Oscillator

The tuned base oscillator is similar to the tuned grid in vacuum tubes (see Figure 11-5). Feedback is supplied by the transformer $T_1$. The common-emitter configuration is utilized.

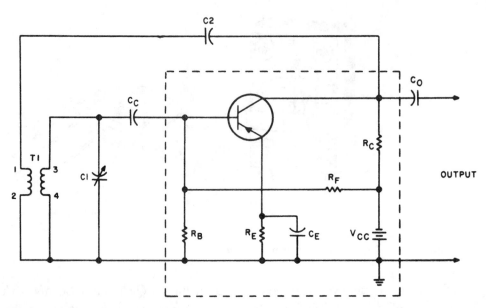

**Figure 11-5.** Tuned base oscillator.

Resistor $R_E$ is the emitter-swamping resistor. Within the dotted line are the transistor's components for making an amplifier. The other resistors within the box are used for biasing the transistor and making it operate properly. The feedback through the transformer makes it an oscillator that will continue to oscillate.

As you can see from looking at the circuit, the LC tank circuit is made up of the transformer secondary and the tuning capacitor $C_1$. Tuning the capacitor can generate a wide range of frequencies. $C_C$ and $R_B$ are $RC$-coupled to bring the output of the tank circuit to the base of the transistor amplifier. $R_E$ is the emitter bias with $C_E$ furnishing emitter bias bypass service. $R_F$ is the base bias resistor, and $R_C$ is the collector load. $C_2$ is the feedback capacitor. $C_2$ and $C_1$ should be approximately equal to the ratio of the output impedance to the input impedance of the transistor. $C_o$ serves as a coupling capacitor to the next stage.

## Clapp Oscillator

You can improve the stability of the Colpitts oscillator with the addition of a capacitor. See Figure 11-6 where the Clapp oscillator has a variable capacitor $C$ in series with the $T_1$ points 1 and 2. The variable capacitor tunes the output of the tank circuit over a wide range of frequencies. The shunting impedance of the $LC$ series combination is at a minimum, thereby making the oscillating frequency comparatively independent of the transistor parameter variations. Compare this with the Colpitts oscillator shown in Figure 11-7.

## Hartley Oscillator

The Hartley oscillator is similar to the Colpitts except for the use of a split inducance instead of the split capacitance to obtain feedback (see Figures 11-8 and 11-9). Both the series-fed and shunt-fed Hartley oscillators are operationally the same. They differ mostly

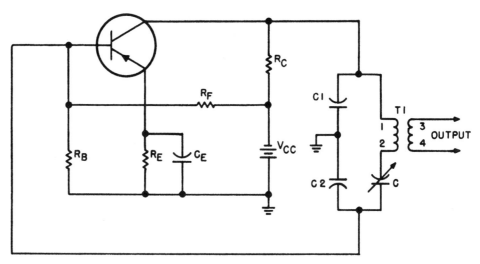

**Figure 11-6.** Clapp oscillator.

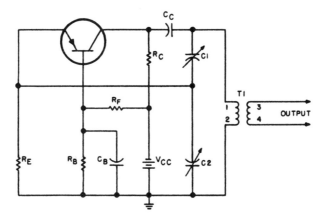

**Figure 11-7.** Colpitts oscillator.

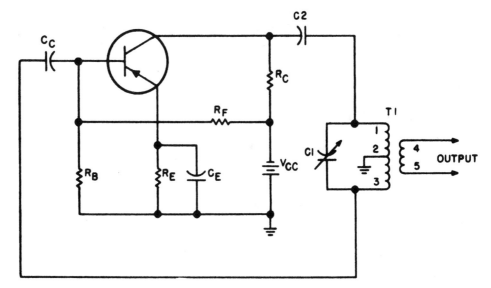

**Figure 11-8.** Shunt-fed Hartley oscillator.

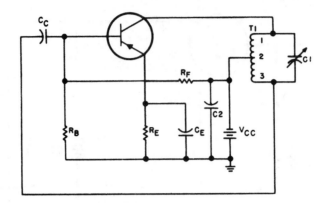

**Figure 11-9.** Series-fed Hartley oscillator.

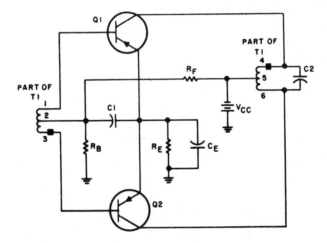

**Figure 11-10.** Push-pull oscillator for higher power output.

in the method of obtaining collector bias. A modified Hartley oscillator which provides greater power output is shown in Figure 11-10. It is called the push-pull oscillator. The tank circuit is made up of part of $T_1$ and $C_2$. Feedback is through the other part of $T_1$ located in the base circuits. After the feedback is accomplished the operation of the circuit is the same as for a push-pull amplifier.

# Crystal Oscillator

The quartz crystal is used to establish the operating frequency of the circuit (see Figure 11-11). This is a crystal-controlled tickler-coil oscillator. This unit uses the series mode of operation of the crystal and functions similarly to the tuned collector circuit or tuned plate circuit in vacuum tubes. Increased frequency stability is ensured by inserting the crystal in series with the feedback path. However, the frequency is essentially fixed by the crystal. You have to change the crystal to obtain a different frequency. Each crystal is tuned to or grounded to its own frequency. Regenerative feedback is through the mutual inductance of the transformer windings. The transformer action provides the necessary 180° of phase shift for the feedback signal. At frequencies above or below the series resonant frequency

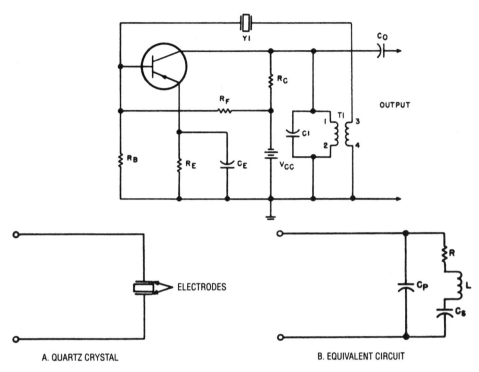

**Figure 11-11.** Crystal-controlled tickler-coil feedback oscillator.

of the crystal, the impedance of the crystal increases and reduces the amount of feedback. This in turn prevents oscillation at frequencies other than the series resonant frequency.

# Colpitts Crystal Oscillator

Figure 11-12 illustrates how the crystal is used to shunt the capacitors $C_1$ and $C_2$. The capacitors are center-tapped, and the center tap is grounded. This causes $C_1$ to feed its voltage back to the base of the transistor 180° out of phase and of the proper amplitude to keep

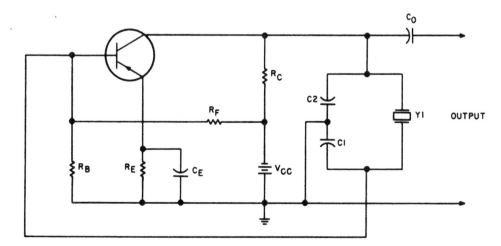

**Figure 11-12.** Colpitts crystal oscillator with collector-base regeneration.

the oscillator going. The circuit is the common-emitter configuration with the feedback supplied from the collector to the base. The resistors provide the proper bias and stabilizing conditions for the circuit. The oscillation frequency of this circuit is determined by the crystal and the capacitors connected in parallel with it.

# Wein-Bridge Oscillator

The Wein-bridge oscillator uses a resistance-capacitance network for the development of a sinusoidal output. The circuit shown in Figure 11-13A and B are identical. Figure 11-13A shows how the bias and feedback is accomplished, and Figure 11-13B indicates how the bridge or diamond-shaped circuit looks when drawn in the shape of the typical bridge circuit.

The two transistors are connected in the common-emitter configuration. The second stage functions as an amplifier and phase inverter from which the feedback signal is taken in proper phase. Regeneration is provided for oscillation and degeneration is provided to obtain frequency stability and distortionless output.

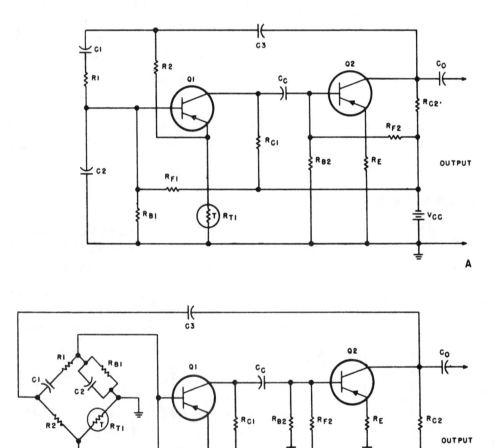

**Figure 11-13.** Wein-bridge oscillator.

The output transistor $Q_1$ is coupled to the input of transistor $Q_2$ through capacitor $C_C$. Capacitor $C_3$ couples a portion of the output of the amplifier stage $Q_2$ to the bridge network to provide the necessary feedback (both positive and negative). The output to the load is coupled through capacitor $C_O$.

# Multivibrators

The free-running (astable) multivibrator is essentially a nonsinusoidal two-stage oscillator in which one stage conducts while the other is cut off until a point is reached at which the stages reverse their conditions. That is, the stage which had been conducting cuts off, and the stage that had been cut off conducts. This oscillating process is normally used to provide a square wave output. Most transistor multivibrator circuits are counterparts of those using electron tubes. For example, the emitter-coupled transistor multivibrator is analogous to the cathode-coupled electron tube multivibrator circuit; the collector-coupled transistor multivibrator (see Figure 11-14) is analogous to the common plate-coupled electron tube multivibrator. Since most multivibrator circuits function similarly, only the collector-coupled transistor multivibrator is discussed here.

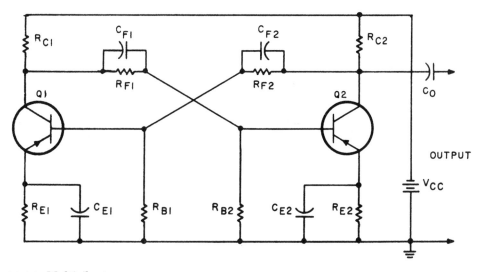

**Figure 11-14.** Multivibrator.

The basic collector-coupled transistor multivibrator of Figure 11-14 is a two-stage resistance-capacitance-coupled common-emitter amplifier with the output of the first stage coupled to the input of the second stage and the output of the second stage coupled to the input of the first stage. Since the signal in the collector circuit of a common-emitter amplifier is reversed in phase with respect to the input of that stage, a portion of the output of each stage is fed to the other stage in phase with the signal on the base electrode. This regenerative feedback with amplification is required for oscillation. Bias and stabilization are established identically for both transistors.

The oscillating frequency of the multivibrator is usually determined by the values of resistance and capacitance in the circuit. In the collector-coupled multivibrator, collector loads are provided by $R_{C1}$ and $R_{C2}$. Base bias for transistor $Q_1$ is established through voltage divider resistors $R_{B1}$ and $R_{F2}$. Base bias for transistor $Q_2$ is established through voltage divider resistors $R_{F1}$ and $R_{B2}$.

Stabilization is obtained with emitter swamping resistor $R_{E1}$ for transistor $Q_1$, and resistor $R_{E2}$ for transistor $Q_2$. Emitter capacitors $C_{E1}$ and $C_{E2}$ are ac bypass capacitors. The coupling capacitor $C_O$ couples the essentially square wave to the next stage. It may be obtained from either of the collectors. If you want a sawtooth waveform output, simply connect a capacitor from the collector to ground to develop the output voltage.

The multivibrator may be modified to produce a sinusoidal waveform. This is accomplished through the connection of a parallel-tuned circuit between the base electrodes of each transistor.

## Blocking Oscillators

Another type of oscillator is the type which conducts for a short period of time and is cut off (blocked) for a much longer period. Blocking oscillators may be either free-running or triggered. Only the free-running type is discussed here. A basic circuit for the blocking oscillator is shown in Figure 11-15. The similarity between this circuit and that of the tickler-coil oscillator should be noted.

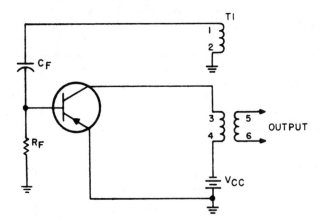

Figure 11-15. Blocking oscillator.

The output waveform is a pulse, the width of which is primarily determined by the winding 1-2. The time between pulses (resting or blocking time) is determined by the time constant of resistor $R_F$ and capacitor $C_F$. The output is coupled through transformer winding 3-4 to the load by transformer winding 5-6. This type is no longer used.

IC's have changed oscillator circuits. Three representative types follow.

# Newer Oscillator Circuits

The crystal oscillator is commonly used in the broadcast industry to keep transmitters on frequency with little or no drift. Modern devices have improved some of the ways basic frequencies are generated for use in many devices.

# RC Phase Shift Oscillator

A popular oscillator using an op-amp is shown in Figure 11-16. This oscillator uses a resistor-capacitor network to determine the oscillator frequency. The op-amp provides the gain required so that the signal fed back through the *RC* phase-shift network provides the required conditions for circuit oscillation. Adjustments to frequency can be made by varying either the resistance or capacitance.

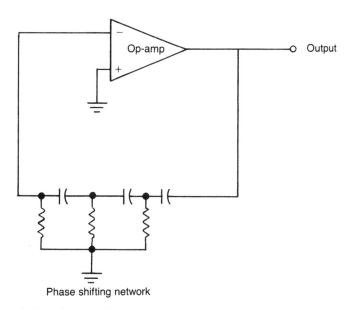

**Figure 11-16.** Phase-shift oscillator using an op-amp.

# Wein-Bridge Oscillator Circuit

Another arrangement for the op-amp and the *RC* phase-shifting circuitry is the Wein-bridge oscillator (see Figure 11-17). The circuitry shown in (A) is the conventional representation for a bridge circuit. However, (B) shows the same circuit drawn differently, but it still has all the same components and values as in the bridge configuration. The resistors and capacitors determine the frequency at which the oscillator operates. Bridge inputs are at points 1 and 2 while the op-amp inputs are from points 3 and 4. The frequency of the output is sinusoidal and can be calculated by using a formula: $f_o = {}^1\!/_2\pi\sqrt{R_1 C_1 R_2 C_2}$.

# Voltage Controlled Oscillator

A 566 IC chip can be used to generate both square-wave and triangular-wave signals with its frequency set or by an external resistor and capacitor; in this case it is $R_1$ and $C_1$. The frequency variation is developed by an applied dc voltage. See Figure 11-18 for a diagram of a voltage controlled oscillator (VCO) that provides an output frequency that is a linear function of a controlling voltage.

A Schmitt trigger circuit is used to switch the current source between charging and discharging the capacitor. A triangular voltage is developed across the capacitor. The square

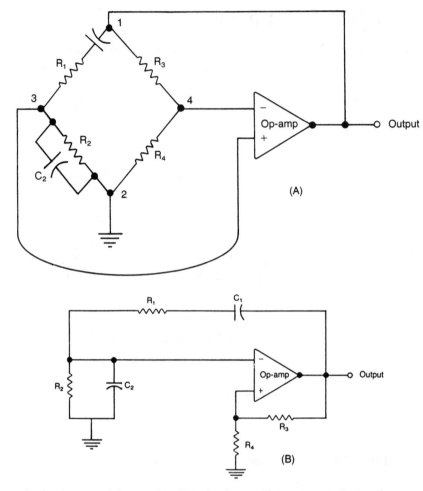

**Figure 11-17.** A. Op-Amp amplifier used in Wein-bridge oscillator circuit. B. Another way to draw the Wein-bridge circuit.

wave from the Schmitt trigger, along with the triangular-shaped wave form are produced as outputs through buffer amplifiers located within the chip or IC (see Figure 11-18). The pin connections for the 566 chip are the same as the numbers on the schematic diagram.

There are other types specially designed for the person with a particular application problem. That is why engineering is such an interesting field: the challenge to design new circuits to do new jobs is always before you.

# Microwaves

Microwaves are superhigh frequencies. They operate in thousands of millions of hertz. The microwave oven you use at home operates on the S band of frequencies as did the older police radar speed meter (mounted on a tripod by the side of the road). Both use the 2450-megahertz (MHz) frequency. With increased use of the S band, the police moved up to the X band which is 10,525 MHz. Since the X band is experimental and also had its limitations, the next higher band was soon used by the police speed detector radar. The K band radar

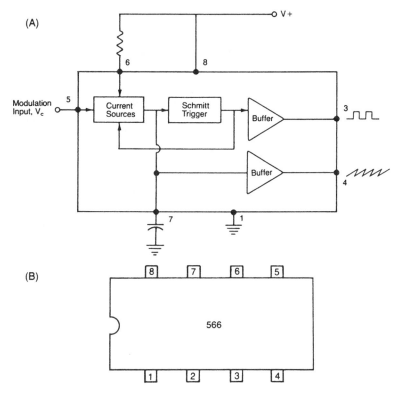

**Figure 11-18.** A. Block diagram and B. pin configuration for a VCO.

operates on 24.15 gigahertz (GHz), or 24,150 MHz. If you would like to see just how short the waves are, take the speed of light (300 million meters per second [m/s]) and divide it by the frequency in hertz. This will give you the distance the wave travels before the next one is ready to take its place. (This works out to be 0.012 422 360 2 m, or 0.491 180 124 3 in.)

There are two basic types of oscillators used in microwaves: the klystron and the magnetron. Each has some unique (up to now) features to examine in order to have a grasp on how they operate to generate these extremely high frequencies.

# The Klystron

The source of the high-frequency energy that a radar unit uses is the klystron. It is a mechanical oscillator inasmuch as the shape and size of the cavities in the unit determine the frequency at which it oscillates.

The principle of bunching of electrons and velocity modulation was discovered in 1935 by two German scientists named Heil and Heil. The principle of the cavity resonator was discovered by William W. Hansen of Stanford University in 1938. The klystron was developed and constructed in 1938 by Hansen and by Russell and Sigurd Varian, brothers. The klystron and the principle of hunching were put together, and the klystron produced a signal that was extremely high in frequency. Figure 11-19 is a breakdown of a klystron. Figure 11-20 shows how the klystron works. The schematic of the single-cavity, reflex klystron and the voltages required for operation are shown in Figure 11-20. The various elements that compose the tube are the cathode, a focusing electrode at the cathode

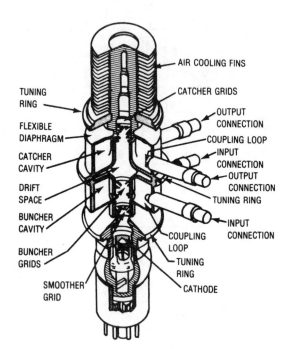

**Figure 11-19.** Cutaway view of a klystron.

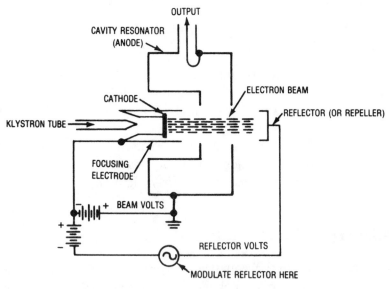

**Figure 11-20.** How the klystron works.

potential, a resonator which also serves as an anode, and a reflector (repeller) that is at a negative potential with respect to the cathode. The combination of the cathode, focusing electrode, and anode beams the electrons through the resonator gap and out toward the reflector. Since the reflector element is negative with respect to the anode, the electrons are turned back toward the anode where they pass through the gap a second time. When the klystron is oscillating, an alternating voltage appears across the gap of the resonator. As electrons pass through the gap, they are either accelerated or decelerated as the voltage across it changes in magnitude with time. This results in accelerated electrons leaving the gap at an increased velocity and decelerated electrons, at a reduced velocity. Because of this, electrons leaving the gap at different parts of the gap voltage cycle take different lengths of time to return to the gap; that is, they have different transit times. As a result, the electrons group together in bunches as they return through the gap. This variation in velocity of the electrons is called *velocity modulation*.

As the bunches pass through the gap, they react with the voltage appearing across the gap. If the bunches pass through the gap at a time in the gap voltage cycle such that the electrons are slowed down, then energy will be delivered to the resonator, and oscillations will be sustained. If the time of transit in the anode-reflector region is such as to cause the bunches to arrive at a time when they will be accelerated by the gap voltage, then energy is removed from the resonator, and oscillating will tend to stop. To generate oscillations at a given frequency and fixed anode voltage, it is necessary to vary the transit time in the anode-reflector space to a suitable value by means of adjusting the reflector voltage: the more negative the reflector voltages, the shorter the transit time. Though frequency of operation is determined primarily by the resonator dimensions, a small change in frequency may be obtained by adjusting either the reflector or anode voltage. This adjustment is known as *electronic tuning*. A frequency change of several percent is possible in some tubes.

At one time the klystron was designed for low-power types of radar units; now there are medium- and high-power units (see Figure 11-21). Today they are used for that as well as for many commercial purposes. They are used as part of the earth stations which beam television programs and telephone conversations to satellites somewhere in the 6-GHz frequency range.

# The Magnetron

This tube is used to produce an extremely high frequency that can be built to very high power levels. It is very much in demand for surveillance radar and other military purposes. It can also be used for commercial purposes.

The magnetron tube is made in any number of sizes, but the basic shape is shown in Figure 11-22. It is a cylindrical brass block with a large hole drilled down the center and eight smaller holes drilled between the center and outer edges. Slots are used to interconnect the holes. The smaller holes are called *cavities*. When the magnetron is operating, electrons take a back-and-forth path along the walls of the cavities as indicated by the arrows in one cavity. Actually a cavity oscillates in a manner similar to a lower frequency coil and capacitor circuit. The walls of the cavity form the inductance, and the capacitance across the cavity opening forms the capacitor.

A cathode is placed down the middle of the magnetron. It has an internal heater wire. A small hook in one of the cavities acts as a pickup loop. This takes the RF energy from the cavity when it is oscillating and feeds it to the transmission line. The waveguide is used in microwaves to handle the movement of the RF energy. A wire is unable to handle the microwaves since they travel on the outside (skin effect) of any conductor. They are contained inside the waveguides which are made to a particular size to accommodate the particular frequency being generated.

In some cases a spark gap is used to generate the pulses needed to get the magnetron to operate or start oscillating. When the tube is pulsed, the cathode is driven negative by

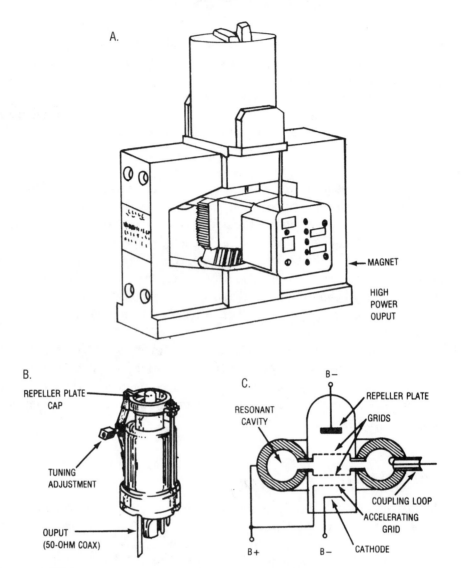

**Figure 11-21.** A. High-power-output and B. medium-power-output klystrons. C. Schematic for a klystron.

10,000 to 20,000 volts (V). This makes the plate relatively positive, and electrons from the hot cathode start moving toward it. However, a strong, external horseshoe-shaped permanent magnet, with its north pole at one end of the cathode and the south pole at the other end, produces an intense magnetic field down the center hole. The electrons are deflected at right angles to the lines of force through which they are passing. This results in an elliptical path for the electrons as they progress toward the anode areas. The positive potential of the anode accelerates the electrons toward it. This is the same as saying that the electrons pick up energy from the difference of potential. As the electrons move past the slots between the anode areas, they induce voltages between the slot faces. This drives currents into oscillation along the surfaces of the cavity walls. In this way the energy of the cathode electrons is transferred to the oscillating currents in the cavities. All the cavities are the same size and oscillate at the same frequency. The magnetrons do get hot and have to be

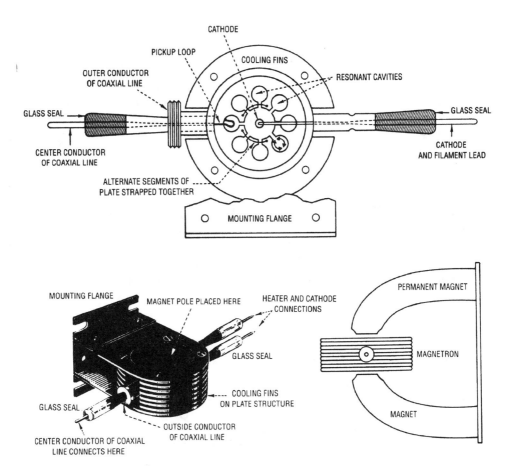

**Figure 11-22.** Magnetron.

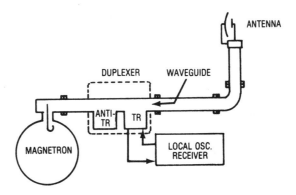

**Figure 11-23.** Magnetron coupled to a radar antenna.

cooled. Air-cooled fins are usually attached to the unit making it appear much larger than its working area really is. Some kilowatt and megawatt units have to be water-cooled.

Figure 11-23 shows how the magnetron is connected to the radar antenna for high-power operation. T/R stands for *transmit-receive*. This allows the radar to send out a signal and then turn off the transmitter for a small fraction of a second to allow the reflected signal to be returned to the receiver without ruining the receiver with too much energy from the magnetron. The waveguide is merely the transmission line or device used to get the energy from the magnetron to the antenna.

# The Gunn Oscillator

A semiconductor device called the Gunn diode operates at lower power levels than the klystron and the magnetron. The device generates microwave energy at the frequencies needed for police radar. The units which are mounted on the dashboard or side window of a police car operate on the X-band frequencies and put out either 100 or 150 milliwatts (mW) of power at 10,525 MHz. The Gunn diode is ideal for this application. It is also used in the hand-held units used to check the speed of baseballs, cars, and any moving thing. The hand-held units operate on 24.15 GHz. They too are very low power devices. Thus, the Gunn diode is best for that particular operation. The power output of the Gunn diode is about 1 watt (W) maximum (see Figure 11-24). Since newer devices are being developed constantly, wattage ratings may be increased at any time.

Newer applications in microwave technology makes it one of the more interesting fields of research. Improvements are announced on a daily basis for military, commercial, and domestic uses of microwaves, some of which are labeled Top Secret.

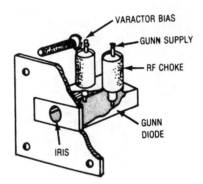

**Figure 11-24.** Note how the Gunn diode fits into this Gunn oscillator designed for use by amateur operators.

# Summary

Oscillators are used to produce desired frequencies. In mechanical oscillators the speed of the generator makes a difference in the output frequency. In electronic oscillators a number of methods are used to produce a desired frequency. The tank circuit consisting of a capacitor and inductor are the older method of producing radio frequencies and audio frequencies. However, other means have been found to produce various frequencies for specific jobs.

Feedback, for instance, is the most important factor that an oscillator needs. It causes the oscillators to continue once started. Feedback aids in a number of ways to keep the oscillator going and to improve its signal output.

Transistor oscillators include the tuned base, Clapp, Hartley, Colpitts, crystal, Wein-bridge, and multivibrator oscillators. The semiconductor Gunn diode serves the same purpose at radar frequencies.

The klystron, magnetron, and Gunn diode are used in microwave applications. They serve as the signal source for these extremely high frequencies. There are many military, commercial, industrial, and domestic uses for microwaves today.

# Review Questions

1. What is an oscillator?
2. Where are oscillators used?
3. How does an *LC* circuit produce a frequency?
4. What is a damped oscillation?
5. What do you need in an oscillator circuit to keep the circuit going once it starts to oscillate?
6. Differentiate between regenerative and degenerative feedback.
7. What is inverse feedback? Where is it used most often?
8. What is the difference between a Hartley and a Colpitts oscillator?
9. How does the crystal oscillator work?
10. What advantage does the crystal oscillator have over the regular *LC* circuit type?
11. What is the difference between the Clapp and Colpitts oscillators?
12. Why is the Wein-bridge oscillator called by that name?
13. What is a multivibrator?
14. What is a blocking oscillator?
15. At what frequencies do microwaves operate?
16. What is a klystron?
17. What is a magnetron?
18. Why would you use a klystron over a magnetron?
19. What is a Gunn diode?

# Chapter 12

# MODULATION AND DEMODULATION

Until the radio came along, long-distance communication was carried on by way of the telegraph and before that the Pony Express. Two long wires were stretched across the nation; at various points taps were attached to run a solenoid that made clicks. A code was devised so that people could read the dots and dashes or clicks. In fact, we still use this type of code in communicating between points with the radio. Amateur radio operators must be able to send and receive 13 words per minute in order to obtain their operator license.

Once the ability to transmit across the ocean and for long distances without wires was established, people were not satisfied to send dots and dashes but wanted to be able to talk with someone many miles away without having an operator in between to interpret the words. However, it was, and still is, rather difficult to transmit the audio frequencies for any distance. Even 200-watt (W) amplifiers can be heard only a few city blocks away from the source. This limitation on the audio led to experiments with the radio carrier wave. They already knew that the radio frequencies would travel great distances, but now they had to put the audio on top of it and then get rid of the carrier and amplify the audio after it arrived at its destination. Much work was done before the process of modulation was perfected.

## Modulation

Let us take a look at the telephone and how it transmits sound along its wires; this will give us a basis for what we need to understand before delving into what is called modulation. The human voice and other sounds can be transmitted over wires. This is accomplished by a device which changes the voice vibrations to current variations. If a single sound is emitted at the transmitting end, the current through the wire varies at the frequency of the sound. In the telephone system, only a variation in current produces sound at the receiver. When no sound is being transmitted, the current is simply some dc value. This means that when the single sound is transmitted, the current through the receiver has the form as shown in Figure 12-1. Since a sound varies at some audio frequency rate, the current varies at the same audio frequency rate.

This means that the direct current (dc) flowing continuously in the circuit is made to carry the sound signal from the transmitter to the receiver. The carrier does not produce any sound at the receiver unless it is varied by a sound at the transmitter. In the telegraph

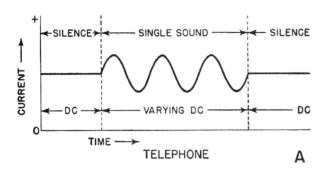 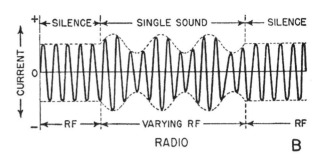

**Figure 12-1.** Telephone and radio signals compared.

system, the dc itself carries information in the form of dots and dashes. This is not true of the telephone system. The sound causing the variation in carrier current flow is known as the *modulating signal*. A carrier on which modulation has been imposed is a *modulated carrier*, which carries information in the form of audio frequency variations, usually produced by voice or music.

A voice-modulated radiotelephone system functions in the same manner as the telephone system. The alternating current (ac) in the transmitting antenna is modulated by the sound at the audio frequency rate. The alternating transmitter current is the carrier. In radio transmission the frequency of the transmitter current is called the *radio frequency* (RF). The frequency of the modulating signal is the *audio frequency* (AF). The radio carrier is referred to as the RF carrier, or carrier wave (CW).

# AM Radiotelephone Transmitter

Before you get too far with the modulation, you have to be able to see how it is useful in a total system. That is where Figure 12-2 comes in. Look at the block diagram which shows how an oscillator is used to produce the RF carrier wave. A buffer amplifier is used to isolate the oscillator from load variations occurring at the power amplifier state. On a parallel line below, you see the speech amplifier (an AF amplifier) that pushes a driver to get the AF frequency up to a level where it can push the modulator. The modulator is one large audio amplifier since it is required to put out about 50% as much in AF power as the power amplifier does in RF power. In larger transmitters these tubes are either air-cooled by fans or, in some instances, water-cooled.

# Amplitude Modulation

The waveform analysis of an amplitude-modulated (AM) amplifier is shown in Figure 12-3A. The carrier signal is applied to the amplifier, and the bias (which varies the gain of the amplifier) is varied by the modulating signal input. The output of the amplifier is an amplitude-modulated carrier. As the carrier signal passes through the amplifier, the gain of the amplifier is increased or decreased. When the amplifier gain is increased, the output is increased, and when the amplifier gain is decreased, the output is decreased. Thus, the amplitude of the carrier signal is varied at the modulating rate.

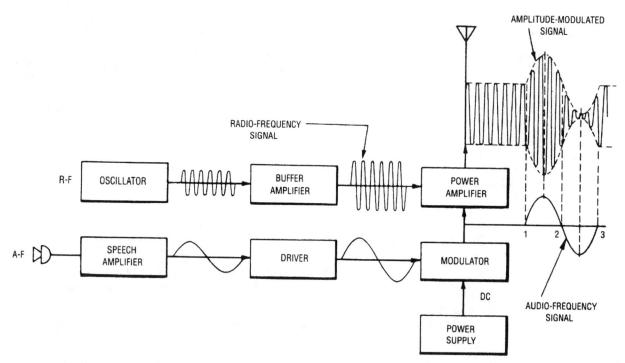

Figure 12-2. AM radio transmitter with waveforms.

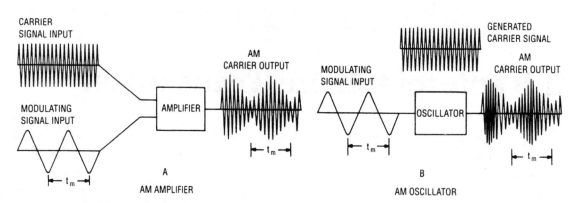

Figure 12-3. Amplifier and oscillator modulation.

A number of methods are used to vary the RF carrier with audio frequencies. In this chapter we will take a look at the modulation aspects without getting into transmitter operation and its circuits (which will be covered in Chapter 13). Figure 12-3B shows the oscillator being modulated. The carrier signal is generated by the oscillator, and the modulating signal varies the bias. Changing the bias of the oscillator changes the gain and the operating point of the transistor. Changing the bias provides amplitude changes in the output of the oscillator.

# Frequency Modulation

In a frequency modulation (FM) wave, the frequency varies instantaneously about the unmodulated carrier frequency in proportion to the amplitude of the modulating signal.

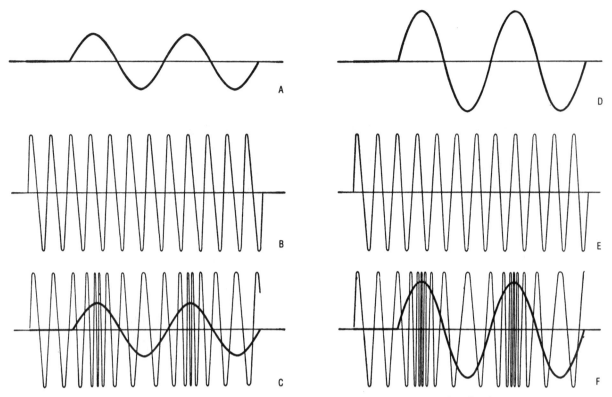

**Figure 12-4.** FM waveforms. Modulating signals with differences in amplitude.

When the modulating signal increases in amplitude, the instantaneous frequency increases; when the modulating signal decreases, the frequency decreases. An RF and an AF carrier are shown in Figure 12-4A and B. When they are combined in the modulation process, the resultant signal is the FM waveform (Figure 12-4C). As the amplitude of the audio signal increases in the positive direction, the modulated wave seems to bunch up, spreading out when the audio signal goes in the negative direction. These changes in the spacing of the modulated wave are caused by instantaneous changes in frequency. When the modulating signal is increased in amplitude (Figure 12-4D), the changes in the spacing of the waveform are proportionally greater (Figure 12-4F). Therefore, the frequency deviation of the modulation wave is directly proportional to the amplitude of the modulating signal. When the audio voltage reaches its peak voltage in the positive direction, the frequency of the carrier is at its highest value above the center value. When the modulating voltage reaches its negative peak, the frequency of the carrier wave is reduced to its lowest value below that of the center carrier frequency. Maximum frequency deviation, therefore, takes place at the peaks of the audio signal.

## Frequency Deviation

FM has some characteristics which should be understood when compared with AM or amplitude modulation. The amplitude of the modulated wave remains constant in FM. Remember, in AM the amplitude varies with the frequency of the audio. See Figure 12-5. In FM, the frequency of the modulated wave varies directly as the amplitude of the

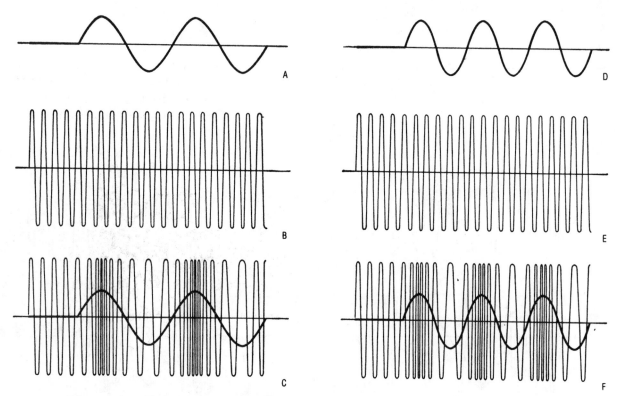

**Figure 12-5.** FM waveforms. Modulating signals with differences in frequency.

modulating signal. The limits of the frequency shift on either side of the carrier are known as the *frequency deviation limits*. In an FM system, the frequency of the modulating voltage determines the number of times per second that the frequency shifts between the deviation limits. The higher the frequency of the modulating signal, the greater the number of times per second the frequency varies between the deviation limits set by the peak amplitude of the modulating signal. The ratio between the maximum frequency deviation and the maximum frequency of the modulating signal is called the *modulation index*.

$$\text{Modulation index} = \frac{\text{maximum frequency}}{\text{maximum frequency of modulating signal}}$$

## Percentage of Modulation

The percentage of modulation of an FM signal cannot be determined in the same manner as an AM signal because 100 percent modulation would mean that the entire carrier varies in frequency from zero to twice the carrier frequency. Percentage of modulation in FM is determined as the percentage of maximum *deviation* incorporated in a transmitter for a particular type of service. For an FM transmitter with maximum deviation of 75 kilohertz (kHz), 100 percent modulation occurs when the transmitter deviates the full 75 kHz. When the deviation falls to 37.5 kHz, the transmitter is being modulated only 50 percent. Such a definition is flexible, of course, and depends on the maximum deviation of the equipment used.

# Phase Modulation

All FM transmitters use either direct or indirect methods for producing FM. The modulating signal in the direct method has a direct effect on the frequency of the carrier. In the indirect method, the modulating signal uses the frequency variations caused by phase modulation. In either case, the output of the transmitter is a frequency-modulated wave, and the FM receiver cannot distinguish between them.

In the phase-modulated (PM), or indirect, FM transmitter, the modulating signal is passed through some type of correction network before reaching the modulator. When comparing the PM to the FM wave, it should be remembered that a phase shift of 90° in the PM wave makes it impossible to distinguish it from the FM wave. Figure 12-6 shows the phase-modulated waveform and how it is produced. The phase modulation will come in handy later when we look at color television and how the colors are transmitted to make them compatible with black-and-white receivers.

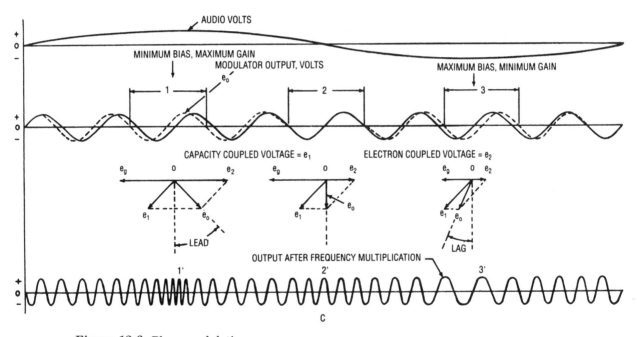

**Figure 12-6.** Phase modulation process.

Figure 12-7 shows the three types of transmitters used for AM, FM, and PM. *Mults* stands for multipliers. These stages take the input frequency and change it to double, triple, or quadruple its original frequency by tuning to a harmonic in the output stage. These harmonics are weaker than the original and must have one or two other amplifiers to boost them up to the level needed to drive the power amplifier. The block diagram simplifies this as one block, and it may represent three or four stages.

Figure 12-8 shows how the phase-modulated carrier is shifted back and forth to produce the resulting PM waveform with the information on it.

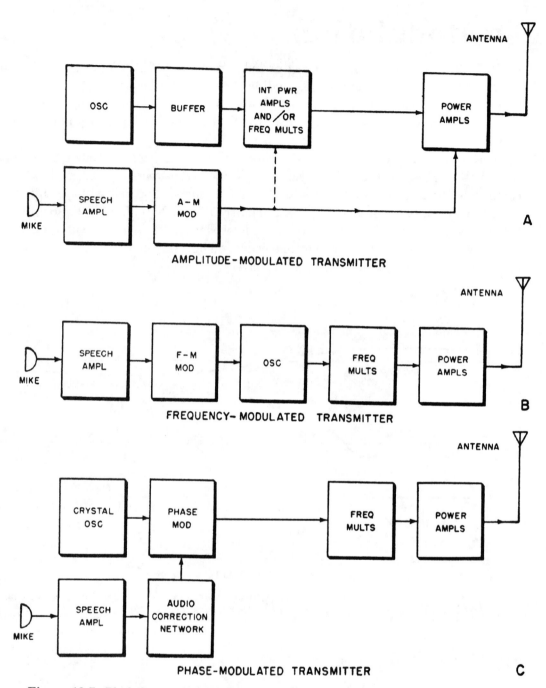

**Figure 12-7.** Block diagrams of AM, PM, and FM transmitters.

# Detectors or Demodulators

Now that we have produced the modulated waveform and it has been sent out over the air, it is time to receive it and convert it back to its original audio form so that the human brain may react to the transmitted information. The carrier wave has served its purpose: to take the message or information from the transmitter antenna to the receiver antenna without the aid of wires. Once the carrier, either AM or FM, has done its job, it is time to decipher

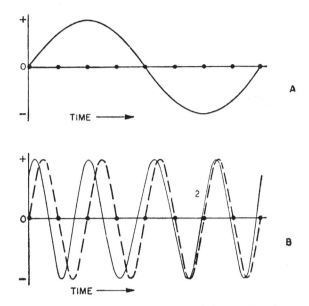

**Figure 12-8.** Phase-modulated carrier with only 1 Hz of modulation signal.

its message. This may be done with a number of different types of demodulators or detectors. The *detector* is a circuit that detects the presence of the audio and separates it from the RF carrier wave.

## Diode Detector

One of the most frequently used detectors for AM is the diode detector. It was a common detector when used in vacuum tube circuits and is used today with semiconductor diodes. The diode detector is one of the simplest and most widely used; it has nearly an ideal resistance characteristic. Diodes have a point of sharp transition between the conducting (forward) and nonconduction (reverse) directions and therefore make good detectors. (See Figure 12-9 for an illustration of the diode circuit.)

Early radio receivers used a crystal detector that was made of galena, a mineral compound of sulfur and lead. The end of a short length of fine wire touched the surface of the crystal and was held against it by the pressure of a spring. This wire was called the *cat whisker*. The wire could be adjusted manually until the best or most sensitive spot on the

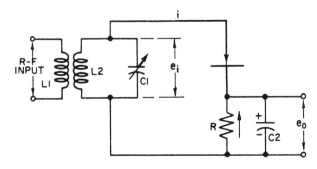

**Figure 12-9.** Diode detector.

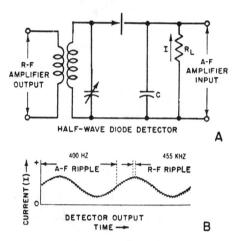

**Figure 12-10.** Half-wave rectification of the AM signal.

crystal was found. The crystal rectified the signal, and the operator heard the signal in earphones. The diode detector has low sensitivity, good linearity, poor selectivity, and a high ability to handle strong signals without overloading.

For a better look at what the diode detector can do and how it works, step by step, look at Figure 12-10. This is the half-wave diode detector. The purpose of the circuit is to convert the modulated RF carrier to a direct current varying at the AF rate of the original modulating sound. The amount of current flow induced in the receiver antenna by the RF signal is small. Thus, several stages of RF amplification often precede the detector. The output of the last RF amplifier usually is a transformer-coupled stage which has the diode connected across it and in series with a load resistor. The load resistor is usually the volume control of the receiver. The diode rectifies the RF by allowing current to flow on every other half-hertz only. This results in unidirectional current flow through the resistor $R_L$.

If an unmodulated RF carrier is applied to the detector, the output is similar to the output of the rectifier with an ac input. The RF is rectified by the diode and filtered by the capacitor. The result is a relatively smooth dc with a ripple varying at the RF rate.

If a modulated RF carrier is applied to the detector, the RF is rectified and filtered as before. However, the resultant dc output has two ripples, shown in Figure 12-10B. One ripple is caused by the RF, and the other by the AF. The RF carrier frequency is 455 kHz, and the single modulating sound is at an audio frequency of 400 hertz (Hz). For all practical purposes this waveform can be considered as dc varying at the AF rate of the original modulating sound.

The output waveform appearing across $R_L$ can be used to drive an earphone. More sophisticated receivers have stages following the detector stage for the purpose of amplifying the audio signal until it is sufficient to drive a speaker.

Figure 12-11 shows how the diode of the semiconductor detector is introduced into the receiver for the purpose of detection. One configuration is for voltage output. The other is for current detection. Current detection is more apropos for transistor operation, whereas voltage detection is more suited for vacuum tubes.

At least four methods can detect the audio in an FM signal: the discriminator, the slope detector, the ratio detector, and quadrature detection make up the four.

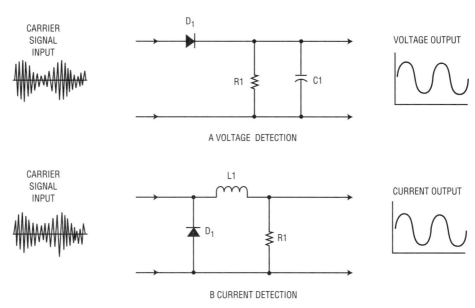

**Figure 12-11.** Semiconductor diode rectifiers.

# Discriminator

A discriminator in an FM receiver performs the same function as a detector in an AM receiver. Figure 12-12 shows a transistorized version of the intermediate frequency (IF) stage and the discriminator.

Amplifier $Q_1$ amplifies the IF signal applied to the discriminator. Resistor $R_1$ is the emitter swamping resistor, and capacitor $C_1$ is an IF bypass. Capacitor $C_2$ and the primary of the transformer $T_1$ form a parallel resonant circuit for the IF signal which is coupled through the transformer to the discriminator. Capacitor $C_3$ couples the IF signal to the secondary of transformer $T_1$ for phase shift comparison. The IF signal, coupled across capacitor $C_3$, is developed across coil $L_1$. Capacitor $C_4$ and the secondary of transformer $T_1$ form a resonant circuit for the IF signal coupled through the transformer. The top half of transformer $T_1$ secondary, diode $D_1$, coil $L_1$, load resistor $R_2$, and filter capacitor $C_5$ form one-half of the comparison network. The bottom half of transformer $T_1$ secondary, diode $D_2$, coil $L_1$, load resistor $R_3$, and filter capacitor $C_6$ form the second half of the comparison network. The audio output of the discriminator circuit is taken from the top of

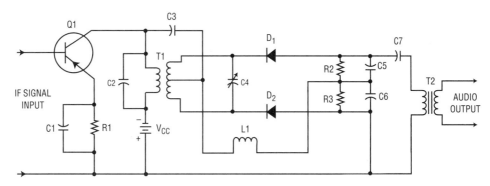

**Figure 12-12.** IF amplifier and discriminator stages.

capacitor $C_5$ and the bottom of capacitor $C_6$. The audio output is coupled through capacitor $C_7$ to the primary of transformer $T_2$. The audio signal, coupled through transformer $T_2$, is applied to the following stage.

# Slope Detector

The main reason for using a slope detector is its ease of operation. It converts the FM signal or frequency changes of a carrier signal into amplitude changes. The amplitude changes can then be detected by an AM diode detector or an AM transistor detector. The input and output waveforms of a slope detector and an AM diode detector are shown in Figure 12-13. The IF signal with frequency deviations is applied to the slope detector $Q_1$. The output of slope detector $Q_1$ and the IF signal with amplitude and frequency deviations is applied to the diode detector $D_1$. The resultant output is an audio signal which is equivalent to the frequency deviations of the IF input signal.

The IF signal coupled through transformer $T_1$ is applied to the base circuit. The resonant circuit consisting of coil $L_1$ and capacitor $C_2$ (tuned slightly off the carrier frequency) develops a large amount of IF signal when the frequency deviation is near the resonant frequency. As the frequency deviation of the IF signal becomes lower than the resonant frequency of the resonant circuit, a smaller amount of IF signal is developed. A large amount of IF signal added to the bias voltage developed across resistor $R_1$ increases the emitter-base bias, and a small amount of IF signal decreases the emitter-base bias. The emitter-base bias is therefore increasing and decreasing as the frequency of the IF signal increases and decreases, respectively. Since the bias of slope detector $Q_1$ changes at the frequency

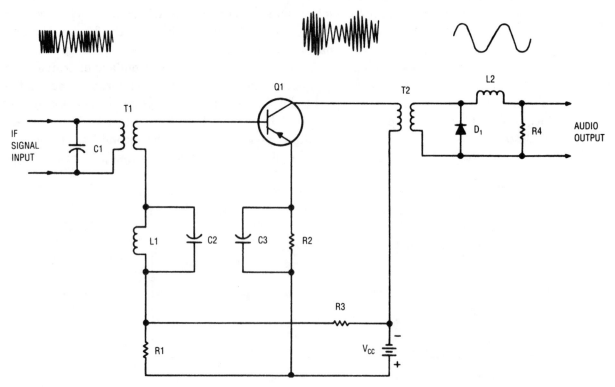

**Figure 12-13.** Slope detector for FM.

deviation rate, the gain also changes at the frequency deviation rate. Thus, the output of the slope detector is an IF signal that is changing in amplitude and frequency. The IF signal applied to the diode detector $D_1$ is rectified, filtered by coil $L_2$, and developed across resistor $R_4$. The output of the current diode detector is an audio signal.

Capacitor $C_1$ and the primary of transformer $T_1$ form a parallel resonant circuit for the IF signal, which is coupled through the transformer to the base circuit of slope detector $Q_1$. Capacitor $C_2$ and coil $L_1$ form a parallel resonant circuit for a frequency slightly higher than the maximum frequency deviaton of the IF signal. Resistor $R_1$ is the base-emitter bias resistor, and resistor $R_3$ is a voltage-dropping resistor. Resistor $R_2$ is the emitter-biasing resistor, and capacitor $C_3$ is a bypass capacitor for the IF signal. Transformer $T_2$ is an output coupling transformer for slope detector $Q_1$. Diode $D_1$ is the AM detector, resistor $R_4$ is a load resistor, and coil $L_2$ is a filter.

# Ratio Detector

Figure 12-14 shows the ratio detector using semiconductor diodes. There are now complete chips which can do the whole operation without discrete components.

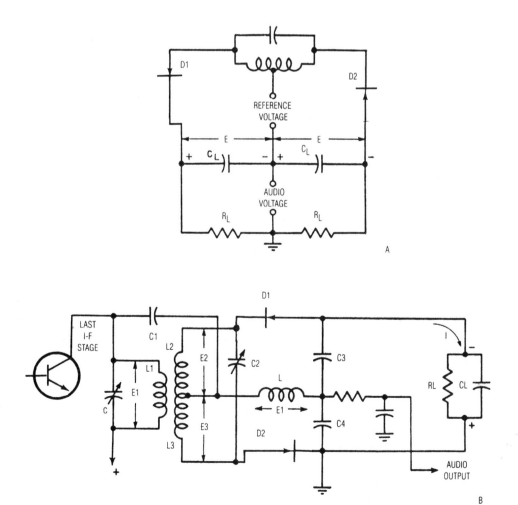

**Figure 12-14.** Ratio detector for FM.

The discriminator requires a limiter preceding it in order to be able to limit the noise and amplitude modulation which may appear on the FM signal fed to the IF stages. The ratio detector, as you can see, can be fed directly from the IF stage without the benefit of an additional stage such as the limiter.

Figure 12-14A shows the simplified version of the ratio detector which can be used for explanation purposes. Figure 12-14B shows how a practical circuit is utilized to produce an output from an FM signal.

The ratio detector splits the rectified voltages in such a way that their *ratio* is directly proportional to the ratio of the applied IF voltages, which vary with the frequency.

When the sum of the rectified voltages from the transformer is maintained at a constant value, the ratio between them must remain constant, and the individual rectified voltages also must be constant. Output, therefore, is independent of amplitude variations in the signal, and no limiter is necessary. A simplified ratio detector circuit is shown in Figure 12-14A. It shows both diodes connected so that their output adds, instead of subtracting as in the discriminator. Capacitors $C_L$ across the load resistors have a large value of capacitance and are charged by the output voltage of the rectifiers. This tends to make the total voltage across the load constant over the period of the time constant $R_L C_L$, since a large capacitor across the combined loads maintains an average signal amplitude that is adjusted automatically to the required operating level. The rectified output must not vary at audio frequency, and the time constant of the capacitor and the load resistors must be great enough to smooth out such changes. This time constant is approximately 0.2 second. The basic phase comparison circuit and the appropriate vector diagram of the ratio detector and the phase discriminator are the same.

In the circuit for a practical ratio detector shown in Figure 12-14B, the voltages $E_1$, $E_2$, and $E_3$ are obtained in the same way as in the discriminator. Therefore, the applied voltage to the diodes also is the same. The diodes are connected in series, and the current through load resistor $R_L$ is always in the same direction. Therefore, $R_L$ acquires the polarity shown when the current flows from the plate of $D_1$ to the cathode of $D_2$. When an unmodulated signal is applied to the primary of the transformer, equal and opposite voltages $E_2$ and $E_3$ are developed across the secondary in respect to the center tap. These voltages are rectified by the diodes, with the output voltage across the load resistor equal to their sum (or, $E_2$ plus $E_3$) and the large capacitor $C_L$ charged to this constant voltage. The time constant of $R_L$ and $C_L$ is long compared with the lowest audio frequency.

Since the voltage across $C_L$ is constant, the sum of the voltages across $C_3$ and $C_4$ must remain fixed. When the carrier frequency shifts with modulation, however, the voltages across the two capacitors $C_3$ and $C_4$ change, but the sum of their voltages stays fixed at the amplitude of the charge on $C_L$. When frequency decreases, $C_4$ acquires a greater charge than $C_3$. When the frequency increases, $C_4$ loses charge to $C_3$. That means the voltage between the center tap and the two capacitors and ground varies as the ratio of the voltages across $C_3$ and $C_4$ — the ratio depending on the instantaneous frequency. A variable voltage whose amplitude depends on the frequency deviation of the carrier consequently can be applied to the audio output. As the rate of variation increases with frequency deviation, the voltage at the center tap changes frequency, producing a higher audio frequency. Any amplitude variation in the input signal to the transformers, no matter where the carrier is in its swing, also tends to change the voltage across $C_3$ and $C_4$. The voltage across the $RC$ network, however, cannot change rapidly enough to follow the amplitude modulations, and the ratio of the voltage across $C_3$ and $C_4$ does not change enough to produce an audio output.

The rectified voltage across the load circuit of the ratio detector adjusts itself to the amplitude of the input signal, and there is no minimum level where amplitude variation still can appear in the output. No matter how weak the signal is, the amplitude variations are removed to some extent by the constant charge on the capacitor. However, if signals of greater strength are tuned in, the charge on the capacitor is increased, and the total voltage across $C_3$ and $C_4$ is increased. Thus, the ratio detector can produce audio output which is proportional to the average strength of the received signal. Ratio detectors can operate with as little as 100 millivolts (mV) of input which is much lower than that required for limiter saturation and less IF gain is required. This receiver also is relatively quiet when no signal is received.

# Quadrature Detection

Another method of detection used for FM is the quadrature detection method. It is necessary to use an IC chip to do the job since it involves amplification and phase shifting of the input IF signal. Figure 12-15 shows two of the types of ICs used for FM and TV IF amplifier, detector, limiter. These 14-pin DIPs are used in most television sets and FM receivers made today.

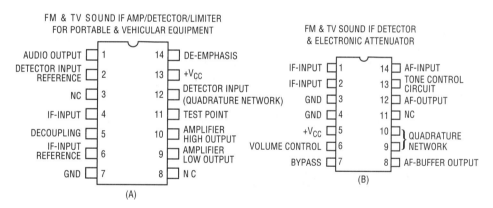

**Figure 12-15.** Quadrature detection is accomplished by 14-pin DIP integrated circuits like the two shown in A and B.

Figure 12-16 shows how the IC is used to produce the quadrature detection of FM with its external components. The IF signal is amplified and limited and then applied to the IC at pin 4. The IF output is amplified and then leaves the chip at pin 10. This signal is also coupled internally to the phase comparator. The capacitor attached to pin 9 serves to couple some of the IF signal to the external quadrature circuit, enclosed in the dotted lines. The center frequency for the quadrature circuit is 10.7 MHz or the IF frequency. The quadrature signal is developed so it will lead the IF signal by 90°. The *leading* quadrature signal is sinusoidal in shape. It is applied to pin 12 of the chip. Pin 12 is the other input to the phase comparator. The IF signal at this point is varying above and below 10.7 MHz in step with the modulation of the transmitter signal. Upward deviation of the IF frequency makes the quadrature circuit become capacitive and the quadrature signal then lags its

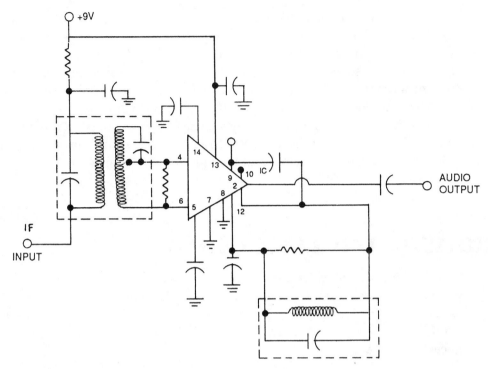

**Figure 12-16.** Circuitry associated with utilizing a 14-pin DIP integrated circuit.

former phase angle. Downward deviation of the IF frequency makes the quadrature circuit become inductive and the quadrature signal leads its former phase angle. The configuration of the phase comparator is such that it compares the phase angle of the quadrature signal with that of the IF signals and develops an output voltage that varies inversely with the phase angle and outputs it at pin 1.

# Other Types of Modulation

Digital communications have accounted for a variety of different types of modulation. These include pulse modulation, digital modulation, digital data processing, and telemetry.

In the early days of radio the carrier wave was turned on and off in order to send dots and dashes. These dots and dashes were coded to represent letters of the alphabet and numbers. Once the operator learned the code it was possible to communicate with other operators who decoded the information and wrote it down in whatever language was being used. However, this was slow and very labor intensive. What was needed was a quicker way of doing the same thing, using the advantages of the coding and not having to wait for an operator to decode it. The answer was forthcoming with the advent of radio telegraphy and the application of pulse modulation, digital modulation, and digital data processing.

Instead of turning the carrier on and off to cause a receiver to receive the dot and dash, the coding that takes place today is digitized. That means the binary system is used to represent the information in coded form. The high or 1 level is used to represent the on situation. This is called the mark. Two 1 s can be used to represent the dash of the old Morse Code. The 0 level can be used to represent spacing between the on conditions (see Figure 12-17).

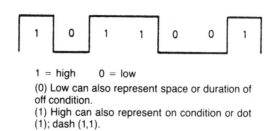

1 = high    0 = low
(0) Low can also represent space or duration of off condition.
(1) High can also represent on condition or dot (1); dash (1,1).

**Figure 12-17.** Binary information in pulse form.

A *bit* is a unit of information required to allow proper selection of *one out of two* equally probable events. By using a 7-bit arrangement it is possible to have a binary result of 2 raised to the seventh power, or 128. The binary code is then used to represent the 26 letters of the alphabet and the numbers as well as other needed signs for punctuation. The period, comma, slash mark, plus and minus, ampersand, dollar mark, percent sign, parentheses, apostrophe, exclamation point, quotation mark, and the number sign are a few, as well as lowercase letters. Figure 12-18 shows how the 7-bit ASCII code is used to represent the needed components for transmission of messages. There are also other codes possible and used for the transmission of messages. Two of those are the 5-bit Baudot code and the ARQ code. The ARQ code is an adaption of the Baudot code and uses 7 rather than 5 bits. It is used as a telegraphic code. The ARQ means *automatic request for repetition*.

# Tone Modulation

It is possible to transmit an audio code on the radio frequency carrier. This tone or audio signal is then interrupted in the form of dots and dashes. The person at the other end does not need to have a receiver with a beat frequency oscillator to beat against the incoming signal in order to produce a difference that can be heard by the human ear.

A two tone modulation system has been developed for the transmission of information. It uses a frequency of 470 Hz to represent the high or mark condition. The other frequency is 300 Hz, and it represents the space condition. At the receiver only the 470 Hz is heard since the received signal is passed through a filter that extracts the 470 Hz.

This type of modulation has its advantages. It only takes a bandwidth of 940 Hz for transmitting the information. The bandwidth is obtained by multiplying $470 \times 2$ since 470 is the highest frequency used for the tone modulation. Three-hundred hertz is used for the 0 point or spacing information and 470 Hz for the dot or dash. That means you can design a system that can transmit one hundred 1-kHz channels on a band that uses only 0.1 MHz.

# Frequency Shift Keying

Frequency modulation or FM can provide a better method of broadcasting coded information. FM has a number of advantages, one of which is the elimination of noise that can cause misinterpretation of messages. Frequency shift keying (FSK) is a form of frequency modulation. The modulating wave shifts the output between two predetermined

| | |
|---|---|
| A | 1 000 001 |
| B | 1 000 010 |
| C | 1 000 011 |
| D | 1 000 100 |
| E | 1 000 101 |
| F | 1 000 110 |
| G | 1 000 111 |
| H | 1 001 000 |
| I | 1 001 001 |
| J | 1 001 010 |
| K | 1 001 011 |
| L | 1 001 100 |
| M | 1 001 101 |
| N | 1 001 110 |
| O | 1 001 111 |
| P | 1 010 000 |
| Q | 1 010 001 |
| R | 1 010 010 |
| S | 1 010 011 |
| T | 1 010 100 |
| U | 1 010 101 |
| V | 1 010 110 |
| W | 1 010 111 |
| X | 1 011 000 |
| Y | 1 011 001 |
| Z | 1 011 010 |
| a | 1 100 001 |
| b | 1 100 010 |
| c | 1 100 011 |
| d | 1 100 100 |
| e | 1 100 101 |
| f | 1 100 110 |
| g | 1 100 111 |
| h | 1 101 000 |
| i | 1 101 001 |
| j | 1 101 010 |
| k | 1 101 011 |
| l | 1 101 100 |
| m | 1 101 101 |
| n | 1 101 110 |
| o | 1 101 111 |
| p | 1 110 000 |
| q | 1 110 001 |
| r | 1 110 010 |
| s | 1 110 011 |
| t | 1 110 100 |
| u | 1 110 101 |
| v | 1 110 110 |
| w | 1 110 111 |
| x | 1 111 000 |
| y | 1 111 001 |
| z | 1 111 010 |
| 0 | 0 110 000 |
| 1 | 0 110 001 |
| 2 | 0 110 010 |
| 3 | 0 110 011 |
| 4 | 0 110 100 |
| 5 | 0 110 101 |
| 6 | 0 110 110 |
| 7 | 0 110 111 |
| 8 | 0 111 000 |
| 9 | 0 111 001 |
| SP | 0 100 000 |
| ! | 0 100 001 |
| " | 0 100 010 |
| # | 0 100 011 |
| $ | 0 100 100 |
| % | 0 100 101 |
| & | 0 100 110 |
| • | 0 100 111 |
| ( | 0 101 000 |
| ) | 0 101 001 |
| * | 0 101 010 |
| + | 0 101 011 |
| , | 0 101 100 |
| − | 0 101 101 |
| . | 0 101 110 |
| / | 0 101 111 |

**Figure 12-18.** The 7-bit ASCII code used for digital when upper- and lowercase letters are needed.

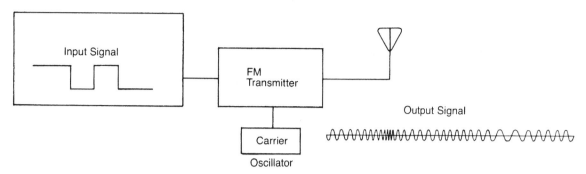

**Figure 12-19.**  Frequency shift keying used to modulate a FM transmitter.

frequencies. These are usually referred to as the mark and space frequencies. This system uses a carrier frequency that is located halfway between the mark and space frequencies and is then modulated by a rectangular wave (see Figure 12-19).

This type of modulation allows for a number of narrow-band FM channels to be placed in a small portion of the spectrum. The wide-band FSK uses a 10- to 20-kHz channel, while the narrow-band utilizes a channel that is less than 10 kHz.

The telephone lines can be used to transmit information using the FSK system. This is one of the better ways to transmit digital information or data over long distance lines.

# Pulse Modulation

Pulse modulation is different from the normal AM and FM inasmuch as the AM and FM uses some part of the modulated wave to vary continuously with the message. In pulse modulation some part of a sample is used and varied by each sample value of the message. The pulses are usually very short in duration so that a pulse modulated wave is off most of the time. This allows the transmitter to operate on a very low duty cycle, and the time intervals between pulses can be modulated with other messages. This capability means a number of different messages can be transmitted on the same channel. The pulse method of sending many messages over the same channel is called multiplexing or time division multiplexing, in this case TDM. This is where the computer can make use of the system by having a number of users utilize a computer at the same time (see Figure 12-20).

There are a number of ways to accomplish pulse modulation. This is not really a modulation system but a technique used to send messages. The three types of pulse modulation are PAM, PWM, and PPM.

PAM is *pulse amplitude modulation*, which means the pulse amplitude is proportional to the modulating signal's amplitude. This technique is similar to using a rotator resembling, in operation only, that of an automobile's ignition system. It takes eight channels and places their inputs into a frequency-modulated transmitter and sends the messages to a receiver that redistributes the eight channels to envelope detectors that process and reconstruct the information (see Figure 12-21). This method does have some problems with noise. Noise problems can be better handled by PWM and PPM since they have constant amplitude pulses.

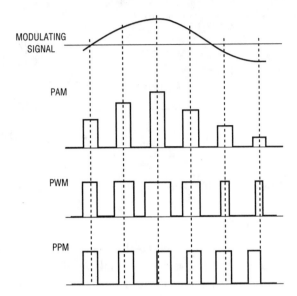

**Figure 12-20.** Comparison of pulse modulation signals.

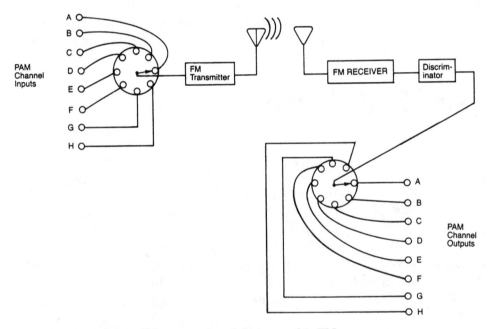

**Figure 12-21.** Eight-channel PAM system of modulation used in FM.

# Pulse Width Modulation

Pulse width modulation (PWM) is different from PAM inasmuch as it is more pulse time modulation (PTM). The time varies instead of the amplitude. It can also be referred to as PLM or pulse length modulation. A phase locked loop can be used in the generation of both PWM and PPM. The phase lock loop 565 IC can be used. By varying the voltage controlled oscillator (VCO) it is possible to cause the phase detector to increase or decrease its output. The output then is a phase-shifted signal that can be applied to a carrier and

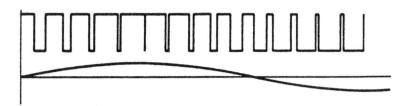

**Figure 12-22.** The processing of PWM waveforms through a comparator and an integrator to produce the desired results.

transmitted. The advent of ICs and the phase lock loop have made it possible to obtain the phase shifting easily (see Figure 12-22).

# Pulse-Position Modulation

Pulse-position modulation (PPM) is generated from PWM. PPM is usually used since it has better noise characteristics. It is possible to generate PPM by inverting PWM (see Figure 12-23). Once inverted the signal is differentiated and produces PPM with a very short pulse. This pulse can be used to modulate an AM carrier. Its improved quality at reception is due to the pulse's very short duration, and the information content is not contained in either the pulse amplitude or width.

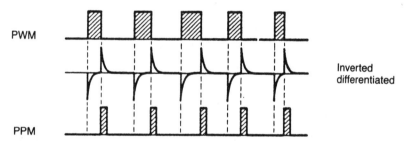

**Figure 12-23.** How the PPM signal is generated by inverting and then differentiating the PWM.

Time division multiplexing or PCM transmission for the telephone has proven its ability to cram more messages into short haul cable than frequency division multiplex analog transmission. PCM is the most noise-resistant transmission system available. Much work is being done with the transmission of data, and other methods of transmission may be developed as newer ICs are made available with the necessary characteristics to handle the requirements of various situations.

## Summary

Modulation is the process of putting an audio frequency (AF) on a carrier wave of radio frequency (RF). The information contained in the AF signal is impressed on the carrier by means of amplitude in amplitude modulation (AM) or in the way of frequency in frequency modulation (FM).

The modulator is a large AF amplifier. It is used to impress the audio voltage on the RF carrier wave in a transmitter.

FM transmitters have a modulation index which is the maximum frequency divided by the maximum frequency of the modulating signal. In FM the frequency of the modulated wave varies directly as the amplitude of the modulating signal. The limits of the frequency shift on either side of the carrier are known as frequency deviation limits.

There are three common methods of modulation: amplitude, frequency, and phase. Frequency and phase modulation are both used in FM.

Demodulation is the process whereby the audio signal is removed from the RF carrier. Demodulation is the process that is handled by the detector circuit. The detector circuit may be a simple diode in the AM receiver, but in the FM receiver it has a number of possibilities. In FM the demodulator may be a discriminator, a ratio detector, or a slope detector. All three work, and each has its advantages and disadvantages.

# Review Questions

1. What is modulation?
2. How does AM differ from FM?
3. What is phase modulation?
4. Where is FM used today?
5. What is a carrier wave?
6. What is frequency deviation?
7. What is the modulation index?
8. What type of detectors can you use for AM?
9. What type of detectors can you use for demodulating FM?
10. Explain briefly how the AM diode detector works.
11. What is a discriminator? Where is it used?
12. What is a ratio detector? Where is it used?
13. What is a slope detector?
14. Name an advantage of the ratio detector over the discriminator.
15. What does a slope detector detect?

# Chapter 13

# TRANSMITTERS, TRANSMISSION LINES, AND ANTENNAS

A transmitter is a device for converting messages, whether in code, computer language, voice, or music, into electric impulses for transmission either on closed lines or through space from a radiating antenna. The various types of transmitters include the amplitude-modulated, the frequency-modulated, and the phase-modulated. Each has a job to do and does it well. There are also radar transmitters that deal with super-high frequencies.

Transmission lines take the signal from the transmitter to the antenna in most instances. Antennas are designed to direct the radiated energy in a path that is predetermined by design. Special designs are available for individual purposes.

## Continuous Wave Transmitter

One type of radiation is the continuous wave (CW), or unmodulated wave. This waveform is like that of the RF current in the tuned tank circuit of a power output stage. In this type of wave the peaks of all the waves are equal, and they are evenly spaced along the time axis. The waveform is sinusoidal. It is designed for use as a code transmitter. Note the keying in Figure 13-1.

## Amplitude-Modulated Transmitters

The function of a transmitter is to take the audio or code from one location to another without the use of wires. In most instances it is used to broadcast information which can, in turn, be picked up by a receiver and decoded or demodulated to obtain the intelligence thus transmitted. Figure 13-2 shows a block diagram of an AM transmitter with waveforms.

The amplitude may be modulated by means of a signal of constant frequency which is varied to send coded messages. This is called *modulated continuous wave* (MCW). It is modulated, but a tone is used to do the modulating.

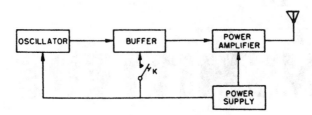

**Figure 13-1.** An MOPA (master oscillator power amplifier) transmitter with keying in the buffer stage.

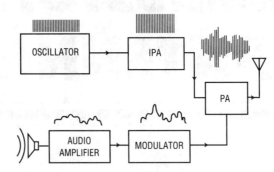

**Figure 13-2.** Block diagram of an AM transmitter with waveforms.

The amplitude-modulated transmitter may also have voice or music impressed on its carrier wave. This is one form of communicating. Those who pick up the radiations are able to demodulate them and listen directly without having an operator decode the message.

There are other types of modulation, frequency modulation (FM), for instance, but we will concern ourselves with amplitude modulation (AM) at the moment.

A CW transmitter, with equal power output, will get greater range than one that is modulated. This results from the fact that all the intelligence is contained in the sidebands, and the fewer the number of sideband frequencies, the greater will be the signal strength in the remaining sideband frequencies. In CW operation the sidebands are not extended very far on each side of the carrier, and all the energy is therefore contained in a narrowband and not wasted in nonessential bands.

# Oscillator

Every transmitter needs an oscillator to establish its operating frequency. The frequency of a transmitter can be stabilized by the use of a crystal oscillator. However, this arrangement would require a large number of crystals to cover many frequency channels used by some communicators. Thus, some oscillators that can be easily changed in frequency are employed. These oscillators would be the ones in which the coil or capacitor in the tank circuit could be adjusted to change the resonant frequency. Ham operators use this type of circuit so that they can change frequencies to those needed to talk with the individual contacted.

Placing the frequency-determining components of the oscillator in a temperature-controlled oven eliminates drift. To ensure further stability, the oscillator is loaded very lightly and isolated by a buffer stage.

# Buffer Stage

The buffer stage is located between the oscillator and the power amplifier in order to keep the oscillator stable. The output of the buffer can be tuned to the second harmonic of the oscillator and then become a frequency doubler. Thus, the output of the stage has twice the frequency of the oscillator. It can also become a tripler or a quadrupler if so tuned to the third or fourth harmonic of the oscillator. Keep in mind that some stages or circuits are not suited for multiplication work. The push-pull stage, for instance, is not suited for even-harmonic multiplication. It is good for odd harmonics only — such as third, fifth, seventh, etc. The buffer may also be called the intermediate power amplifier (IPA). If this is the case, it usually is not a doubler, tripler, or quadrupler. Stages preceding it will be frequency multipliers, whereas the IPA is strictly a power amplifier, a little less powerful than the final power amplifier (PA).

# Power Amplifier

Figure 13-3 shows the location of the power amplifier (PA) in the transmitter stages. Note that the driver is an IPA or amplifier of the doubled frequency of the oscillator that was multiplied by the multiplier stage.

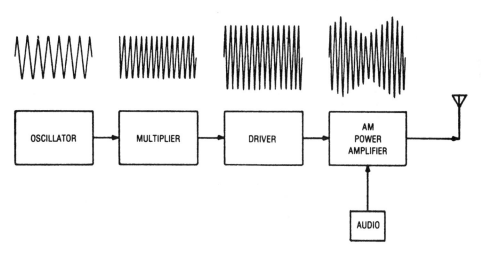

**Figure 13-3.** Block diagram of AM transmitter and waveforms.

The buffer stage isolates the oscillator from the varying load caused by modulation or keying. The PA's purpose is to increase the magnitude of the RF current and voltage by the resonant action of its output tank circuit. The power amplifier is usually operated as class C. Note that the PA is connected to the antenna. Figure 13-4 shows the complete AM transmitter. Remember that we covered the audio section or modulation earlier in Chapter 12.

The amplitude modulation is also used in television for transmitting the picture information and its synchronous signals.

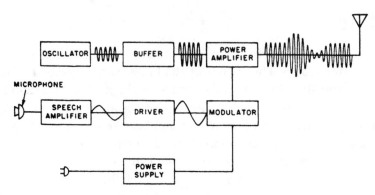

**Figure 13-4.** AM transmitter with audio section included.

# Frequency-Modulated Transmitters

The various direct and indirect methods for producing frequency modulation (FM) involve changing either the frequency or phase of an oscillator in accordance with some modulating signal. In the direct method, the modulating signal is injected into a modulator whose output varies the frequency of the oscillator in accordance with the original modulating signal. In the indirect method, the modulating signal is passed through a correction network to a phase modulator. The correction network changes the phase of the modulation in such a manner that, when the output of a crystal oscillator is passed through the modulator, the oscillations are frequency-modulated in accordance with the modulating signal. See Figure 13-5 for a block diagram of the FM transmitter so you can see what takes place from the audio input to the antenna.

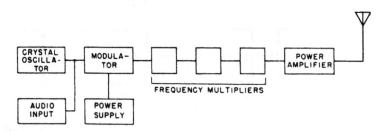

**Figure 13-5.** FM transmitter block diagram.

## Frequency Multiplication

It is extremely important that the transmitter be transmitting at its designated frequency. To achieve maximum frequency stability, the oscillator is operated at relatively low frequencies, and the lower the frequency, the more stable the oscillator. This signal output means that the center frequency of the FM signal output of the modulator-oscillator section is lower than the carrier frequency desired for transmission. To raise the FM signal to the correct frequency, it is passed through a series of frequency multipliers. Each stage of frequency multiplication raises the frequency of the signal input by some multiple of the fundamental frequency. When the input to the frequency multiplier is an FM signal, the multiplier produces an increase in the frequency deviation. The oscillator-modulator and frequency multiplier sections of the transmitter are operated at low power levels, and the output of the final multiplier is too weak to be transmitted. A power amplifier similar to those in the AM transmitter acts as the final stage in the FM transmitter to build up the signal to the power level desired.

In FM transmitters, frequency multiplication of the FM signal performs two functions. It increases the frequency of the signal to the value desired for transmission, in this way acting the same as a frequency multiplier in an AM transmitter. It also increases the effective frequency deviation of the FM signal.

# Frequency Deviation

The FM signal from the oscillator-modulator section has a center frequency of $f_c$ and a frequency deviation of $\Delta f$ caused by the modulating signal. The FM signal therefore varies from a maximum of $f_c - \Delta f$. For example, with an oscillator whose unmodulated output frequency is 100 kilohertz (kHz), a certain audio signal causes this frequency to swing between 95 and 105 kHz, and the frequency deviation therefore is ±5 kHz.

If this FM signal is impressed on the grid of a tube or the base of a transistor which is operating as a doubler, the center frequency of the doubler output is twice $f_c$, or 200 kHz. Since the multiplier doubles any frequency appearing at the grid or base, when the FM signal is deviated to 95 kHz, the output frequency is 190 kHz. When the FM signal is 105 kHz. the output is 210 kHz. The multiplier output therefore varies from 190 to 210 kHz, and the deviation is 10 kHz. By doubling the frequency at its input, the multiplier also has doubled the frequency deviation. The amount of multiplication used depends on the frequency to which the signal must be raised and the amount of frequency deviation desired. The greater the deviation, the greater is the bandwidth of the FM signal transmitted.

# Power Amplifier

The requirements for an FM power amplifier are somewhat different from those for AM, in which the power amplifier is usually the stage in which the modulation is introduced. Therefore any losses that take place during the modulation process must be dissipated in the power amplifier stage. Since the FM power amplifier has no connection with the modulation process, the only losses that are involved are those inherent in the transistor (or tube) and circuit when amplifying an unmodulated carrier.

# Uses of FM Transmitters

FM transmitters are used for the broadcast band on 88 to 108 megahertz (MHz). This gives static-free reception of music and voice communications or entertainment. The FM transmitter is also used on airplanes for ground-to-air and air-to-ground as well as air-to-air communications. The Army uses FM transmitters and receivers to set up temporary communications links between the battlefield and headquarters. Any number of uses for FM have been found, and more are being utilized all the time. Television uses FM for transmitting the sound portion of the channel.

# Transmission Lines

The output of a transmitter must be transferred from the power amplifier to the antenna in order to be useful. This is the job of the transmission line. The transmission line is used to conduct or guide the energy from the transmitter to the load. A transmission line must be capable of handling the power output of the transmitter and must possess low effective resistance to keep power losses at a minimum.

There are five types of transmission lines to be discussed in this chapter. They are the parallel two-wire, the twisted pair, the shielded pair, the concentric coaxial line (rigid and flexible), and waveguides. The use of a particular line depends, among other things, on the applied frequency, the power-handling capabilities, and the type of installation.

## Two-Wire Open Line

Figure 13-6 shows a parallel two-wire line. This line consists of two wires that are generally spaced from 2 to 6 in. apart. This type of line is most often used for power lines, rural telephone lines, and telegraph lines. It is sometimes used as a transmission line between the antenna and the transmitter or between the antenna and the receiver. An advantage of this type of line is its simple construction. The principal disadvantages are the high radiation losses and noise pickup due to the lack of shielding. Radiation losses are produced by the changing fields that are produced by the changing current in each conductor. Some of these lines of force will radiate from the transmission line in much the same manner as energy is radiated from the sun.

Another type of parallel line is the twin lead or two-wire ribbon line often seen in the installation of television receiver antennas (Figure 13-7). As you can see, it is essentially

BAR INSULATED

**Figure 13-6.** Parallel two-wire transmission line.

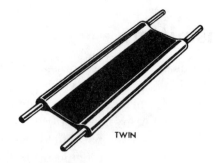

TWIN

**Figure 13-7.** Two-wire ribbon transmission line.

the same as that in Figure 13-6. It does, however, have a uniform spacing ensured by imbedding the two wires in a low-loss dielectric, usually polyethylene. Since the wires are imbedded in a thin ribbon of polyethylene, the dielectric space is partly air and partly polyethylene.

# Twisted Pair

The twisted pair line is shown in Figure 13-8. As the name says, the line consists of two insulated wires, twisted to form a flexible line without the use of spacers. It is not used for high frequencies due to the high losses that occur in the rubber insulation. When the line is wet, the losses increase greatly.

**Figure 13-8.** Twisted pair transmission line.

# Shielded Pair

The shielded pair line (Figure 13-9) consists of parallel conductors separated from each other and surrounded by a solid dielectric. The conductors are contained within copper braid tubing that acts as a shield. The assembly is covered with a rubber or flexible composition coating to protect the line from moisture or mechanical damage. Outwardly, it looks much the same as the power cord of a washing machine or refrigerator.

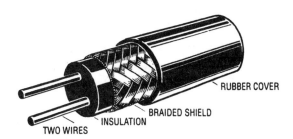

**Figure 13-9.** Shielded pair transmission line.

The main advantage to the shielded pair is that the conductors are balanced to ground. That is, the capacitance between the cables is uniform throughout the length of the line. This balance is due to the grounded shield that surrounds the conductors with a uniform spacing along the entire length. The copper braid shield isolates the conductors from stray magnetic fields.

# Coaxial Lines

There are two types of coaxial lines. The rigid, or air, coaxial line and the flexible, or solid, coaxial line are used where two concentric conductors are needed (see Figure 13-10). The

CABLE WITH WASHER INSULATOR

**Figure 13-10.** Air coaxial transmission line.

rigid type consists of a wire mounted inside of, and coaxially with, a tubular outer conductor. In some uses the inner conductor is also tubular. The inner conductor is insulated from the outer conductor by insulating spacers, or beads, at regular intervals. The spacers are made of pyrex, polystyrene, or some other material possessing good insulating characteristics and low loss at high frequencies.

This line minimizes line losses that occur because of radiation. The electric and magnetic fields in the two-wire parallel line extend into space for relatively great distances and radiation losses occur. No electric or magnetic fields extend outside the outer conductor in a coaxial line. The fields are confined to the space between the two conductors, thus the coaxial line is a perfectly shielded line. Noise pickup from other lines is also prevented.

This line has several disadvantages: (1) It is expensive to construct, (2) it must be kept dry to prevent excessive leakage between the two conductors, and (3) although high-frequency losses are somewhat less than in previously mentioned lines, they are still excessive enough to limit practical length of the line.

The condensation of moisture is prevented in some cases by the use of an inert gas — such as nitrogen, helium, or argon — pumped into the line at a pressure of from 3 to 35 pounds per square inch (lb/in.$^2$). The inert gas is used to dry the line when it is first installed, and a pressure is maintained to ensure that no moisture enters the line.

The flexible coaxial line is preferred by most people today (see Figure 13-11). It has been improved until it has the ability to handle large amounts of power without the need to keep the line pressurized. Polyethylene is used to fill in the coaxial line and keep the center conductor evenly spaced from the shield or outer conductor. Polyethylene is unaffected by sea water, gasoline, oils, and other liquids that may be found around transmitter sites. High-frequency losses due to the use of polyethylene, although greater than the losses would be if air were used, are lower than the losses resulting from the use of most other materials.

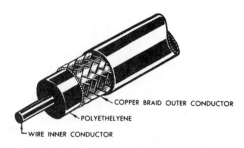

COPPER BRAID OUTER CONDUCTOR
POLYETHELYENE
WIRE INNER CONDUCTOR

**Figure 13-11.** Flexible coaxial transmission line.

# Waveguides

The waveguide is classified as a transmission line. It is designed for the higher frequencies and is always used with radar frequencies. It does transmit its energy down the line a little differently than the other lines. The common types of waveguides are the cylindrical and the rectangular (see Figure 13-12). Waveguides are discussed in detail in Chapter 17.

Waveguides are temperamental. If you dent one or allow solder to run inside, it may cause a loss of energy to the antenna. Anything left in the waveguide can cause arc overs and a breakdown of the current-carrying properties of the guide.

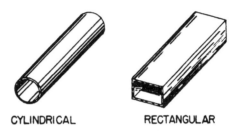

CYLINDRICAL          RECTANGULAR

**Figure 13-12.** Cylindrical and rectangular waveguides used for transmission lines at microwave frequencies.

# Antennas

An antenna is a conductor that is so constructed that it can radiate energy or receive it. Or it can both transmit and receive.

Antennas can be classified as receiving antennas and transmitting antennas. A transmitting antenna converts electric energy into electromagnetic waves that radiate away from the antenna at speeds near the velocity of light. A receiving antenna converts electromagnetic waves which it intercepts into electric energy and applies this energy to electronic circuits for interpretation.

The atmosphere slows the radio frequency waves slightly. However, for all practical purposes it can be said that radio waves travel *at* the speed of light.

Radiation of electromagnetic energy is based on the principle that a moving electric field creates a magnetic field, and conversely, a moving magnetic field creates an electric field. This creative process is called *propagation* and is used to describe the manner in which electromagnetic waves travel through space.

The electric ($E$) and the magnetic ($H$) fields make up the electromagnetic radiation and are perpendicular to each other and to their direction of motion.

When radio frequency currents flow through an antenna, radio waves are radiated in all directions. They spread out in much the same manner that waves spread out on the surface of a pond into which a stone has been thrown. Antennas which radiate in all directions are referred to as *omnidirectional*. Being electromagnetic in nature, radio waves can be directed so that radiation occurs in specific directions or is concentrated into a narrow beam. Such directional antennas form an important part of point-to-point communication systems and radar systems.

# Dipole Antenna

The *Hertz* antenna is a half-wavelength in physical length. That means the wavelength ($\lambda$) lambda is found by dividing 300 million by the frequency of the transmitted or received signal. The 300 million represents the speed of light which is 300 million meters per second (m/s). You get an answer that is in meters or part thereof. If you want to find the length of the antenna in feet, divide 982,080,000 by the frequency in hertz. Then if you want the half-wavelength, divide the answer by 2.

The Hertz antenna is predominantly used on frequencies that are above 2 MHz. The antenna is too long for frequencies lower than 2 MHz. The 2-MHz antenna is 300 million divided by 2 million to get 150 meters (m). Take the 150, multiply it by 39.54 to convert to inches, and divide by 12 to get feet. Since the 2-MHz full wavelength would be 494.25 ft., the Hertz antenna would be half of that, or 247.125 ft., long. As the frequency gets lower, the antenna gets longer. Figure 13-13 shows the voltage and current distribution on a Hertz antenna.

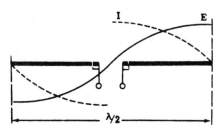

**Figure 13-13.** The Hertz antenna, or half-wave.

Figure 13-14 shows the *Marconi* antenna. The Marconi antenna is a quarter-wavelength. It is shorter than the Hertz and can be used on lower frequencies. Ground acts as the other half of the quarter-wavelength Marconi. The word dipole means two poles. Look at the Hertz antenna, and it is obvious that there are two poles. With the Marconi, however, the ground makes up the other pole or half of the antenna.

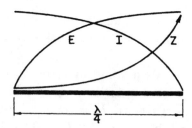

**Figure 13-14.** The Marconi antenna, or quarter-wave.

When you figure the true length of the antenna, you have to make up for the influence of the buildings and trees and other things which may be near the antenna and cause it to be electrically longer than need be physically. Therefore, if the antenna is longer than half-wave, it should have a 0.95 factor utilized in figuring the actual length. Since the wave travels slower in the antenna than in air, this 5 percent loss is compensated for by multiplying the answer so arrived at previously by 0.95.

If the antenna is not of the proper length, the source will see an opposition other than the pure resistance offered under perfect conditions. The source may see an impedance that will look like a capacitive circuit or an inductive circuit, depending on whether the antenna is shorter or longer than the specified wavelength. A Hertz antenna slightly longer than a half-wavelength will act like an inductive circuit, and an antenna slightly shorter than a half-wavelength will appear to the source as a capacitive circuit. Any two-wire open line longer than a quarter-wavelength appears electrically as a quarter-wavelength section with an additional section of open-circuited transmission line attached to it. The open section, which is capacitive in itself, will have its characteristics inverted and appear to the source as an inductive circuit. Compensation for the additional length can be made by cutting the antenna down to proper length, or by tuning out the inductive reactance by adding capacitance in series. This added $X_C$ will completely cancel the inductive reactance, and the source will then see a pure resistance, providing the proper-size capacitor is used. If the antenna is shorter than the required length, the source end of the line will appear capacitive. Therefore simply add inductance to make up the difference.

## Propagating a Wave from an Antenna

An antenna produces an $E$ and an $H$ field. The $E$ field is perpendicular to the $H$ field (Figure 13-15). The magnetic field is called the $H$ field. The electric field is called the $E$ field. Since the current and voltage that produced these $E$ and $H$ fields are 90° out of phase, the fields will also be 90° out of phase.

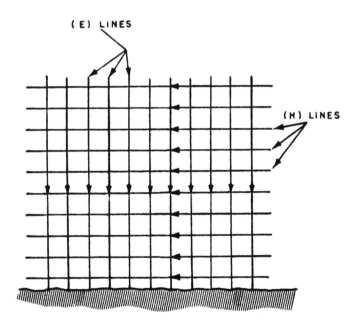

**Figure 13-15.** Instantaneous cross section of a radio wave.

## Radiation Patterns

Theoretically, a vertical dipole in free space has no vertical radiation along the direct line of its axis. However, it may produce a considerable amount of radiation at other angles measured to the line of the antenna axis. Figure 13-16C shows a vertical cross section of

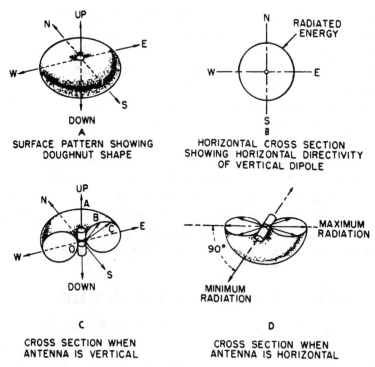

**Figure 13-16.** Radiation pattern of a dipole.

the radiation pattern of Figure 13-16A. The radiation along *OA* is zero, but at another angle, represented by angle *AOB*, there is appreciable radiation. At a greater angle *AOC*, the radiation is still greater. Because of this variation in field strength pattern at different vertical angles, a field strength pattern of a vertical half-wave antenna taken in a horizontal plane must specify the vertical angle of radiation for which the pattern applies.

Figure 13-16D shows half of the doughnut pattern for a horizontal half-wave dipole. The maximum radiation takes place in a plane perpendicular to the axis of the antenna and crossing through its center. A polar diagram representing the radiation pattern of a horizontal dipole is shown in Figure 13-17.

# Radio Waves

The ground wave is very important in radio communications. When a radio wave leaves a vertical antenna, the field pattern of the wave resembles a huge doughnut lying on the ground with the antenna in the hole at the center. Part of the wave moves outward in contact with the ground to form a ground wave, and the rest of the wave moves upward and outward to form the sky wave (see Figure 13-18). The ground and sky portions of the radio wave are responsible for two different methods of carrying the message from the transmitter to the receiver. The ground wave is both for short-range communication at high frequencies with low power, and for long-range communications at low frequencies with very high power. Daytime reception from most nearby commercial stations is carried by the ground wave. The sky wave is used for long-range, high-frequency daylight communication. At night, the sky wave provides a means for long-range contacts at somewhat lower frequencies.

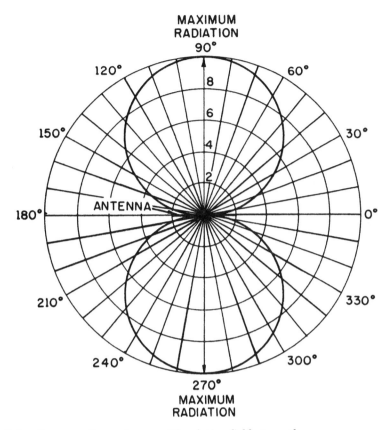

**Figure 13-17.** Polar diagram of an antenna with relative field strength.

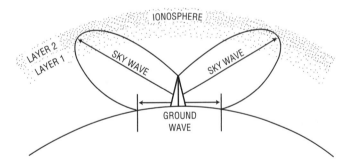

**Figure 13-18.** Ground and sky waves.

# Ground Wave

The *ground wave* is made up of two parts: a surface wave and a space wave. The *surface wave* travels along the earth's surface.

The *space wave* travels in the space immediately above the earth's surface in two paths: one path travels directly from the transmitter to the receiver, and the other follows a path in which the space wave is reflected from the ground before it reaches the receiver. Since the space wave follows two paths of different lengths, the two components may arrive in or out of phase with each other. Thus, as the distance from the transmitter is changed, these two components may add, or they may cancel. Neither of these component waves is affected by the reflecting layer of the atmosphere (called the *ionosphere*) high above the earth's surface (see Figure 13-18).

The space wave part of the ground wave becomes more important as the frequency is increased or as the transmitter and receiver antenna height is increased. When the transmitting and receiving antennas are both close to the ground, the space wave components cancel. This is true because the ground reflected component is shifted 180° in phase upon reflection, it has the same magnitude as the direct component, and it travels a path of approximately the same length as that of the direct components. Thus the surface wave part of the ground wave is responsible for most of the daytime broadcast reception.

As it passes over the ground, the surface wave induces a voltage in the earth, setting up eddy currents. The energy to establish these currents is absorbed from the surface wave, thereby weakening it as it moves away from the transmitting antenna. Increasing the frequency rapidly increases the attenuation so that surface wave communication is limited to relatively low frequencies.

Since the electrical properties of the earth along which the surface wave travels are relatively constant, the signal strength from a given station at a given point is nearly constant. This holds true in nearly all localities except those that have distinctly rainy and dry seasons. There the difference in the amount of moisture causes the soil's conductivity to change.

## Sky Wave

That part of the radio wave that moves upward and outward and that is not in contact with the ground is called the *sky wave*. It behaves differently from the ground wave. Some of the energy of the sky wave is refracted (bent) by the ionosphere so that it comes back toward the earth. A receiver located in the vicinity of the returning sky wave will receive strong signals even though it is several hundred miles beyond the range of the ground wave (Figure 13-19).

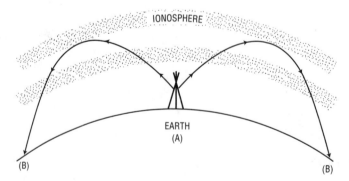

**Figure 13-19.** Ionospheric refraction of the sky wave.

## Ionosphere

The ionosphere plays an important part in radio communications. That is why a closer look is called for at this point. It is found in the rarefied atmosphere approximately 40 to 350 mi above the earth. It differs from the other atmosphere in that it contains a much higher number of positive and negative ions. The negative ions are believed to be free electrons. The ions are produced by ultraviolet particle radiations from the sun. The rotation of the earth on its axis, the annual course of the earth around the sun, and the development of sun spots all affect the number of ions present in the ionosphere, and these in turn affect the quality and distance of radio transmission.

The ionosphere acts as a conductor, and absorbs energy in varying amounts from the radio wave. The ionosphere also acts as a radio mirror and refracts (bends) the sky wave back to the earth (see Figure 13-19).

The ability of the ionosphere to return a radio wave to the earth depends upon the angle at which the sky wave strikes the ionosphere, the frequency of the transmission, and the ion density.

The radio wave may be refracted many times between the transmitter and the receiver (see Figure 13-20). In this example the radio wave strikes the earth at location A, with sufficient intensity to be reflected back to the ionosphere and there to be refracted back to the earth a second time. Frequently, a sky wave has sufficient energy to be refracted and reflected several times, thus greatly increasing the range of transmission. Because of this so-called multiple-hop transmission, transoceanic and around-the-world transmission is possible with only moderate power.

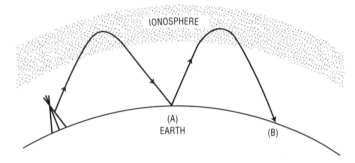

**Figure 13-20.** Multiple refraction and reflection of a sky wave.

# Fading

*Fading* is a term used to describe the variations in signal strength that occur at the receiver during the time a signal is being received. There are several reasons for fading, some of which are easily understood, while others are more complicated.

One probable cause is a direct result of interference between single-hop and double-hop transmissions occurring simultaneously from the same source. If the two waves arrive in phase, the signal strength will be increased, but if the two waves arrive in phase opposition (180° out of phase), they will cancel each other, and the signal will be weakened.

Interference fading also occurs where the ground wave and sky wave come in contact with each other. This type of fading becomes severe if the two waves are approximately equal in strength. Fluctuations in the sky wave with a steady ground wave can cause worse fading than sky wave transmission alone.

Variations in absorption and in the length of the path in the ionosphere are also responsible for fading. Occasionally, sudden disturbances in the ionosphere can cause complete absorption of all sky wave radiations.

Receivers located near the outer edge of the skip zone are subjected to fading as the sky wave alternately strikes and skips over the area. This type of fading sometimes causes the received signal strength to fall to nearly the zero level.

# Blackouts

Frequency blackouts are closely related to certain types of fading, some of which are severe enough to completely blank out the transmission. Changing conditions in the ionosphere shortly before sunrise and shortly after sunset may cause complete blackouts at certain frequencies. The higher-frequency signals pass through the ionosphere while the lower-frequency signals are absorbed by it.

Turbulent conditions in the ionosphere (Ionospheric storms) often cause radio communication to become erratic. Some frequencies will be completely blocked out, while others may be reinforced. Sometimes these storms develop in a few minutes, and at other times they require as much as several hours to develop. A storm may last several days.

# Summary

The continuous wave (CW) transmitter covers a longer distance than the MCW or voice modulation transmitter. It is used primarily for sending code. Every transmitter must have an oscillator to set up the fundamental frequency on which it operates. Frequency multipliers and intermediate power amplifiers are inserted where needed before the final amplifier stage is connected to the antenna.

Frequency-modulated transmitters modulate the oscillator while the amplitude-modulated transmitters modulate the output stage for better-quality speech or music. The FM transmitter has a frequency deviation caused by the swing of the audio modulation. Since the power amplifier of the FM transmitter is not modulated, it can be designed for maximum power output of the modulated signal it received from previous stages.

The transmission lines are used to connect the transmitter to the antenna. They include the two-wire open line, twisted pair, shielded pair, coaxial lines, and waveguides.

Each frequency band has its own special uses. These uses depend on the nature of the waves — surface, sky, or space — and on the effect that the sun, the earth, the ionosphere, and the earth's atmosphere have upon them.

Waveguides are used for microwaves and are discussed in Chapter 17. Coaxial cables today are usually composed of a solid core instead of the hollow core filled with a pressurized gas.

In most instances an antenna can be used for both transmitting and receiving. An antenna is a conductor that is so constructed that it can radiate energy or receive it. Radiation of electromagnetic energy is based on the principle that a moving electric field creates a magnetic field, and, conversely, a moving magnetic field creates an electric field.

The Hertz antenna is half-wavelength. The Marconi antenna is the quarter-wavelength. A dipole antenna is one with two poles. An antenna produces an $E$ and an $H$ field. The $E$ field is the electric field, and the $H$ field is the magnetic field. They are perpendicular to one another. As mentioned earlier in the chapter, when a radio wave leaves the antenna, part of it is radiated along the surface of the earth and is called the ground wave, while the other part is radiated upward and is called the sky wave. Each of these — ground wave and sky wave — has its own particular characteristics.

The *ionosphere* is a layer of ionized gases which surrounds the earth. Radio waves may hit the ionosphere and be absorbed, they may penetrate it, or they may be reflected or refracted (bent) by it. It all depends on the angle of incidence and the frequency of the radio wave energy. Fading and blackouts of certain frequencies can be traced to the action of the ionosphere.

It is difficult to establish fixed rules for the choice of a frequency for a particular purpose. Only some general statements can be made as to which frequency bands are best suited to the type of transmission to be made. However, only after experience with various frequencies under many different conditions can the appropriate choice be made.

# Review Questions

1. What is the function of a transmitter?
2. What is the purpose of a transmission line?
3. What is the purpose of an antenna?
4. What is the difference between FM and AM?
5. Why is an oscillator needed in a transmitter?
6. What is a buffer?
7. What does a frequency multiplier do?
8. What is a doubler?
9. Why do you need frequency multipliers?
10. What is frequency deviation?
11. What is the broadcast band for FM?
12. What is the difference between a twisted pair and a two-wire open line?
13. How is a flexible coax constructed?
14. Where are waveguides used?
15. What is a dipole antenna?
16. What is a Hertz antenna?
17. What is a Marconi antenna?
18. What is the formula for finding wavelength?
19. What is the ionosphere?
20. Define ground wave.
21. Define sky wave.
22. What causes fading?
23. What is meant by frequency blackouts?
24. What causes frequency blackouts?

# Chapter 14

# RECEIVERS

The other half of the radio frequency (RF) communications consists of a receiver. The transmitted wave is of very little use unless it is received and decoded to present the information, music, or speech as intended. This is where the receiver does its half of the job in a communications system.

## Antennas

Antennas have already been mentioned and discussed in transmitters. The reciprocity principle can be applied here since the transmitting antenna can also be used as a receiving antenna. When a transmitted wave passes a receiving antenna, it induces a current in the antenna as a result of electromagnetic induction. This current varies with the frequency of the wave. This weak current can then be amplified and processed to produce the intelligence it contains.

A number of different stages are designed for various purposes in a receiver. It is the intent of this chapter to cover the stages and how they operate to produce the output in the form of speech, music, or other intelligence.

Requirements for receiving antennas are much less severe than those for a transmitter. If you are not using the same antenna that the transmitter used, you can use some rather simple ones for receivers since they use very low signal strengths. Antennas for receivers may include a straight piece of wire, coils of wire, or basic dipoles. They can also be a little more complicated when it comes to bringing in very weak signals from long distances. The parabolic reflector in the form of a receiving dish is used in satellite communications and in home reception of satellite television.

A stacked array, or dish, can become more directional and enable the operation of a receiver on very weak signals (see Figure 14-1). Figure 14-2 is a loop antenna used for simple amplitude modulation (AM) reception on the broadcast band, i.e., 535 to 1605 kilohertz (kHz). A dipole for frequency modulation (FM) enhances reception if the signal is weak or if stereo is to be picked up. A stereo signal requires a stronger signal fed into the amplifier than the mono FM signal. Figure 14-3 shows an FM dipole antenna that aids in bringing in weaker stations. In most instances, point-to-point communications on FM need a stacked array.

## Transmission Lines

Getting the signal from the antenna is no problem in most small portable AM receivers or in FM sets that operate on local stations, but the signal from the antenna to the receiver,

**Figure 14-1.** Dish antenna for satellite TV reception.

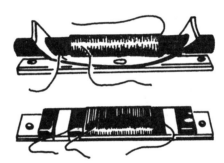

**Figure 14-2.** Typical antenna coils used as antennas for AM.

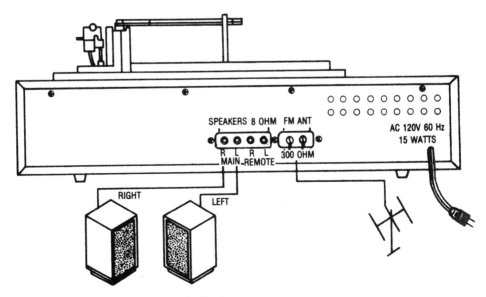

**Figure 14-3.** Antenna hookup for outside dipole.

in the case of very high frequency (VHF) and ultrahigh frequency (UHF), can be a problem. Signal loss from a number of sources may occur, mainly from the transmission line.

In some instances the antenna is on the rooftop, and the receiver is downstairs or in the basement or in any number of locations from the site of the antenna. That calls for a transmission line. The idea of the line is to make the setup as efficient as possible so that whatever gain was obtained by the antenna is not lost in transmission to the receiver. However, every transmission line has resistance. It also has inductance and capacitance all along the length of the line. A transmission line has its own characteristic impedance due to its type and physical characteristics including length.

The characteristic impedance of a transmission line is independent of the line's length. The usual impedances (Z) are 52 and 72 ohms (Ω) for the coaxial type and 300 Ω for the twin-lead line used for TV receivers. The characteristic impedance of a line is important for matching purposes. If the impedance of the load connected to the line is different from the characteristic impedance, some of the energy applied to the line will be reflected from the load back to the source. This is a loss in power. However, if the load impedance is the same as the line's characteristic impedance, all the energy applied to the line from the antenna will be absorbed by the load (the receiver). The common impedances of transmission lines are chosen to fit the TV receiver and its input terminals. Most TV receivers have both 72- and 300-Ω input impedances.

## Standing Waves

When impedance for the load and the line does not match up, the reflected wave has a tendency to get in the way of efficient movement of the antenna output on its way to the receiver (see Figure 14-4). The reflected wave and the applied wave are on the line at the same time. This means *interference*. The reflected wave goes from the load to the source while the applied wave is moving in the opposite direction from the antenna or source to the receiver or load. These two waves combine at some point along the line to produce a third wave which is stationary on the line. This is called a *standing wave*. There are

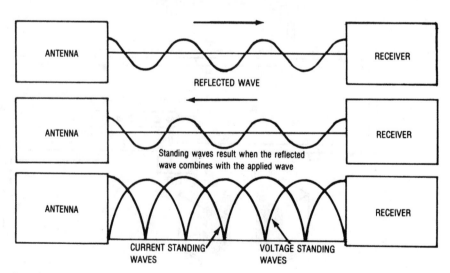

**Figure 14-4.** Standing waves.

standing waves of voltage and standing waves of current. Both types are out of phase since the current and voltage waves are reflected out of phase from the load.

## Standing Wave Ratio

The transmission line has a standing wave ratio (SWR) when there is a mismatch in impedance between the line and the load. This is the ratio of the maximum effective value of the standing waves to their minimum effective value. The SWR is also equal to the ratio of the characteristic impedance of the line to the impedance of the load, or vice versa. For example, if the line has a characteristic impedance of 300 Ω and the load impedance is 25 Ω, the SWR is 300/25, or 12. You can see, therefore, that the higher the SWR, the greater is the mismatch between the line and the load. An SWR of 1 indicates that there are no standing waves on the line, or 300/300 = 1, and all the power is absorbed by the load.

Transmission lines should have low SWRs. There are a number of ill effects from standing waves. Standing waves can reduce the power-handling capacity of a line. They do this by producing very high voltage points that can cause insulation breakdowns in the case of transmitters, and they can cause a reduction in the power being received by the load. They can also radiate as an antenna and rob power from the source. We have already taken a closer look at transmission lines in the chapter on transmitters. They are much more important in transmitters than in receivers and need to be matched to the transmitting antenna. That is why citizens band transceivers needed a meter in most cases to check for SWR so that damage would not be done to the output transistors by standing waves.

# Amplitude-Modulated Receivers

In this discussion we will limit our study to a six-transistor AM receiver that is most often purchased in a discount store for less than $10. Communications receivers have some special circuits that are not important to the basics of AM reception at this time.

## AM Receiver

The block diagram of an AM receiver will give us some idea of how the entire unit operates, then we will take it stage by stage to obtain the circuit functions. See Figure 14-5 for a block diagram of the AM receiver.

Note how the signal is injected from the antenna into the RF amplifier. This is the case in most communications and more expensive sets. If you have a standard six-transistor type, there will be no RF amplifier. The mixer will be the first stage to view. However, in the interest of getting a better look at the receiver, we will take a closer look at the RF amplifier when we analyze the circuit functions later in the chapter.

The RF amplifier takes the antenna signal and selects it. The broadcast AM band covers a wideband of frequencies [535 to 1605 kilohertz (kHz)]. This wideband of frequencies has to be beaten down to one in order for selectivity to be improved. This is where the principle of heterodyning comes in.

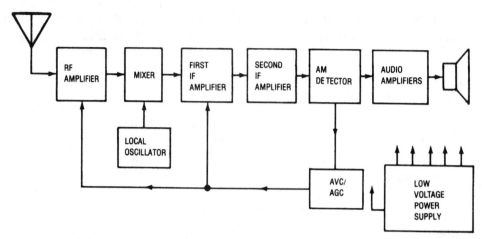

**Figure 14-5.** Block diagram of an AM receiver with an RF stage.

Heterodyning takes place in the mixer stage. This is where two frequencies (in this case the incoming frequency from the antenna, which is selected by the first tuning circuit) are mixed or beaten against a local oscillator frequency. The local oscillator puts out a non-modulated frequency. However, the incoming frequency is modulated. The beating of the two frequencies produces the sum and difference frequencies, and the original two are also present. All four frequencies are now modulated in the collector tank circuit which is the output of the mixer stage. Once the proper frequency [455 kHz is the usual *intermediate frequency* (IF) selected for AM] is selected, it is passed on to the IF amplifier because of the signal loss during the previous processing. There may be a minimum of two IF amplifiers to amplify the signal, each time there is a tuned circuit tuned to only 455 kHz with ±5 kHz for sidebands.

After the IF with the sidebands or modulation is amplified to the desired level, it is put through the detector stage where the modulation is taken from the RF carrier and passed on to the audio amplifiers. This audio amplification causes the signal to be of the proper level to drive a speaker.

The automatic volume control (AVC) or automatic gain control (AGC) is part of the detector stage and feeds back a strong signal bias voltage when the signal is too strong and cuts down on the amplification of the signal before it can be heard by the human ear as being too loud. (This too will be discussed in detail later.) The power supply is needed to furnish the proper voltages for the operation of the various stages.

# The RF Amplifier

The first stage in an AM receiver is the RF amplifier. The major assignment of the RF amplifier is to amplify the incoming signals from the antenna after a particular one has been selected. Four characteristics of the RF stage are very important in any receiver. It should have the proper ability to amplify or produce the required gain; it should have a low noise output since the signal itself will have a certain amount of noise depending on the signal-to-noise ratio of the received signal; it will need to be selective since it has the entire band to select from; it will also need to have a linear characteristic throughout the entire band of AM frequencies.

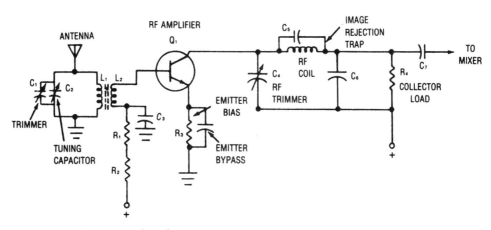

**Figure 14-6.** Typical RF stage found in receivers.

Figure 14-6 is a typical RF amplifier used in AM radios. As we mentioned before, not all small portable transistorized AM radios have the RF amplifier. The RF signal is coupled from the antenna to the tank circuit consisting of $C_2$ and $L_1$. Across $C_2$ is another capacitor which is also adjustable. It is a trimmer. They are called *trimmers* if in parallel and *padders* if in series. The tank circuit is chosen carefully so that the capacitor and inductor will work together to tune in the whole broadcast band. The inductor $L_2$ is really a secondary of an RF transformer, and it couples the signal to the base and emitter of the transistor. The incoming signal placed on the base of $Q_1$ and the capacitor $C_3$ is such that its capacitive reactance at these broadcast frequencies makes it a virtual short circuit. This effectively places the RF signal on the emitter. $C_3$ shorting to ground of the RF signal causes it to pass through the emitter bypass capacitor to be applied to the emitter of $Q_1$.

$R_3$ serves as the emitter stabilizer or emitter bias resistor. The emitter bypass capacitor is needed to keep the dc bias clean. The bypass capacitor is of such size as to allow the RF to bypass the resistor and in combination with $C_3$ puts the RF signal on the emitter of the transistor. The base bias is provided by $R_1$ and $R_2$. The collector obtains its proper dc polarity by going from the (+) connection through the collector load resistor and through the RF coil back to the collector.

The RF coil and the capacitor across it make up an image rejection trap that, along with $C_4$ can be tuned to take out the frequencies not close to the broadcast band. $R_4$ serves as the load for the collector and develops the transistor's output, which is then coupled through $C_7$ to the next stage, the mixer.

Keep in mind that the RF amplifier stage is not a common stage for less expensive, portable radios used for local reception. It is used in military equipment, ham radios, or advanced communications receivers and in the more demanding situations where special equipment is needed for receiving weak signals.

# The Oscillator

The oscillator stage is needed whenever heterodyning is used to improve the selectivity of a receiver. It produces the beat frequency that is used to beat against the incoming frequency to produce the IF.

You have already looked at oscillators in an earlier chapter. From what you have already learned, you can see that the oscillator used in Figure 14-7 is a Hartley since it uses tapped inductors for the tank circuit. The transistor $Q_1$ is used as an amplifier, and it is made to oscillate since a path of feedback is available. The feedback path is through the $RC$ coupling produced by $C_4$ and $R_3$. $R_3$ also serves as a source of emitter bias for the transistor. $C_5$ is part of an $RC$ coupling network that takes the oscillator signal and injects it on the emitter of the mixer as seen in Figure 14-8. The oscillator tank circuit is made up

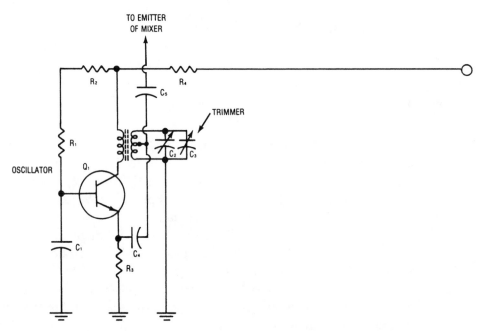

**Figure 14-7.** Oscillator stage found in receivers.

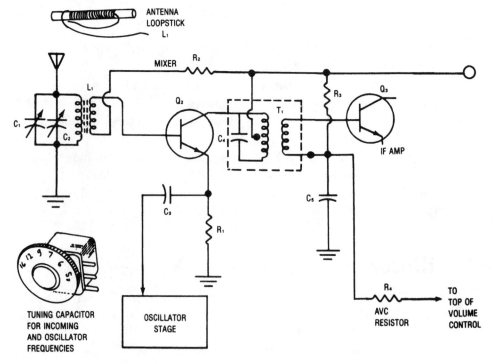

**Figure 14-8.** Mixer stage found in receivers.

of a tapped coil and $C_2$ and $C_3$. $C_2$ is connected on a shaft with the tuning capacitor in the mixer stage. When one moves, so does the other. The oscillator has an operating range of roughly 1005 to 2055 kHz. When used on the broadcast band, the desired IF is 455 kHz. As you can see, the oscillator *tracks above* the incoming signal, which means that it operates at a higher frequency than the incoming signal. $C_1$ serves to place the base at RF ground, and $R_1$ and $R_2$ serve a part of the bias network for base bias. $R_4$ makes sure the correct voltage is applied to the collector of $Q_1$.

# The Mixer

The purpose of combining an RF with an oscillator frequency to produce an IF is called *mixing*, or *frequency conversion*. The two basic methods of frequency conversion used with electron tubes are also used with transistors. The first method, where a transistor combines an oscillator frequency and a radio frequency, is called a *mixer*. In the second method, only one transistor functions as an oscillator and a mixer and is known as a *converter*. When a transistor is used as a mixer or converter, it is operated on the curved portion of the dynamic transfer curve. Under these conditions when two frequencies are applied to the transistor input, four frequencies are produced in the output. Two of the output frequencies are the original frequencies that were present in the input. Another one of the output frequencies is a frequency that is equal to the sum of the original frequencies. The remaining frequency that is presented in the output is a frequency that is equal to the difference of the two original frequencies in most superheterodyne receivers, but only the difference frequency is of interest, and all other frequencies must be filtered out. The difference frequency used in a receiver is referred to as the *intermediate frequency* (IF).

Figure 14-8 shows the mixer stage used in most inexpensive radio receivers. Figure 14-9 shows how the two frequencies are mixed to produce four.

Figure 14-8 shows how the input signal is placed directly on the base of the mixer by mutual induction. The coil $L_1$ is the antenna on a ferrite rod and is really a transformer rather than a coil. The tuning capacitors $C_1$ and $C_2$ are in parallel, with $C_2$ serving as the trimmer. The trimmer is adjusted at the high end of the band to enable better selectivity.

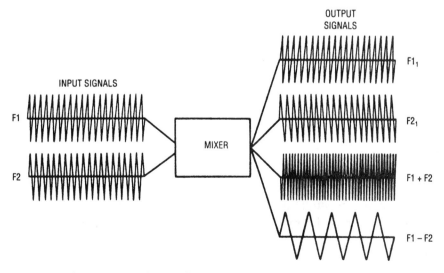

**Figure 14-9.** Input and output waveforms of a mixer.

$C_3$ and $R_1$ make up the $RC$ coupling for the oscillator signal being placed on the **emitter.** $T_1$ is the first IF can. It tunes the IF frequency (455 kHz) and rejects the others. **The** secondary of $T_1$ couples the signal to the first IF amplifier stage. This means the **tank** circuit in the collector circuit of the mixer is very critical since it determines the **frequency** selected for use as the IF. It must also be tuned broad enough to allow the sidebands **to pass.**

## The IF Amplifiers

There is usually a minimum of two IF amplifiers. (See Figure 14-10 for an example **of two** typical RF amplifiers used as IFs.) The major function of the IF amplifier is to **give the** signal good linear amplification through the range of frequencies it will be handling. **The** main objective of the amplifier after its use as an amplifier is to provide selectivity. **By** using a couple of additional tuned circuits tuned to the same frequency, it is **possible to** become very selective. That is one of the reasons for using superheterodynes **rather than** some earlier type of receiver. The 455-kHz signal has been chosen as the input to the **first** IF, and the output of that stage is also tuned. The 455-kHz signal is fed to the **second IF** and amplified further so that it is sufficient to be rectified by the detector to **provide an** audio signal that is not too noisy. The automatic volume control (AVC) serves as part **of the** base bias for the first IF stage. (More about AVC later.)

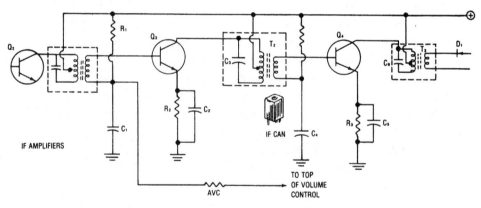

**Figure 14-10.** IF amplifiers used in a receiver.

In most AM receivers the bandwidth is about 3 kHz. This severely limits **the** quality of music heard from the speakers. That means the IF cans have to be **tuned to** 455 kHz ± 3 kHz.

## The Detector

You have already read about the detector in the chapter on modulation and demodulation.

To recover the audio intelligence that has been broadcast on the AM band, the 455-kHz signal, which has the modulation on it from the incoming signal, is demodulated **by the** detector stage. The diode in Figure 14-11 that is labeled $D_1$ serves as the detector. **The** diode serves as a half-wave rectifier. The RF filter consisting of $R_1$, $C_1$, and $C_2$ takes **care**

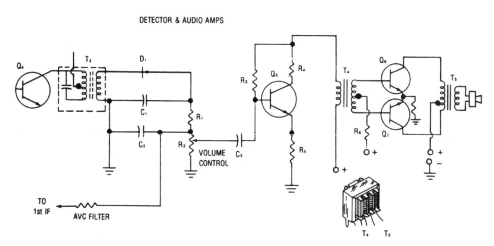

DETECTOR & AUDIO AMPS

**Figure 14-11.** Detector and audio amps for an AM receiver.

of the radio frequencies and keeps them from the volume control. The low impedance path for the RF is through the filter capacitors. The low impedance path for the audio is through the volume control. The rectified signal that produces the audio across the volume control is very weak and needs further amplification if it is to be used to drive a speaker. If you wish, you can check the output at this point by using a headset across the volume control.

# The Audio Stages

The audio stages are shown in Figure 14-11. Note how the signal from the volume control is coupled through $C_3$ to the base of $Q_5$ which serves as the first audio amplifier. The output of the first audio is through the audio transformer $T_4$. This output is coupled to the bases of $Q_6$ and $Q_7$ by an audio transformer. The transformer has a signal that is (+) on one side and (−) on the other. This means that the push-pull amplifier stage has the positive signal on one base and the negative signal on the other. Since these are NPN transistors, they operate when a positive signal is applied to the base. The signal with the positive base conducts. When the signal changes on the next half-cycle, the other transistor conducts for it will have the positive signal applied. The output of the push-pull amplifier is sufficient to drive the speaker.

# Automatic Volume Control

Now, let us back up for a minute and look at the automatic gain control, or as it is sometimes called the *automatic volume control* (AVC). This stage is used to adjust the amplification of the RF and IF stages. The strength of the input signal determines the loudness of the speaker. Therefore, controls must be placed on the RF and IF stages if we want to control the volume. If the signals that reach the antenna are weak, the signal amplitude reaching the detector will also be weak, producing a low output volume. If strong signals are received, they will create overload signal conditions in the amplifiers and cause distortion. To correct this, a feedback signal is used to adjust bias on the first IF amplifier.

Note where the AVC filter connection is made between resistors $R_1$ and $R_2$ of

Figure 14-11. The resistor works with capacitor $C_1$ of Figure 14-10 to form a filter for the voltage fed back to the first IF stage. When the rectified signal that appears across the volume control is applied to $C_1$, it charges $C_1$ to the level of voltage applied across the volume control. If this is a strong signal, it puts a relatively higher charge on the capacitor. If the signal happened to decrease, the amount of voltage dropped across the volume control would decrease, and if the signal increased, it would charge the capacitor again to the level of the voltage across the volume control. This is a *rippling* voltage since it is mainly the audio component of the rectified wave and must be filtered. That is where the proper selection of the AVC filter resistor is important. It has a tendency to keep the capacitor from discharging back through the volume control until another change in the voltage comes along with the varying signal strength. As the signal voltage swings negative, it causes the capacitor to be charged with a negative polarity at the top of it. This negative polarity causes the transistor used for the first IF to conduct less since it needs a positive polarity to operate properly. As it cuts the amplification down in that stage, it also reduces the voltage that appears across the volume control. This takes place so quickly that you cannot even hear the action.

# AM Radio on a Chip

It is now possible to place the entire AM radio on one integrated circuit chip. This can be done with one 14-pin DIP as shown in Figure 14-12. This particular chip does not contain the audio output stages. That means another chip of whatever audio power desired can be connected to cause this to operate as a radio. Note the antenna, oscillator coil, and IF cans are mounted externally. The dotted lines with numbers in circles make up the chip with its pins being identified for external connection purposes. This chip operates well with 12 volts DC so it is ideal for use in automobile radios.

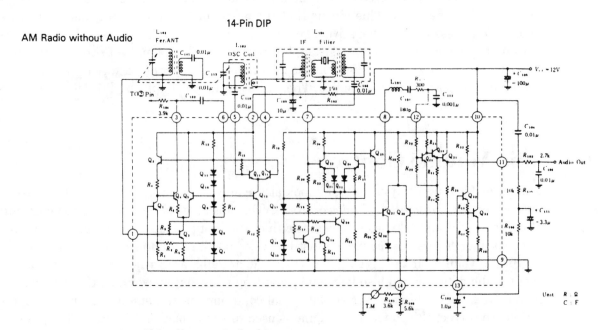

**Figure 14-12.** AM radio on a single chip.

# Frequency-Modulated Receivers

The FM receiver is not as simple to understand as the AM. The method used to produce FM and then to produce stereo is somewhat more complicated than the AM receiver. There are some rather interesting devices in the way integrated circuits are used to decode the stereo signal and to produce the output sufficient to drive a couple of speakers. In fact, the whole realm of FM electronics has changed markedly in the past ten years and is going to change even more rapidly as the impact of the newly manufactured sets reaches the market. We will have improved sensitivity and selectivity as well as noise reduction.

Keep in mind that the FM band is not used only for wideband transmission of music for home receivers. Narrowband FM where only the voice is used can be made to serve a number of purposes. For instance, the commercial broadcast band covers 88 to 108 megahertz (MHz) with 200-kHz channel sidebands. Television audio signals use 50-kHz channel sidebands at 54 to 88 MHz, 174 to 216 MHz, and 470 to 890 MHz. The narrowband amateur radio channels are at 29.6 MHz, 52 to 53 MHz, 146 to 147.5 MHz, 440 to 450 MHz, and in excess of 890 MHz for experimental purposes.

There is a reason for not using FM on frequencies lower than 30 MHz. The earth's ionosphere introduces phase distortion to FM signals at frequencies below the 30-MHz point. The line-of-sight method of transmission is used for those signals above 30 MHz because of nature's way of not reflecting these signals and allowing them to continue through the ionosphere. Since the earth curves, it limits the communications range of FM to about 80 mi maximum. This is especially true of the narrowband public service channels that operate on 108 to 175 MHz, right in between channels 6 and 7 on TV. The narrowband is also assigned to frequencies in excess of 890 MHz. Output power from a TV FM transmitter is about 50 kilowatts (kW), while the amateurs use some walkie-talkies with only 100 milliwatts (mW) of power.

Of course, one of the main advantages of FM over AM is its ability to transmit music with very little noise, or static, being heard in the receiver. This was one of the selling points in its favor when it was introduced and then dropped and introduced once again.

## The RF Amplifier

The antenna is usually a piece of wire where the signal is very strong, but a folded dipole is needed if any type of signal strength is desired at the front end of the receiver. The first two stages are called the front end of the receiver, which usually includes the RF amplifier and the mixer.

Figure 14-13 shows a block diagram of the FM receiver with the stages needed for reception of the standard monaural signal.

The RF amplifier, or preselector, performs the same function in the FM receiver as it does in the AM receiver, that is, it increases the sensitivity of the receiver. Such an increase in sensitivity is often a practical necessity in fringe areas. However, the gain of the IF stages is relatively much greater, perhaps 100 times that of the preselector, since the chief advantage of the superheterodyne lies in the uniformity of response and gain of the IF stages within the receiver band. The principal functions of the RF stage are to discriminate against undesired signals and to increase the amplitude of weak signals so that the signal-to-noise ratio will be improved.

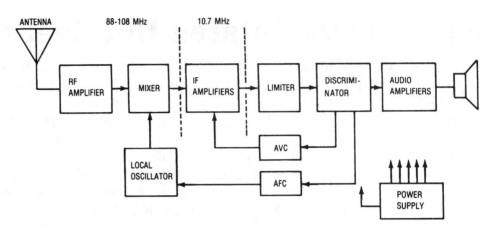

**Figure 14-13.** Block diagram of a basic FM receiver.

# The Mixer

Mixing the incoming FM signal with the set's oscillator frequency is basically the same in principle, but the frequencies are changed. FM for home use employs the 10.7-MHz IF. The beating together of the incoming frequency and the oscillator frequency produces the 10.7-MHz IF. The oscillator tracks 10.7 MHz above the incoming frequency. As the tuning capacitor is moved to select the frequency desired, the two capacitors are connected on the same shaft, and they tune the tank circuit for the desired incoming frequency and the oscillator at the same time. This keeps the two frequencies 10.7 MHz apart. Some drifting of oscillator frequency occurs, and that is why AFC or automatic frequency control is needed. AFC keeps the oscillator on frequency. (AFC will be discussed later in this chapter.)

The output of the mixer stage is 10.7 MHz with the modulation of the incoming signal selected by the tuner. This signal is passed on the IF strip where it is amplified.

One of the problems with any oscillator is its ability to become a transmitter if given the proper-length antenna. This happens when the FM set does not have an RF amplifier preceding the mixer. The oscillator signal can be reradiated and produce interference in TV sets and other FM sets nearby. Another reason for using the RF stage is the ability of the FM signal to be relatively noise-free even at low signals around 1 microvolt ($\mu$V). This is possible primarily because FM inherently produces less noise in its signals. It takes about 30 $\mu$V for a mixer on an AM set to operate without destroying the signal altogether. Since the FM set requires a signal of at least 20 $\mu$V, its mixer stage noise level does not destroy the intelligence on the signal. Therefore, an RF amplifier stage is needed to bring the 1-$\mu$V signals up to at least 20 $\mu$V before it goes to the mixer (see Figure 14-14). The RF amplifier also reduces the image frequency problem that exists in most sets.

# The IF Amplifiers

The IF amplifiers are important here since they take the signal from the mixer and amplify it sufficiently for detection. They also increase the selectivity of the set since the heterodyning principle produces the IF for the purpose of allowing it to be amplified by stages which are tuned slightly off their normal 10.7 MHz. This detuning produces a wide enough bandwidth to allow the modulation (up to 200 kHz) to pass through the IF cans.

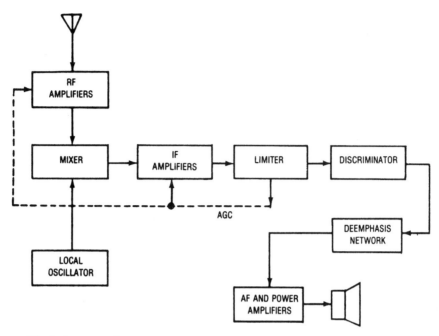

**Figure 14-14.** Block diagram of an FM receiver.

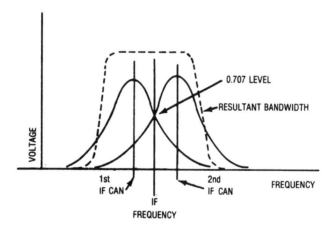

**Figure 14-15.** Stagger-tuned IFs produce a broadband pass.

The IF cans in the IF stages are stagger-tuned to produce a wide enough band pass for the music to pass through (Figure 14-15). This means that there are usually three IF stages in an FM receiver.

# The Limiter

A limiter puts out a signal with a constant amplitude. This eliminates any noise that may be riding on the incoming signal. A good limiter is essential to the proper operation of an FM receiver (see Figure 14-16). The limiter stage also provides automatic gain control (AGC) since its signals are from minimum value up to maximum value constant in amplitude. This provides a constant input level to the discriminator. Figure 14-17 shows how the

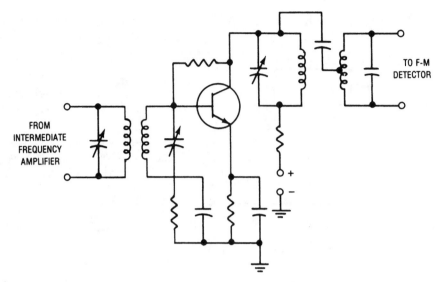

**Figure 14-16.** Limiter circuit.

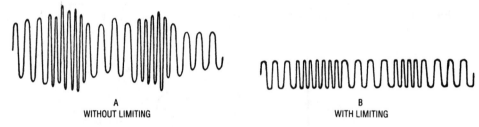

A
WITHOUT LIMITING

B
WITH LIMITING

**Figure 14-17.** FM signal before and after limiting.

input signal and the output signal from the limiter compare. Keep in mind that the modulation is on the frequency swing, not the amplitude.

Limiters require an input signal voltage of at least 1 V. This is why the IFs are used to boost the signal from the antenna (which may be about 1.5 $\mu$V) to this level. The sensitivity of an FM set refers to how much input signal is required to produce a specific level of quieting. This is normally 30 decibels (dB). A good-quality receiver will have a sensitivity of 1.5 $\mu$V with a background noise 30 dB down from the input signal level of 1.5 $\mu$V. Most IF stages in today's receivers are made in integrated circuits that have built-in limiting action.

# The Discriminator

The translation of the FM variations into audio is the function of the discriminator, ratio detector, or slope detector. The detection takes place at the 10.7-MHz level of the IF. The IF has the modulation ($\pm$75 kHz) on it and must be separated to obtain good audio for the audio stages.

The Foster-Seely discriminator is the most popular one used in FM. The newer integrated circuit (IC) technology causes the ratio detector and the Foster-Seely to be one of the older techniques used in FM demodulation. These circuits are shown and analyzed in the chapter on modulation and demodulation.

The ratio detector does not need a limiter stage since it does not respond to the variations in amplitude that exist in the FM signal. The ratio detector responds only to the frequency changes in the input signal. You have to look closely at the ratio detector to see that the two diodes are reversed, but in the discriminator they are both the same.

The slope detector has decided limitations when used on the wideband FM and is therefore limited in application to very inexpensive types of receivers. An AM receiver can be adjusted to receive FM when the slope characteristic of the last IF's tank circuit is detuned. (See Figure 14-18 for a transistorized circuit of the slope detector.) It operates by tuning the last-tuned circuit of the limiter so that the frequency of the IF (10.7 MHz) falls in the middle of the most linear region of the response curve (Figure 14-19). The increasing and decreasing of the FM signal from 10.7 MHz to its sidebands cause deviations above and below the 10.7 IF. The slope detector therefore changes these variations into amplitude variations.

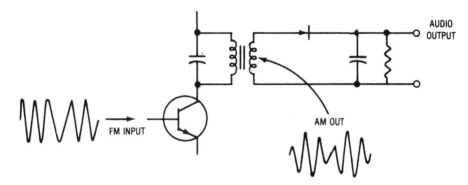

**Figure 14-18.** Slope detector for FM demodulation.

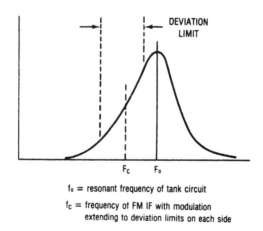

$f_o$ = resonant frequency of tank circuit

$f_c$ = frequency of FM IF with modulation
extending to deviation limits on each side

**Figure 14-19.** Portion of curve used for slope detection of FM.

# Automatic Frequency Control

In order for the receiver to operate properly, the local oscillator that beats against the incoming frequency must be stable in its output frequency to produce the 10.7 MHz needed for the IFs. One way of keeping an oscillator frequency stable is to keep the voltages applied to the transistor (or vacuum tube) constant. This source of voltage correction can be obtained from the output of the audio stage. Take this dc produced by the audio signal and feed it back to the local oscillator to keep it on frequency.

The automatic frequency control (AFC) may cause the receiver to miss some weak stations when tuning. That is why the *defeat switch* is used. FM and AFC are the switch labels used to designate this condition. The FM or defeat position means weaker stations can be tuned in and then the AFC turned on to capture or hold the stations.

Newer receivers do not use the AFC system of control since the problem of keeping an oscillator over 100 MHz from drifting is no longer the problem it once was. The newer chips and phase-locked loops have automatic correction for frequency drift.

Once the signals have been detected, the audio is coupled to the audio amplifiers. In the AM-FM sets the same audio section takes care of both the AM and FM outputs. However, they are switched with an external source usually mounted so that the AM, FM, and stereo are marked on the switch. FM stereo calls for two speakers and two amplifiers for the audio.

# FM Stereo

Most of today's FM receivers are made with stereo. For the receiver to receive and decode this stereo signal, several stages have to be added to the regular monaural FM receiver.

Figure 14-22 shows a typical block diagram for such a receiver. The FCC authorized FM stereo in 1961. This made it possible for home receivers to obtain the complete information on records and tapes that were already available. Stereo uses two separate signals to produce a spatial dimension to the music or speech. This also called for another channel to be added to the FM single-channel transmissions. Stereo high fidelity requires two channels of 30-Hz to 15-kHz signals to modulate the carrier frequency in such a way that the receiver can separate them and reproduce the outputs in a left and right speaker (see Figure 14-20).

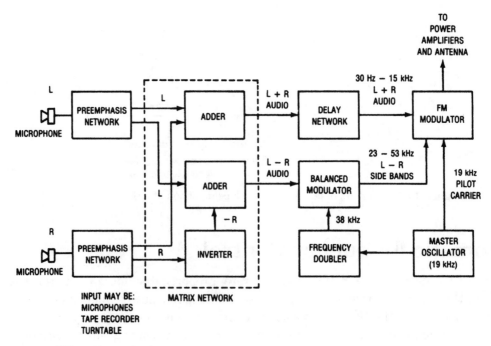

Figure 14-20. FM stereo transmitter.

More efficient use of the 200-kHz bandwidth was the answer to the stereo problem. This was done by the process of multiplexing. *Multiplexing* is the simultaneous transmission of two different signals on one carrier. It is also possible to broadcast more than two under the right conditions. The FCC approved a compatible system for stereo broadcasts. This means that the stereo and the FM monaural signals can be received on receivers that normally receive the monaural. Or, the stereo receiver can receive the monaural signal and reproduce it properly.

# Stereo Signal

Figure 14-21 shows how the composite signals are impressed on one carrier. The sum of the left and right (L + R) signals extends from 30 Hz to 15 kHz, just as the full audio signal is used to modulate the carrier in standard FM broadcasts. The left-minus-right (L − R) channel extends from 23 to 53 kHz. Note the placement of the 19-kHz pilot subcarrier on the line. This 19-kHz pilot subcarrier is included in the composite stereo signal that modulates the transmitter. Thus, two different signals are multiplexed together by placing them in two different frequency ranges.

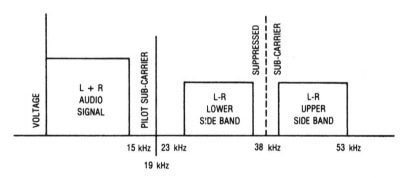

**Figure 14-21.** Modulating signals for an FM stereo transmitter.

Figure 14-20 shows briefly how the stereo signal is generated and then applied to the FM modulator for transmission.

1. Right and left channels are picked up by two separated microphones with preemphasis for the high frequencies.
2. Both signals are applied to the matrix network. The network inverts the right channel. This makes it a −R signal, and then it combines or adds L and R to provide the L + R signal. The two outputs are still 30-Hz to 15-kHz audio signals here.
3. The L − R signal and a 38-kHz subcarrier signal are then applied to a balanced modulator. This modulator suppresses the carrier but provides a double sideband signal at its output. The upper and lower sidebands extend from 30 Hz to 15 kHz above and below the suppressed 38-kHz carrier or from 23 to 53 kHz. This is obtained by subtracting 15 kHz from the 38 kHz to obtain 23 kHz and adding 15 kHz to the subcarrier (38 kHz) to get 53 kHz.
4. This means that the L − R signal has been changed from audio up to a higher frequency so as to keep it separate from the 30-Hz to 15-kHz L + R signal.

5. The L + R signal is slightly delayed so that both signals are applied to the FM modulator in time phase due to the slight delay that was encountered by the L – R signal in the balanced modulator.
6. The 19-kHz master oscillator signal is applied directly to the FM modulator.
7. The 19-kHz oscillator signal is also doubled in frequency to 38 kHz so that it can be inserted into the balanced modulator.

Now, let us look at the receiver to see how it handles this suppressed subcarrier and puts the signals together to produce stereo.

# Stereo Demodulation

The FM receiver and the FM stereo receiver are the same up to the discriminator stage. At this point the stereo signal must be detected and processed properly to add the missing channel.

Take a look at Figure 14-22 to see the common stages and those needed for obtaining the extra channel. The output of the discriminator is 30 Hz to 15 kHz, or the audio

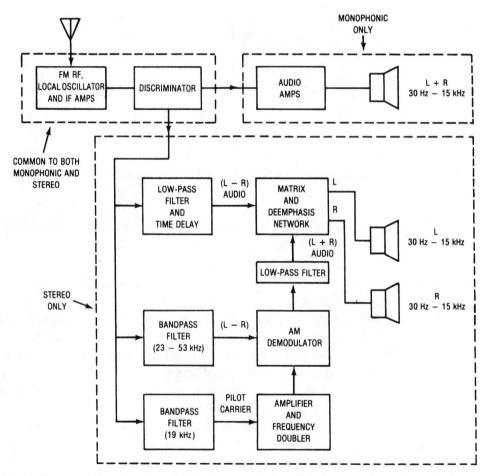

**Figure 14-22.** Comparison of stages needed for mono and stereo reception.

frequencies normally transmitted by the monaural transmitter, but there is also the 19-kHz subcarrier and the 23-kHz to 53-kHz (L – R) signal. Remember that the 30- to 15-kHz signal is called the L + R signal. Keep in mind that the 19 kHz is above the human hearing range and the audio amplifiers are not capable in most instances of amplifying it.

The standard monaural receiver reproduces the 30-Hz to 15-kHz (L + R) signal and is not aware of the other frequencies. That makes it compatible with stereo signals; since it does not have the stages to reproduce it, the other channel is ignored, and the L + R signal is sufficient to broadcast what is normally heard through one channel or speaker.

To take a closer look at what takes place after the signal goes from the discriminator to the stereo stages, see Figure 14-23.

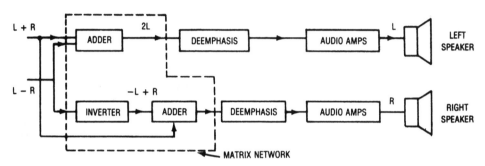

**Figure 14-23.** Processing of the two stereo signals to drive speakers.

Filters separate the three components of the stereo signal.

1. The L + R signal is obtained through a low-pass filter.
2. The L + R signal is delayed so that it reaches the matrix network in step with the L – R signal.
3. The 23- to 53-kHz band-pass filter allows the L – R signal with double sidebands to pass.
4. The 19-kHz band-pass filter allows the pilot carrier to pass to the frequency doubler where it is changed to 38 kHz.
5. The 38 kHz is the suppressed carrier for the L – R signal.
6. The AM demodulator generates the sum and difference outputs of which the 30- to 15-kHz (L – R) signal is selected by the low-pass filter.
7. The L – R signal is translated back to audio (with a frequency range of 30 to 15 kHz), and this signal plus the L + R signal are applied to the matrix and deemphasis network.
8. The matrix has an adder which adds L + R and L – R to produce 2L.
9. The L – R is also applied to an inverter which makes it a –L + R signal. This inverted signal (–L + R) combines to produce 2R.
10. Then the two individual signals for the left and right channels are deemphasized and individually amplified by audio stages to drive speakers.

# FM Radio on a Chip

It is now possible to place the entire FM radio on one or two integrated circuit chips. This can be done with two 14-pin DIPs as shown in Figure 14-24. These chips do not contain

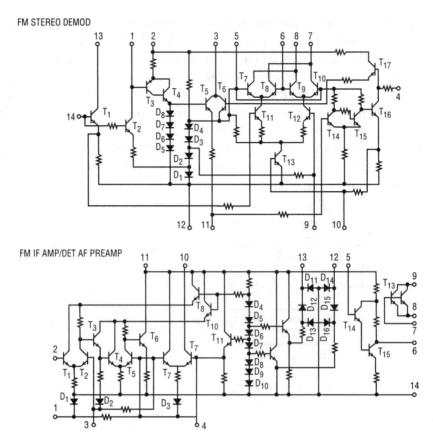

**Figure 14-24.** FM stereo on a chip.

the audio output stage but do have the preamplifier for the audio. That means another chip of whatever audio power desired can be connected to cause this to operate as a radio. This arrangement for FM stereo can be utilized when making an AM-FM stereo receiver for a car radio. It is easy to manufacture and needs no maintenance once it is installed properly with the correct power source. Note how the series of diodes are used for detection in both of the chips.

# Other FM Users

There are other users of FM. The FCC allows some stations to multiplex music on a channel with no commercials, and the station owner can charge customers for the music they use as background for workers in plants, on elevators, and in offices.

This multiplexing system is referred to as the subsidiary communication authorization (SCA). It usually employs a 67-kHz carrier and a ±7.5-kHz (narrowband) deviation. Decoders, usually using the NE 565 IC chip which is a phase-locked loop, do the demodulation and filtering. However this is beyond the realm of this book.

Keep in mind that television uses FM for the audio portion of the program. The TV FM is not broadcast as regular FM on 88 to 108 MHz. The FM standard deviation is limited to ±25 kHz to conserve on the bandwidth. FM uses ±75-kHz deviation. The television receiver is able to produce better quality sound if it has a better audio amplifier and speaker system.

We have already mentioned the use of FM transmission by people who wish to communicate for distances less than 80 mi. The amateur uses these frequencies for short distances; the walkie-talkie is limited to 100-mW output, and this further reduces its range. They come in handy for people enjoying the outdoors and for those who must work in two different locations (such as putting up a TV antenna on the roof and checking its performance inside the house). This type of communication provides comparatively low noise levels and is satisfactory for many purposes.

# Summary

A number of different stages are designed for various purposes in a receiver. Requirements for receiving antennas are much less rigid than those for a transmitter. Transmission lines present a problem in TV antenna installations. Matching of impedance of the antenna, transmission line, and the load is important for good-quality reception. The characteristic impedance of a transmission line is independent of the line's length. The usual impedances are 52, 72, and 300 $\Omega$. Standing wave ratio (SWR) is determined when there is a mismatch in impedance between the line and the load. This ratio is the maximum effective value of the standing waves to their minimum effective value. The SWR is also equal to the ratio of the characteristic impedance of the line to the impedance of the load, or the opposite.

The superheterodyne receiver is used in AM, FM, and television. Heterodyning is the beating together of two frequencies to obtain four. Then one is selected, and tuned circuits are adjusted to the selected frequency to make sure no others get past. This reduces the possibilities of receiving two stations at once. The intermediate frequency (IF) is the one picked out of the four produced by heterodyning.

The AM radio may or may not have an RF amplifier. The other stages are the mixer, oscillator, IF amplifiers, detector, audio amplifiers, and the speaker. Various other circuits are involved in the proper operation of an AM radio. The automatic volume control (AVC) aids in keeping the output constant, and the feedback circuits improve the quality of the output signal.

FM receivers rely upon a discriminator, ratio detector, or slope detector to separate the music or speech from the carrier wave. In the case of stereo where there are two channels, the setup requires a subcarrier, a pilot carrier, a matrix, and some sum and difference channels. The complexity of the FM stereo signal becomes clear once a block diagram is examined of the whole system.

Television receivers use both the AM and FM systems. The picture information is broadcast on AM while the FM is only a few megahertz away on an FM signal. TV receivers have both an AM and an FM receiver. The specifics of separation of signals was covered in a previous chapter.

There is a narrowband FM for use in communications systems. The wideband FM that is used for commercial broadcasting uses very wide sidebands to get the full audio range of music to the listener. FM operates best above 30 MHz. This means it will have line-of-sight transmission, and so maximum distance is about 80 mi. The Federal Communications Commission regulates the transmission of all radio frequencies in the United States.

# Review Questions

1. Why do you need a transmission line for receivers?
2. What is the SWR?
3. What causes standing waves?
4. Why do some AM radios need an RF amplifier stage?
5. Why do most AM radios eliminate the RF amplifier?
6. What is the function of a mixer?
7. What is meant by local oscillator?
8. What is an IF?
9. What type of device is used for a detector?
10. How is AM different from FM?
11. Why does FM need a limiter?
12. What is the function of a discriminator in FM?
13. Why is a limiter needed in FM receivers?
14. What type of FM detection does not need a limiter?
15. What is a subcarrier?
16. What is a pilot carrier?
17. What is meant by multiplexing?
18. How does AVC operate on an AM receiver?
19. What is the purpose of AFC on an FM receiver?
20. What is meant by narrowband FM, and where is it used?
21. What does the name Foster-Seely refer to?
22. When was FM stereo authorized by the FCC?
23. What type of modulation does television use for the picture information?
24. What type of modulation does television use for the sound information?
25. What does FM standard deviation compare to in an AM receiver?

# Chapter 15

# TELEVISION

elevision has an interesting history. It was developed over a period of years by a number of individuals who contributed one or two ideas, and then others would build on those ideas and put them together to produce a picture. Others would produce the scanning devices, and still others a means of getting them from one place to another. At the receiving end it was difficult to display the results in a series of neon tubes or selenium disks. Until the cathode ray tube (CRT) came along, picture quality was very poor.

Mechanical scanning was invented by Paul Nipkow in 1884. Vladimir Zworykin, a Russian who studied at the St. Petersburg Technical Institute, came up with the CRT and made television as we know it today possible. Actually, 1926 was a banner year for television. That was the year John L. Baird of England successfully transmitted halftone pictures by means of a mechanical scanning system. A few years later Bell Telephone Laboratories produced a system using mechanical scanning and transmitted a picture from New York City to Whippany, New Jersey. The picture was displayed on 2000 neon glow tubes and was of poor quality. In 1928 the first television program was broadcast from Station WGY in Schenectady, New York. Regular broadcasting was done in England during 1929 to 1931.

A number of basic elements must be considered in setting up any television system:

- Camera and scanning
- Sync signals
- Video amplifiers
- Audio amplifiers
- FM transmitter for audio
- AM transmitter for video

## Camera Tubes

A number of different camera tubes were tried before the ones available today, which can work in low levels of light, were produced. The oldest was the iconoscope developed in 1933 by Zworykin. The iconoscope tube had an electron gun with a magnetic beam deflection yoke located in the handle (see Figure 15-1). The photosensitive mosaic consisted of silver globules, sensitized with cesium and deposited on one side of a mica sheet; the other side had a metallic backing. The mica was the dielectric of the capacitor formed by the silver deposit and the metal plate. When light struck the silver surface, it produced photoemission at each of the silver globules. An electron beam scanned the surface and struck each globule that was charged by the light striking it. The electrons from the scan replaced the

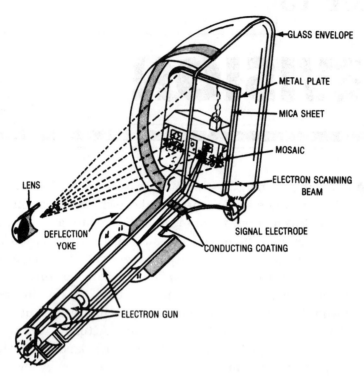

**Figure 15-1.** The iconoscope.

electrons lost by photoemission. This caused a proportionate current pulse to flow in the signal plate. The amplitudes of the pulses from the signal plate current would then represent the relative brightnesses of the mosaic elements as the beam scanned across them.

The iconoscope had some very strong light requirements and was replaced by the image orthicon (see Figure 15-2). That tube worked basically the same with some improvement in light levels. It was bulky, temperamental, and costly. It was the standard tube used in television studios for a number of years. However, another type of camera tube was the vidicon. It was smaller, simpler, and more rugged than the image orthicon. Its operation is somewhat different also since its target depends on photoconductivity rather that photoemission.

The vidicon works by having an image focused through a transparent conductive film. The film acts as the signal electrode. The image is impressed on the photoconductive target. The target is biased slightly positive. When the layer is not exposed to light, such as in darkness, it acts as an insulator. The electron gun provides a beam. The beam is slowed by the wall coating and screen to a moderate velocity and deflected magnetically. Once the scanning beam strikes the back of the target, it neutralizes the charge on the target. This causes the target to give up just enough electrons to make up for those that leaked through the partially conductive coatings since the last scan. As the brightness of the target area increases, so does the conductivity of the target. This produces a larger leakage current and a greater number of electrons taken from the beam. This action produces a burst of current at the signal electrode in proportion to the brightness of the spot being scanned.

More sensitive tubes are being developed. It is now possible to make color television programs at night using available light. The tubes have become more rugged for use in portable home cameras. They also use less light to obtain an acceptable picture.

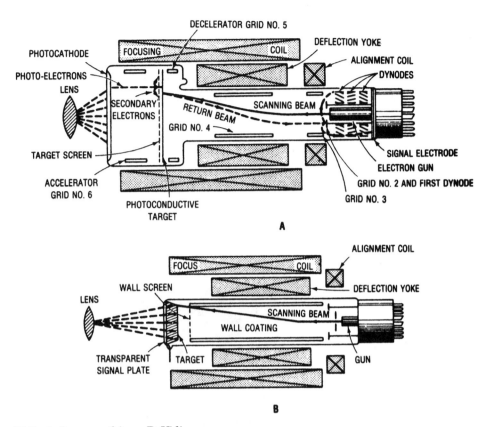

Figure 15-2. A. Image orthicon. B. Vidicon.

# Scanning

In order to have the camera tube operate, it has to be scanned. The beam of light which scans the target area is swept back and forth and up and down. The rate at which it is moved from left to right and from top to bottom of the area will determine what type of picture is reproduced at the receiver.

The Electronics Industries Association standard for the United States is 525 lines horizontally (see Figure 15-3). The 525 lines are produced as the beam scans from top to bottom of the image area. This scanning is done in one-thirtieth of a second. This scan

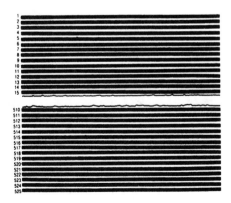

Figure 15-3. The 525 scan lines for a TV picture.

from top to bottom is called *one frame*. However, in the interlaced method, the sweep is from top to bottom in one-sixtieth of a second. This means that the odd-numbered lines are swept from top to bottom, and then the beam returns to sweep the even-numbered lines from top to bottom. The two fields make up a frame.

This type of scanning, interlaced, has a tendency of reducing the flickering effect produced by the succession of pictures. The receiver uses the same scanning technique. That is why the sync signals are produced: to keep the transmitter scanning rate in step with the scanning rate of the receiver's picture tube. If they are not in step, the picture will become scrambled. If the receiver's vertical scan is out of step with the transmitter's vertical scan, the picture on the tube at the receiver will roll slowly. If the horizontal scans are out of sync, the picture will become diagonally slanted and will not be recognizable. As you can see, the synchronization of the camera tube scan and the receiver tube scan is very important (see Figure 15-4).

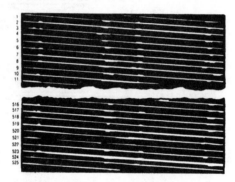

**Figure 15-4.** Scanning pattern for interlacing.

There are 60 fields each second. The vertical sweep frequency is 60 hertz (Hz). The horizontal sweep frequency is 15,750 Hz. It is based on the fact that 30 frames times 525 lines equals 15,750. The sweep voltages are in a sawtooth waveform.

The odd-numbered lines are scanned first and then the even-numbered. During the odd-numbered scanning the even-numbered lines are not illuminated, and so they appear black on the screen. Thus, you actually see only half of the lines at one time, that is, 262.5 lines make up a field (see Figure 15-5). The beam has to travel from the bottom of the screen (midway as you can see from the scan pattern for the odd fields) back to the top left

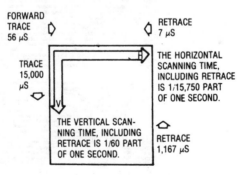

**Figure 15-5.** Scanning frequencies, both vertical and horizontal.

side. (We are actually talking about the picture produced by the picture tube and being watched from the normal viewing position.) The beam traces a line of elements from the left to the right. When the beam reaches the right-hand edge, the magnetically deflecting field for the horizontal direction reverses quickly. This step would move the beam back across the face of the tube, drawing a right-to-left line, except that the beam is momentarily turned off during the retrace by a blanking signal. At the same time, the magnetic deflecting field in the vertical direction has increased to move the starting point on the left down one notch. The blanking signal is turned off and the next left-to-right line is drawn. This pattern repeats until the bottom right corner of the image is reached. Then the beam is deflected sharply back to the top edge, as well as to the left side, during blanking. The picture is scanned along all the even-numbered lines first. This provides the odd field and then the even-numbered lines produce the second field. The two fields together make up one complete picture or frame. You actually see only about 480 lines since the time it takes to retrace from the bottom to the top is such that a few of the lines are not presented. Because the frames consist of two fields, the field rate is 60 per second. This allows the use of 60-Hz power line frequencies for synchronization. The size of the picture is 1.33 times as wide as it is high ($3 \times 4$ is the ratio of height to width). These are, incidentally, about the same proportions that the 35-mm film has. In Europe they use a frame rate of 50 per second since they have a 50-Hz power line frequency. In order to get the same-quality picture at the receiver, they use 15,625 Hz for the horizontal frequency. Therefore, 25 frames times 625 lines equals 15,625 Hz.

# TV Transmitter

A television transmitter is really two in one (see Figure 15-6). The picture information (called video) is transmitted with its synchronization signals, both horizontal and vertical (see Figure 15-7), by way of an amplitude-modulated (AM) transmitter. The audio information is transmitted by a frequency-modulated (FM) transmitter. They both use the same antenna. The band of frequencies which contains both picture and audio is 6-megahertz (MHz) wide (see Figure 15-9).

Since the audio transmitter was discussed in a previous chapter, let us take the time here to concentrate on the video transmitter and how it differs from an AM transmitter.

The critical part of the video transmitter is the sync generator. It provides all the critical timing functions of both the local camera's sweep circuits and the mixer for transmission to the receiver's sync circuits. The output of the camera is amplified and mixed, with the sync signals forming the composite video signal (see Figure 15-8). The rest of the video transmitter is like any other AM transmitter, except that it operates in the TV frequency band and has a sideband filter to suppress the lower video sidebands.

## Bandwidth

As you can see, there is a lot of information in the video signal. On each side of the video is a guard band to prevent interference with the audio and to guard against interfering with adjacent channels (see Figure 15-9). On the right-hand side of the line, which designates *picture carrier frequency*, you see the upper sideband that has the video information.

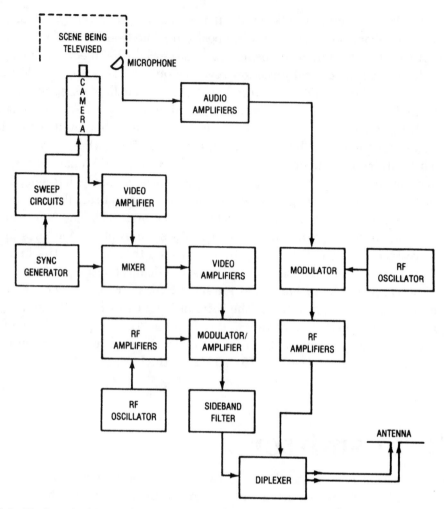

**Figure 15-6.** Black-and-white TV transmitter block diagram.

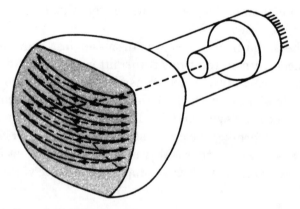

**Figure 15-7.** Scanning pattern for a TV tube.

The left side of the line designates the vestige of the lower sideband. Vestigial sideband modulation is used in the TV transmitter since it takes less of a bandwidth than if both sidebands were transmitted. The most common way to obtain this is by generating both sidebands and then removing the undesired portion of the lower sideband by means of a

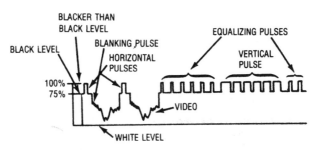

**Figure 15-8.** Television signal waveform, black and white.

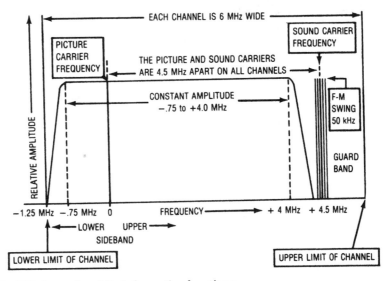

**Figure 15-9.** The TV channel and its information locations.

filter network that is inserted in the transmission line that connects the transmitter with the antenna.

If single sideband modulation were used, it would call for a more expensive receiver. This way, the vestige of the lower sideband makes the receiver think it is all there, and it will operate without requiring the additional oscillators for the single sideband. If you look at Table 15-1, you will see how many channels are available in both the VHF and UHF bands. Note that breaks in the band allocated for TV allow for other operations. For example, there is only a 4-MHz break between channels 4 and 5, but there is a large break in the frequencies between channels 6 and 7. In fact the FM broadcast band is inserted here at 88 to 108 MHz. If you detune channel 6 slightly, you can pick up FM broadcasts on the low 88-MHz end where educational FM stations are operating.

# Television Channel Reallocation

Very few TV stations use channels 70 through 83 on the high end of the UHF band. The U.S. government's Federal Communications Commission (FCC) decided to reallocate these frequencies for land mobile service, or for cell-phone use. All TV stations on channels 70 through 83 will be allowed to continue and will be renewed. No new TV stations will be assigned to this end of the band. See Table 15-2.

## TABLE 15-1

## TELEVISION CHANNEL FREQUENCIES, MHz

**Group 1**

| | Channel | Frequency |
|---|---|---|
| | No. | Limits |
| | | 54 |
| P 55.25 / S 59.75 | 2 | |
| | | 60 |
| P 61.25 / S 65.75 | 3 | |
| | | 66 |
| P 67.25 / S 71.75 | 4 | |
| | | 72 |
| | | 76 |
| P 77.25 / S 81.75 | 5 | |
| | | 82 |
| P 83.25 / S 87.75 | 6 | |
| | | 88 |
| | | 174 |
| P 175.25 / S 179.75 | 7 | |
| | | 180 |
| P 181.25 / S 185.75 | 8 | |
| | | 186 |
| P 187.25 / S 191.75 | 9 | |
| | | 192 |
| P 193.25 / S 197.75 | 10 | |
| | | 198 |
| P 199.25 / S 203.75 | 11 | |
| | | 204 |
| P 205.25 / S 209.75 | 12 | |
| | | 210 |
| P 211.25 / S 215.75 | 13 | |
| | | 216 |
| | | 470 |
| P 471.25 / S 475.75 | 14 | |
| | | 476 |
| P 477.25 / S 481.75 | 15 | |
| | | 482 |
| P 483.25 / S 487.75 | 16 | |
| | | 488 |
| P 489.25 / S 493.75 | 17 | |
| | | 494 |
| P 495.25 / S 499.75 | 18 | |
| | | 500 |
| P 501.25 / S 505.75 | 19 | |
| | | 506 |
| P 507.25 / S 511.75 | 20 | |
| | | 512 |

**Group 2**

| | Channel | Frequency |
|---|---|---|
| | No. | Limits |
| | | 512 |
| P 513.25 / S 517.75 | 21 | |
| | | 518 |
| P 519.25 / S 523.75 | 22 | |
| | | 524 |
| P 525.25 / S 529.75 | 23 | |
| | | 530 |
| P 531.25 / S 535.75 | 24 | |
| | | 536 |
| P 537.25 / S 541.75 | 25 | |
| | | 542 |
| P 543.25 / S 547.75 | 26 | |
| | | 548 |
| P 549.25 / S 553.75 | 27 | |
| | | 554 |
| P 555.25 / S 559.75 | 28 | |
| | | 560 |
| P 561.25 / S 565.75 | 29 | |
| | | 566 |
| P 567.25 / S 571.75 | 30 | |
| | | 572 |
| P 573.25 / S 577.75 | 31 | |
| | | 578 |
| P 579.25 / S 583.75 | 32 | |
| | | 584 |
| P 585.25 / S 589.75 | 33 | |
| | | 590 |
| P 591.25 / S 595.75 | 34 | |
| | | 596 |
| P 597.25 / S 601.75 | 35 | |
| | | 602 |
| P 603.25 / S 607.75 | 36 | |
| | | 608 |
| P 609.25 / S 613.75 | 37 | |
| | | 614 |
| P 615.25 / S 619.75 | 38 | |
| | | 620 |
| P 621.25 / S 625.75 | 39 | |
| | | 626 |
| P 627.25 / S 631.75 | 40 | |
| | | 632 |
| P 633.25 / S 637.75 | 41 | |
| | | 638 |

**Group 3**

| | Channel | Frequency |
|---|---|---|
| | No. | Limits |
| | | 638 |
| P 639.25 / S 643.75 | 42 | |
| | | 644 |
| P 645.25 / S 649.75 | 43 | |
| | | 650 |
| P 651.25 / S 655.75 | 44 | |
| | | 656 |
| P 657.25 / S 661.75 | 45 | |
| | | 662 |
| P 663.25 / S 667.75 | 46 | |
| | | 668 |
| P 669.25 / S 673.75 | 47 | |
| | | 674 |
| P 675.25 / S 679.75 | 48 | |
| | | 680 |
| P 681.25 / S 685.75 | 49 | |
| | | 686 |
| P 687.25 / S 691.75 | 50 | |
| | | 692 |
| P 693.25 / S 697.75 | 51 | |
| | | 698 |
| P 699.25 / S 703.75 | 52 | |
| | | 704 |
| P 705.25 / S 709.75 | 53 | |
| | | 710 |
| P 711.25 / S 715.75 | 54 | |
| | | 716 |
| P 717.25 / S 721.75 | 55 | |
| | | 722 |
| P 723.25 / S 727.75 | 56 | |
| | | 728 |
| P 729.25 / S 733.75 | 57 | |
| | | 734 |
| P 735.25 / S 739.75 | 58 | |
| | | 740 |
| P 741.25 / S 745.75 | 59 | |
| | | 746 |
| P 747.25 / S 751.75 | 60 | |
| | | 752 |
| P 753.25 / S 757.75 | 61 | |
| | | 758 |
| P 759.25 / S 763.75 | 62 | |
| | | 764 |

**Group 4**

| | Channel | Frequency |
|---|---|---|
| | No. | Limits |
| | | 764 |
| P 765.25 / S 769.75 | 63 | |
| | | 770 |
| P 771.25 / S 775.75 | 64 | |
| | | 776 |
| P 777.25 / S 781.75 | 65 | |
| | | 782 |
| P 783.25 / S 787.75 | 66 | |
| | | 788 |
| P 789.25 / S 793.75 | 67 | |
| | | 794 |
| P 795.25 / S 799.75 | 68 | |
| | | 800 |
| P 801.25 / S 805.75 | 69 | |
| | | 806 |
| P 807.25 / S 811.75 | 70 | |
| | | 812 |
| P 813.25 / S 817.75 | 71 | |
| | | 818 |
| P 819.25 / S 823.75 | 72 | |
| | | 824 |
| P 825.25 / S 829.75 | 73 | |
| | | 830 |
| P 831.25 / S 835.75 | 74 | |
| | | 836 |
| P 837.25 / S 841.75 | 75 | |
| | | 842 |
| P 843.25 / S 847.75 | 76 | |
| | | 848 |
| P 849.25 / S 853.75 | 77 | |
| | | 854 |
| P 855.25 / S 859.75 | 78 | |
| | | 860 |
| P 861.25 / S 865.75 | 79 | |
| | | 866 |
| P 867.25 / S 871.75 | 80 | |
| | | 872 |
| P 873.25 / S 877.75 | 81 | |
| | | 878 |
| P 879.25 / S 883.75 | 82 | |
| | | 884 |
| P 885.25 / S 889.75 | 83 | |
| | | 890 |

P — Picture carrier frequency    S — Sound carrier frequency

### TABLE 15-2

### CELLULAR PHONE FREQUENCIES

| Frequency Range | Old TV Channels |
| --- | --- |
| 825–845 MHz | See UHF TV channels 74 to 76 |
| 870–890 MHz | See UHF TV channels 81 to 83 |

# Color Television

The monocolor, or black-and-white (B/W), television was first on the scene. Once it was developed, people were looking for better pictures and with color. Thus, the millions of TV sets that were purchased for B/W reception also had to be able to receive the color-transmitted programs in black and white. This produced a problem or two since the bandwidth had already been assigned and filled with information for black and white and the sound. There simply was not any space to add color signals. However, a few people had ideas that would make color and black and white compatible.

Color television relies on the principle that any visible color may be reproduced by using the proper combinations of the three primary or chromatic colors: red, blue, and green. All systems of transmitting color television signals use some method of analyzing the colors of the scene being scanned in terms of these three colors and of converting the information into electric signals. Figure 15-10 shows how color separators work to produce the color TV picture. Three monochrome cameras look at each scene through the same lens. The color of the scene is then divided into red, green, and blue light by the dichroic mirrors. A mirror of this type is a plate of glass coated with a thin, metallic layer. It will then reflect one of the primary colors while allowing the others to pass through. The camera has three outputs corresponding to the three primary colors. The picture is transmitted by converting these signals into other signals that correspond to the brightness, hue, and saturation of the scene.

In color television the brightness signal, called the *luminance signal*, is represented by the letter Y. Keep this in mind since the Y signal shows up again in the discussion of color receivers. Signals that correspond to the hue and saturation are called the *chrominance signals*. The process of combining the outputs of the cameras to form chrominance signals is called *encoding* and is accomplished in a circuit called a *matrix*.

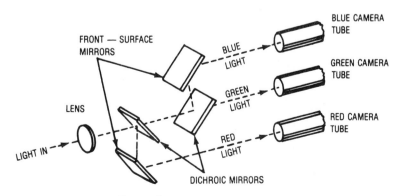

**Figure 15-10.** Color separator system for a color camera.

There is no difference between the luminance, or Y signal, and the monochrome camera signal. It is all that is needed to produce a B/W picture. The luminance signal is made up of 30 percent of the output of the red camera, 59 percent of the output of the green camera, and 11 percent of the output of the blue camera. These are roughly the rates or percentages that the human eye responds to the various colors.

The luminance signal is part of the signal that is amplitude-modulating the transmitter. It appears as the composite waveform. The chrominance information must be transmitted in some other way. The chrominance information from the encoder can be used to modulate both the amplitude and phase of a subcarrier. The subcarrier itself is suppressed and only its sidebands are transmitted. The frequency of the subcarrier was picked so that it could be added to the signal without increasing the bandwidth required for transmission. Take a look at the following to get some ideas as to how it works.

# Frequency Interleaving

The monochrome signal was analyzed, and it was found that the information transmitted was done so in bunches that were clustered together with spaces in between. The separation of the bunches amounted to an amount equal to the horizontal line frequency (see Figure 15-11). It was found that only about 54 percent of the available bandwidth was used. The color subcarrier frequency was chosen, which was an odd multiple of one-half of the horizontal line frequency. That made it possible for the color modulation to fit exactly between the clusters of the monochrome signal (see Figure 15-11B).

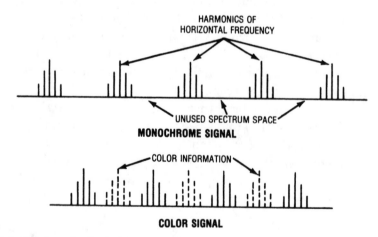

**Figure 15-11.** Interleaving of color information in unused spectrum space of black and white.

A subcarrier of 3.579545 MHz (usually referred to as the 3.58-MHz subcarrier signal) was chosen because it was high enough that the color variations did not show up as noise on the luminance signal. And this frequency does not cause objectionable beats with the sound carrier. Since this is not *exactly* one-half of the original monochrome horizontal line frequency, the horizontal and vertical scanning frequencies were changed slightly for color transmission. The horizontal was changed from 15,750 to 15,734.26 Hz, and the vertical was changed from 60 to 59.94 Hz. The receiver does not notice the difference.

# Color Signal

The color signal looks like Figure 15-12. The sync part of the signal is the same for color as it is for black and white. The only difference is that the video portion has the 3.58-MHz signal added. The amplitude of the video signal determines the saturation of the color being transmitted. The phase of the component corresponds to the hue or tint. The phase of a signal is in reference to the phase of the subcarrier oscillator at the transmitter. The bursts of eight cycles of 3.58 Hz on the back porch of the signal are used to synchronize the phase of the subcarrier oscillator in the receiver with the reference phase at the transmitter.

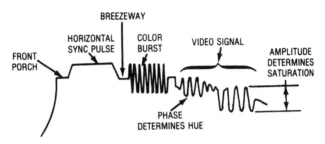

Figure 15-12. Color TV waveform, video portion. Note location of color burst.

# RF Signal

The radio frequency signal as transmitted from the antenna looks like Figure 15-13. All the information is located in this signal to produce color and sound and to keep the picture from rolling or turning into a scrambled mess. Figure 15-14 shows the simplified block diagram of a color television transmitter, and so you can see how the signal was produced from the camera and the color information added along the way to the transmitting antenna. Keep in mind that the subcarrier is not transmitted but has to be put back in the signal when the receiver gets it. The receiver has a crystal that will put the 3.58 MHz back into the signal and cause the phase differences to be retrieved. (More about this later in the discussion of the TV receiver.) The 3.58-MHz pulses are transmitted to keep the receiver oscillator going at the same rate as the transmitter.

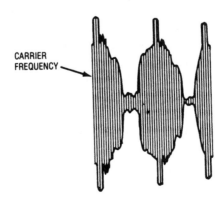

Figure 15-13. Color TV waveform in radio frequency as broadcast.

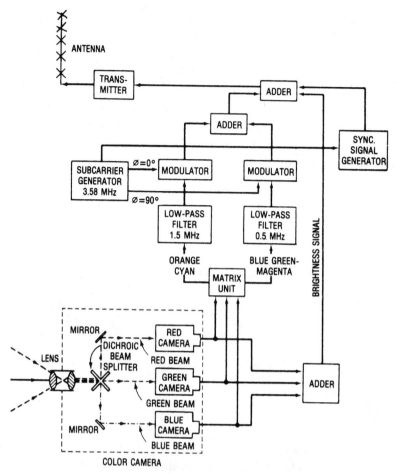

**Figure 15-14.** Simplified block diagram of color TV transmitter.

# The Television Receiver

There is no difference between the television receiver and the AM and FM receivers. (These were discussed in detail in the preceding chapter.) Now we will take a look at the color receiver to see how it works in principle.

Figure 15-15 shows the B/W television set in block diagram form. The audio is separated at the video detector and sent to an FM section which demodulates the signal and produces the sound. The picture information is sent to the cathode ray tube where it is displayed as the signal dictates. In order to keep the picture synchronized properly, sync signals are picked from the incoming signal and fed to the deflection coils around the neck of the tube. These signals cause the beam of electrons in the picture tube to be moved back and forth and up and down by the sawtooth waveform fed into them. The scanning of the phosphorous coating on the picture tube causes it to glow and not glow according to the intensity of the beam. If the beam is not too bright, it will show as a gray. This produces a white-and-gray more so than a black-and-white picture.

Now that you have taken a quick look at the B/W television receiver, we shall modify the receiver slightly to enable it to receive the color signal and produce a picture and sound. Keep in mind that the color and the B/W technology had to be compatible in order for the Federal Communications Commission to approve the system back in the 1940s and early 1950s.

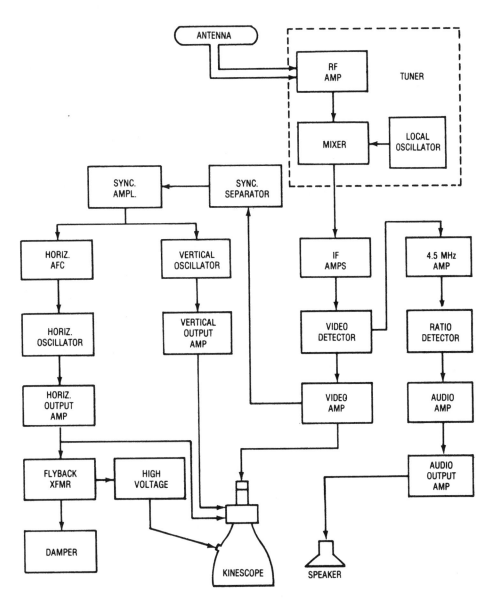

**Figure 15-15.** Block diagram of a black-and-white TV receiver.

In order to understand clearly how the color picture is produced, we had better back up and take a look at some information on color and how the eye perceives it. In order for color television to be developed, there were three things about color that had to be understood and incorporated into the signal to be broadcast: (1) color has hue, or chroma, (2) color has saturation, or purity, and (3) color has brightness. *Hue* is the basic characteristic that gives color its name, such as orange or purple. *Saturation* tells you how much the color is diluted with white. For instance, red can be seen in tomatoes and blood, while the less saturated color would appear as pink in strawberries or anything with a lower intensity of red. *Brightness* refers to the brilliance of the color. Brilliance is a whole range of colors from white to dark grays to black. These three characteristics of color have to be taken into consideration for the purposes of reproducing a color television picture.

Now, take a look at other colors that will have to be understood before a device can be designed to reproduce all the colors known today. White, of course, is not a color, it is the *presence* of *all* the colors. Therefore, it can be reproduced if all the colors are mixed together. Black is not a color either: It is the *absence* of *light*. Gray is a weak white. Browns are reds, oranges, and yellows of low brightness. When a beam of light is broken up by a prism, there are six colors that emerge: red, orange, yellow, green, blue, and violet. They are called the spectral colors. By mixing the red and the violet, purple and the magentas can be formed from the various proportions of red and violet. These are called the nonspectral colors.

When you take three colors — red, green, and blue — and project them until they overlap, some interesting combinations result (see Figure 15-16). The overlap of the red and green produces a yellow area. The red and blue-violet mix to form a magenta. The blue-violet with green combinations produce an aqua color which is called cyan in the printing business. By adjusting the brightness of the various lamps, you can obtain any color. This is the additive process of producing color. Keep in mind that the chrominance signals that are produced by the camera and transmitted by the color TV station are called orange-cyan and blue-green-magenta.

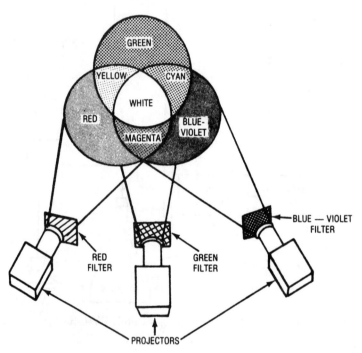

**Figure 15-16.** Mixing the primary colors in an additive mode.

# Phase Modulation

The subcarrier generated at the studio and modulated with the various outputs from three sources of color is filtered out before the TV signal is broadcast. It must be regenerated at the receiver by an oscillator. The oscillator has a tendency to drift if not kept at a perfectly stable temperature. This is why the 3.58-MHz burst is transmitted from the transmitter to keep the oscillator operating properly in the receiver. The two chrominance signals are blended with the subcarrier, but they are 90° out of phase with one another. Once the subcarrier is removed at the transmitter, it leaves only these two sidebands with the chrominance signals 90° apart. These become part of the sidebands of the regular carrier frequency assigned to the channel. They are placed on the regular carrier frequency by the process of phase modulation. Thus, the B/W receiver will ignore them completely since it has no way of recognizing their existence.

The amplitude of the combined chrominance signal carries the saturation information, and the phase angle carries the hue (see Figure 15-12). The sync signals and the luminance signal along with the picture information are broadcast over the assigned frequency for the channel. The receiver picks them up and starts the process of demodulation and of putting the information where it will be used to produce a color picture.

# The Color Picture Tube

The color picture tube has three guns that emit electrons in the end of the tube. There are three cathodes and three filament windings. The filaments heat up the tube (this is the red glow at the base of the tube) and boil off the electrons. A series of plates properly placed inside the neck of the picture tube, called a kinescope, accelerate the electrons toward the phosphorous coating on the front of the tube. Two coils are used to do the horizontal deflecting of the beam of electrons, and two vertical deflecting coils are used to deflect the beam up and down (see Figure 15-17).

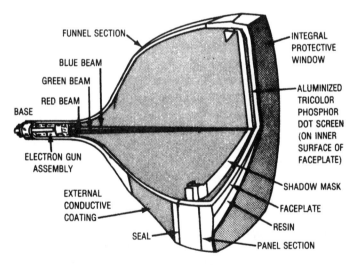

**Figure 15-17.** Cutaway view of the color picture tube. (RCA)

Another name for the picture tube is the *kinescope*. The kinescope made for the reproduction of color pictures is a rather complex device which needs some explanation before you can understand how it reproduces all the colors needed for any picture. Figure 15-18 shows the way the phosphors are placed on the inside of the tube. The inside face of the screen is coated with three different phosphors. A phosphor glows when struck with an electron beam. The phosphors are put onto the screen in a three-dot pattern. The shadow mask is placed near the phosphor-coated screen. The mask is located between the screen and the guns. The mask has about 250,000 holes in it. Each hole is there for a purpose. There is a red, a blue, and a green dot on the screen for each hole in the shadow mask. Thus, there are 750,000 dots on the screen.

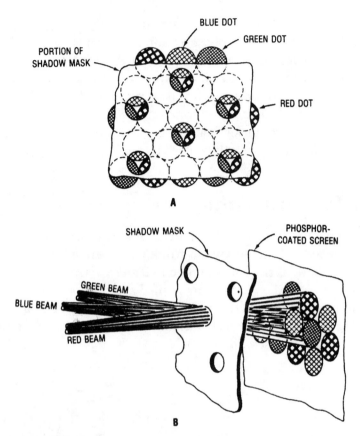

**Figure 15-18.** A. Shadow mask aligned with the color dot pattern. B. The three beams passing through the shadow mask.

A convergence electrode is operated at about 11,000 volts (V) to cause the red, the blue, and the green beams to go through the same shadow mask hole. Once the beams are through the shadow mask, the blue hits the blue dot, the red hits the red dot, and the green hits the green dot in any given pattern (see Figure 15-19B). Now, how does this produce color?

The dots are too small to be seen by the human eye. If the same intensity of beam hits all three dots, they produce a white dot. However, if the red and green are the only ones hitting their dots within the pattern of that hole, a yellow is produced. This goes on for all the various combinations of colors that were discussed earlier in the production of the color signal. The color signal then is displayed on the phosphors of the tube in accordance with what they were in the studio or where the camera saw them.

# LCD Television Pictures

The liquid crystal display has been used for a number of years for such things as calculator displays and signs.

One design developed in London at the Imperial College of Science and Technology has a flat panel with glass optical waveguide strips in vertical columns covered by a glass panel

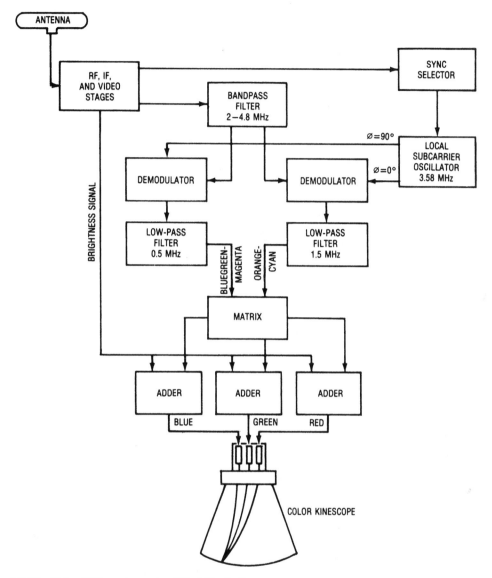

**Figure 15-19.** Color TV receiver, simplified block diagram.

with rows of nearly invisible electrodes (see Figure 15-20). The LC material is sandwiched between the strips and glass sheet. Three colored light emitting diodes (LEDs) at the bottom edge of each strip are fired in a horizontal sequence by a red, green, and blue video signal. Simultaneously, a shift register circuit applies ac pulses to the horizontal electrodes one row at a time. The electric field at each electrode-strip crossover realigns the LCs, allowing light to escape. LED selection here creates pure red, green, and blue picture elements plus white (three LEDs fired at once).

# Receiver Stages

Figure 15-19 shows the block diagram of the color TV receiver. In it you will notice some differences from that of the B/W receiver.

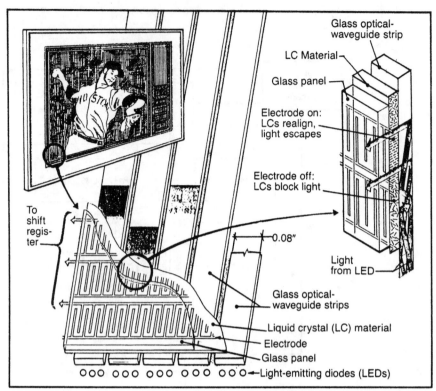

Courtesy of Popular Science and Eugene Thompson

**Figure 15-20.** Liquid crystal display TV picture. (Reprinted from Popular Science with permission © 1986 Times Mirror Magazine, Inc.)

All the stages shown in Figure 15-19 are utilized to produce the color picture. The radio frequency (RF), intermediate frequency (IF), and video stages that are normally found in a B/W receiver are located here also. However, from there the signal is taken into different stages to be processed for color components and sync signals. The sync signals are taken off, and the 3.58-MHz bursts that appear during the horizontal retrace time are used by the automatic frequency control (AFC) circuit to check and compare with the receiver's 3.58-MHz crystal oscillator and to keep it in phase with that which is broadcast with the color information. The subcarrier (3.58 MHz) is again added to the sidebands that contain the color information. This allows the color information to be extracted. The subcarrier oscillator stage puts out two signals for the two modulators, and the signals are 90° out of phase with one another. A low-pass filter of 0.5 MHz then puts out the blue-green-magenta signal, while the other low-pass filter of 1.5 MHz produces the orange-cyan signal. These two signals are fed into the matrix. After demodulation, the R – Y (red minus luminance) and B – Y (blue minus luminance) signals are matrixed to produce a G – Y (green minus luminance) signal. All three signals are then applied to the picture tube where the positive Y signal from the video amplifier cancels all the –Y components. Keep in mind now that the third signal, or brightness, is added after the matrix. From the matrix the signals are fed to the individual adders. The adders add the brightness signal (a positive Y signal), and this removes the –Y signals. That in turn produces a blue, green, and red to drive the individual color guns in the base of the kinescope.

When the tint or hue control is adjusted, it changes the phase angle slightly so that the exact color match can be obtained.

# High Definition Television

High definition television (HDTV) is a type of *digital* television that produces extremely sharp images. HDTV uses from 20 to 1000 scan lines, each of which carries a greater amount of detail than an ordinary line. HDTV provides a picture approximately four times as sharp as standard television sets. This type of scanning system is called progressive. It offers greater clarity than the conventional interlaced 2 : 1 type of television because it scans all the lines at one time in every frame or picture. The screen on HDTV is wider and narrower than the standard 3 × 4 aspect ratio of the regular television set. For example, the 3 × 4 ratio is similar to the 35 mm camera film's picture ratio that was used as a basis for establishing the original standard for television many years ago.

HDTV was first broadcast in the United States as early as 1998. The HDTV receivers were, and still are, very expensive in comparison to the traditional set. Some parts of Europe started broadcasting HDTV in the late 1990s and limited use began in Japan in 1989.

# TV Receiver Antennas

The folded dipole antenna is usually recommended for outside reception. It has a 300-ohm (Ω) impedance that makes it a wideband receptor, wide enough to handle the 2 through 13 channel frequency allocations. The dipole is about one-half wavelength (see Figure 15-21). The folded dipole is bidirectional, the major lobes are at right angles to the elements. Dipoles may be stacked for higher gain in areas of weak signals. Outside antennas usually have directors in front of the dipole and reflectors in the back of it. The directors make the antenna a little more directional, and the reflector rejects the unwanted waves from the back of the antenna. The directors are shorter than the dipole or driven element, while the reflectors are longer in physical size than the driven element. Figure 15-22 shows the proper way to ground and mount a rooftop antenna. The UHF channels respond better to a loop antenna (see Figure 15-23). In some cases where a weak UHF signal is present

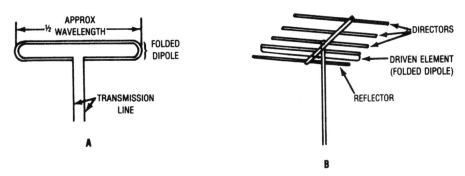

**Figure 15-21.** A. Basic folded dipole. B. Typical single-bay yagi with five elements.

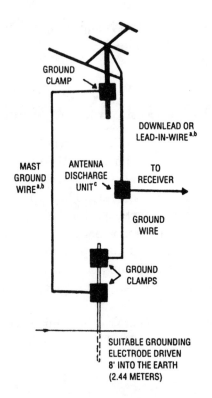

a Use No. 10 AWG copper or No. 8 AWG aluminum or No. 17 AWG copper-clad steel or bronze wire, or larger as ground wires for both mast and lead-in.

b Secure lead-in wire from antenna to antenna discharge unit and mast ground wire to house with stand-off insulators, spaced from 4 feet (1.22 meters) to 6 feet (1.83 meters) apart.

c Mount antenna discharge unit as close as possible to where lead-in enters house.

**Figure 15-22.** Proper installation for rooftop TV receiver antenna.

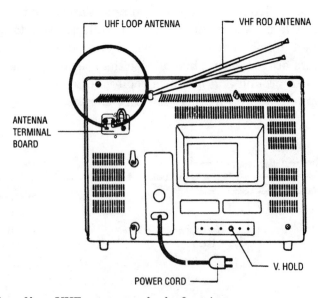

**Figure 15-23.** Location of loop UHF antenna on back of receiver.

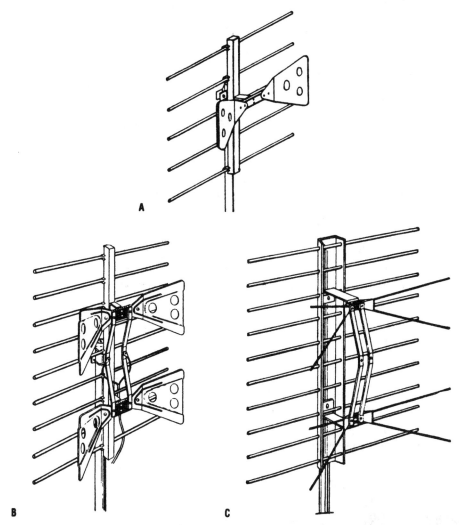

**Figure 15-24.** Bow-tie UHF antenna with a sheet reflector. B. Stacked bow tie with sheet reflector. C. A two-bay UHF antenna using a sheet reflector.

and the color information is lost, it may be to your advantage to obtain a bow-tie UHF antenna (see Figure 15-24). These antennas usually have a screen wire backup with two to four bow-tie shaped antenna elements mounted one on top of the other with proper spacing. Stacking the driven elements has a tendency to double the signal each time another element is added.

## Summary

Television, as we know it, depends on a scanning technique to produce the picture. Mechanical scanning was developed as far back as 1884 in Russia. Vladimir Zworykin came up with a cathode ray tube which caused a rapid development of television. A number of different camera tubes were tried before a reliable one was developed. The iconoscope, the vidicon, and the image orthicon were but three types of camera tubes developed. The camera tube converts the images into electric impulses.

The TV picture is made up of 525 lines in the United States and 625 in Europe. There are 30 pictures per second in the United States and 25 in Europe. We use the interlacing technique for producing the picture in two parts — one scan is the odd-numbered lines and the second scan produces the picture with even-numbered lines. The vertical sweep frequency required for our television picture is 60 Hz, while the horizontal frequency is 15,750 Hz, and in Europe it is 15,625 Hz.

The television transmitter is really two in one. The picture information (video) and its sync pulses are transmitted on an amplitude-modulated (AM) transmitter while the sound or audio is transmitted on a frequency-modulated (FM) transmitter. The carrier frequency is modulated with upper and lower sidebands, but most of the lower sideband is removed before being transmitted. This produces vestigial sideband modulation. This vestige of a lower sideband causes the receiver to work as if it received all the sidebands. It also calls for a less expensive receiver than if only single sideband transmission were made.

Color television relies on the principle that any visible color may be reproduced by using the proper combinations of the three primary or chromatic colors, red, blue, and green.

Color television uses frequency interleaving in order to obtain color without going over the sideband limits for each channel. This means a subcarrier is used to carry the color information and is then suppressed before transmission. At the receiver the circuitry puts the subcarrier back and detects the sidebands and uses them to produce the color signals that produce the color picture. The subcarrier operates at exactly 3.579545 MHz.

The amplitude of the combined chrominance signal carries the saturation information and the phase angle carries the hue. The sync signals and the luminance (brightness) signals along with the picture information are broadcast over the assigned frequency for the channel.

The color picture tube has three guns, one each for red, green, and blue. The three guns are aimed so that the three beams will go through a hole in the shadow mask at the same time and hit the three color dots on the screen with varying intensities as the picture information dictates. Liquid crystal displays (LCD) are now being developed for a flat picture.

The folded dipole is the usual choice for color television receivers since it has a broad-band tuning capability and high impedance. Since it is bidirectional, directors and reflectors have to be added to make it directional. UHF frequencies utilize the closed loop for picking up signals locally. If the UHF signal is weak, a rooftop antenna made up of arrays of bow ties can be used to boost signal strength.

HDTV, high definition TV, is a type of *digital* television that produces extremely sharp images. It uses from 20 to 1000 scan lines, each of which carries a greater amount of detail than an ordinary TV line. The picture is about four times as sharp as an analog TV set. The screen is different than regular television. It is wider and longer than the 3 × 4 aspect ratio of the older TV standard. HDTV was first broadcast in the United States in 1998. HDTV receivers are very expensive when compared with analog television sets.

# Review Questions

1. Who invented the cathode ray tube?
2. Where was the first TV picture transmitted in the United States?
3. Name three types of TV camera tubes.
4. What is another name for the picture tube?

5. What is a vidicon?
6. How many lines do pictures on a TV set have?
7. What is the frame frequency of U.S. television?
8. What is the field frequency of U.S. television?
9. What is the difference between frame and field?
10. What is meant by scanning in a receiver?
11. What is the subcarrier frequency for color television?
12. What is the horizontal frequency used for the United States?
13. What is the bandwidth of TV in the United States?
14. What is vestigial sideband modulation?
15. What are the percentages of red, blue, and green in the luminance signal?
16. What is meant by the term frequency interleaving?
17. On what type of transmitter is sound transmitted for television in the United States?
18. On what type of transmitter is video transmitted for television in the United States?
19. What is the difference between brilliance and luminance?
20. What are the spectral colors?
21. Where is phase modulation used in color TV?
22. What is a phosphor?
23. What is a shadow mask used for in color TV?
24. What does the tint or hue control do in color receivers?
25. What does HDTV stand for?
26. What type of antenna is used for rooftop use with TV receivers?
27. What is the bow-tie antenna used for in television?

# Chapter 16

# LASERS AND FIBER OPTICS

The laser is a device which uses radiant energy. It produces a narrow, highly intense beam of light which can be focused over long distances. In the communications field the laser could carry more than 100,000 telephone calls at once. Present wire capacities stop at about 6000.

The beam could handle 160 television programs. The limit today is 10. This is using the microwave systems on a broadband point-to-point transmission. The main use of the laser is not communications (except for telephone use). Its varied applications include medicine, metalworking, and even the carving of resistors until they are the precise resistance needed.

A maser is similar to a laser. The *laser* receives its name from *light amplification by stimulated emission of radiation*. The maser was used as an experiment for the amplification of microwave frequencies. This attempt to find an amplifier of these extremely high frequencies led to the development of light amplification.

The laser is used as the light source for fiber optics communications systems. The laser can be used to communicate directly by simply transmitting a modulated beam through the atmosphere. This can lead to problems since dust, fog, rain, and clouds can get in the way of the beam. This is why the fiber optics route was selected.

## Types of Lasers

There are three different types of useful lasers. They are designated in terms of the active materials used in their manufacture. The materials used are solid state, gas, and semiconductor.

### Light

All lasers use light, but the *type* of light used is important here. All light, remember, is in the form of electromagnetic radiation. Ordinary light, such as that produced by a light bulb, consists of electromagnetic radiations of different frequencies. Visible light is electromagnetic radiation at $430 \times 10^{12}$ to $730 \times 10^{12}$ hertz (Hz) — far above the frequency of microwaves.

The visible light from a lamp actually consists of several different colors that appear to the eye as white. Such light is referred to as incoherent. *Incoherent light* is a random, out-of-phase emission of energy. It diffuses rapidly and cannot be precisely or accurately transmitted without a significant loss of energy. In short, it scatters and loses its ability to be seen after a short distance. Figure 16-1 shows the incoherent light being scattered.

Light generated by the laser is *coherent* light. It consists of electromagnetic radiations that are in phase (see Figure 16-2). Because of this characteristic, coherent light does not diffuse appreciably and can be transmitted over long distances without significant loss of energy.

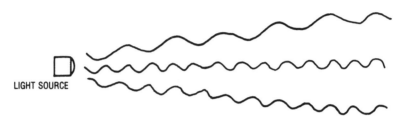

LIGHT SOURCE

**Figure 16-1.** Incoherent light beam.

LIGHT SOURCE

**Figure 16-2.** Coherent light beam.

Figure 16-3 shows how the ruby rod (solid state laser) is used to produce a coherent light for use as laser beams. Many types of different materials have been successfully stimulated to exhibit laser action. The brilliant red, green, blue, and yellow beams seen at laser light shows are produced by gaseous materials. In terms of importance to electronic communications using fiber optics, the semiconductor injection laser is the only useful type.

The devices used in communications are members of the light-emitting diode (LED) family. LED lasers are created by current being injected into a diode laser. When it is below a critical value (the threshold), the diode behaves just like an LED and emits a relatively broad spectrum of wavelengths in a wide radiation pattern. As the current increases to the threshold, however, the light narrows into a distinct beam and is confined to a very narrow spectrum. Lasing action begins, and the device then functions as a laser.

Special types of semiconductor diodes are made to withstand a continuous operation. The stripe geometry of the DH laser (Figure 16-4) illustrates the point. Gallium arsenide (GaAs) and aluminum gallium arsenide (AlGaAs) are used. This allows the light-emitting PN junction region to be sandwiched between two or more semiconductor layers that confine the generation and emission of light to the junction region. This combination of events will allow for a high-efficiency operation at low threshold currents. These devices generate several milliwatts of laser light and are used for communications through optical fibers and for video disk readouts. They are also used in printing systems. This type of laser is called a DH from *d*ouble *h*eterojunction. Heterojunction is another term for combining two dissimilar semiconductor materials.

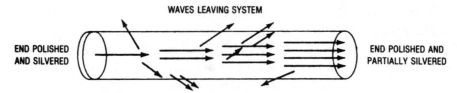

WAVES LEAVING SYSTEM

END POLISHED
AND SILVERED

END POLISHED AND
PARTIALLY SILVERED

The ends of a ruby rod are flattened (so they are parallel),
and silvered to form mirrors. The mirror at one end is made
to reflect only part of the light so that when there is a
buildup in energy between the mirrors, the beam can escape.

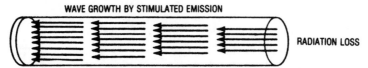

WAVE GROWTH BY STIMULATED EMISSION

RADIATION LOSS

Soon after the chromium atoms in the ruby crystal are pumped by a flash
lamp to a higher energy level, they drop to another level and stimulated
emission takes place. Waves moving at angles to the crystal's axis leave the
system, but those traveling along the axis grow by stimulated emission of
photons.

LASER BEAM

The parallel waves are reflected back and forth between the mirrors and the wave system grows in
intensity. A pale red glow indicates a certain amount of light being lost at the mirror, but beyond a
critical point, the waves intensify enough to overcome this loss and an intense red beam flashes out
of the crystal's partially silvered end.

**Figure 16-3.** Forming laser beams in a ruby rod.

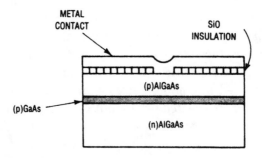

METAL
CONTACT

SiO
INSULATION

(p)AlGaAs

(p)GaAs

(n)AlGaAs

**Figure 16-4.** DH laser made by using stripe geometry.

# Uses for Lasers

There are many uses for lasers. Some have been used extensively, and others are still
experimental. The boundaries and limitations are created only by your imagination. This
is a good field for those who like to experiment and produce new products.

Some applications for lasers include:

1. Industrial welding.
2. Distance measuring equipment and alignment for sewer lines.

3. Surgical procedures such as welding detached corneas in the eye and the evaporation of tumors.
4. Military applications include the smart bomb which follows a laser beam to its target, the death ray, and the shooting down of missiles while still airborne or in space.
5. Producing holograms such as three-dimensional photography.
6. Pickup devices for video disk playback both for picture information for television programs and disks for audio sounds.
7. Communications systems.

# Fiber Optics

For years people have used fiber optics for various craft pursuits. They have made interesting light displays such as shown in Figure 16-5. Other creations include fiber optics displays, image pipes, and the magic wand with multicolor light tips. Plastic and glass rods were put to many decorative uses and made people take notice of the fiber's ability to transmit light without losing it along its sides. Then, more serious applications were found for the magic fibers.

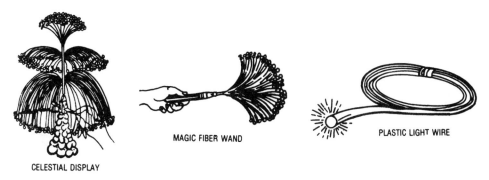

CELESTIAL DISPLAY        MAGIC FIBER WAND        PLASTIC LIGHT WIRE

**Figure 16-5.** Using fibers for bending light beams.

Doctors used the magic fibers to put light into places that could not be easily examined before because of inability to illuminate the area. Cadillac used fiber optics to show the driver whether the headlights, tail lights, and brake lights were on. But a truly dramatic application by the telephone companies brought about a revolution in communications through the use of the magic fibers. They are capable, with the use of a laser, of transmitting enormous amounts of information on one little strand of fiber. The little fiber replaces two strands of wire and can do many times more than the wire could.

## Refraction and Reflection of Light

Methods for controlling light have been devised. Take a look at Figure 16-6. In the early days of electric lighting, the relatively low light output was commonly used without shielding the eyes from direct rays or redirecting the light into useful zones. Improvements in electric lamps brought about the development of larger light sources with higher light

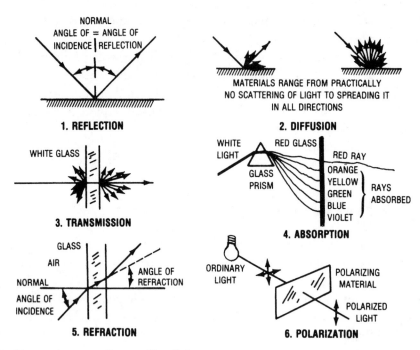

**Figure 16-6.** Six usual means of controlling light.

output. Such lamps were immediately put into use in an attempt to satisfy the demands for more light. At the same time, however, it became evident that lamps in the field of view must be shielded in order to reduce their brightness and minimize glare. Reflectors and other light-control media were very useful in this respect, also giving better distribution of light. The six ways of controlling light are shown in Figure 16-6, but we shall concentrate on refraction and reflection.

Refraction causes the light wave to be bent (see Figure 16-7). The speed reduction and the refraction are different for each wavelength. Take a look at the prism and how it bends the light beam and separates the colors or wavelengths. The amount of bend provided by refraction depends on the refractive index of the two materials. Total reflection for angles greater than the critical angle is of critical importance in fiber optics communications.

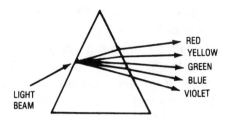

**Figure 16-7.** Refraction of light through a prism.

# Optical Fibers

There are two types of optical fibers: plastic and glass. Fibers used in communications systems may be either glass or plastic. The fiber usually has a protective coating around it (see Figure 16-8). Coated fibers are referred to as *cables*. The core or central part of the cable is the light-handling material.

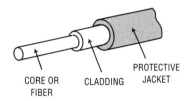

**Figure 16-8.** Construction of a single fiber used in a communications system.

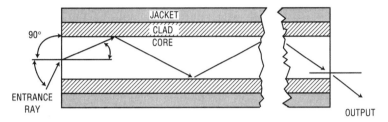

**Figure 16-9.** Reflection of a light beam inside a fiber.

Figure 16-9 shows how the fiber causes the light to be bounced back and forth inside without escaping. If you get total reflection inside the core, the propagation of light down the fiber moves along as shown in Figure 16-9. Propagation is the result of continuous reflection at the core or clad interface so that the ray bounces down the length of the fiber by the process of total internal reflection.

# Light Loss in Fibers

All things have some kind of loss, and fibers are no exception. A number of factors contribute to the loss, or attenuation, of light in a fiber. The most important factors are (1) flaws in the core consistency, (2) imperfections at the core and clad interface, (3) connector couplings, and (4) impurities in the core composition. Certain frequencies are affected by the impurities, and all frequencies are affected by the others.

# Types of Fibers

Three types of fibers are used in the communications field: (1) all-plastic or what is referred to as plastic core and plastic coating, (2) glass core and plastic cladding, and (3) glass core with glass coating or cladding. The all-plastic fiber is rugged and often used when continuous flexing and rough treatment is evident. The plastic clad silica (PCS) is really the glass fiber with a plastic coating or cladding. While the all-plastic fibers have the disadvantage of high attenuation compared to glass cores, the PCS fibers provide the good attenuation characteristics of glass core and are reasonably rugged. The all-glass fibers offer low attenuation and are the ones most easily mated to other cables, light sources, or detectors. However, they are not rugged and are susceptible to attenuation when exposed to outside light.

Fiber communications use only one fiber. All-plastic fibers are bundled for use in low-frequency, short-distance runs. Multiple-fiber cables can be used to enable multiple light sources to send information to a number of detectors. Since the single fibers can handle such

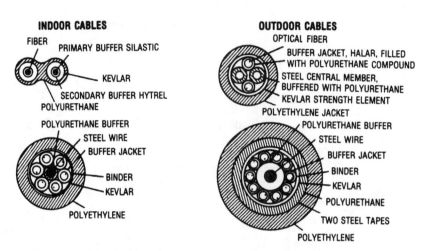

**Figure 16-10.** Fiber optics cables.

wide bandwidths multiplexing is often used and more than one signal is transmitted on a single fiber.

Fiber optics cables are shown in Figure 16-10. Note how the outdoor cables have more protection. These are the four cables available for use in telecommunications. Fiber optics are replacing coaxial cables because of the high loss in signals in the coax. The optical fiber loss is about 0.1 decibels per kilometer (dB/km).

# Fiber Optics Communications System

So far we have looked at lasers and the fibers used in the telecommunications system. Now let us look at the total system (see Figure 16-11).

Light transmitted down the glass fiber by being reflected off its sidewalls can be compared to the movement of microwaves down a *waveguide*. (Waveguides are discussed on

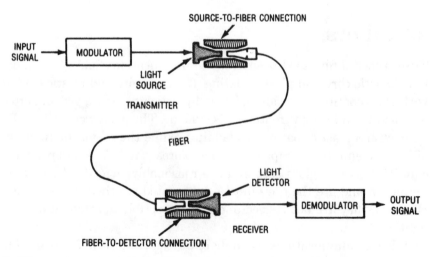

**Figure 16-11.** Fiber optics communications system.

page 300.) However, there are certain advantages to the fiber optics method of transmission over that of the microwave and its waveguides. These include:

- No cross talk
- Lower signal loss
- Lower cost
- Wide system sidebands
- Lighter weight and smaller size
- No interference from electromagnetic waves
- Safe to use
- Rugged
- Helps conserve copper

Since the fiber is made of glass and glass is made of sand, there is no limit to the material available for use in the making of fibers. The connections are important since they are a source of light loss. Take a look at Figure 16-12 to see how these connections are made.

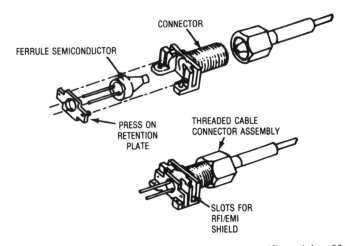

**Figure 16-12.** Latest standard devices used for fiber optics systems. (Copyright of Motorola, Inc. Used by Permission.)

# Sources of Light

In order for the system to operate, there must be a source of light. The light is modulated by the speech or data input and then is detected at the end of the line by a detector.

The laser DH light source is used as well as the high-radiance LED. The diode laser is preferred for moderate to wideband systems: 50 to 300 megahertz (MHz). It has certain advantages, such as fast response time of less than 1 nanosecond (ns) (0.000 000 001 second [s]). It is also used where the fiber is very small in diameter because it is easily coupled into the small fibers. The expected lifetime of the DH diode is 100,000 to 1 million hours (h) at room temperature.

LEDs are cheaper and need no temperature stabilizing. They have a slightly longer expected lifetime of 1 to 10 million hours. They fail gradually, which is not the case with the DH diodes.

The LED emits light as a result of the recombining of electrons and holes. It is a purely semiconductor device. The LED is a PN junction diode. When it is forward-biased, the minority carriers are injected across the junction. Once they cross, they combine with the majority carriers and give up their energy. The energy released is about equal to the material's energy gap. This process is called *radiative*. It does not happen with all materials, especially such materials as silicon. The LED used here is made of gallium arsenide (GaAs), which does radiate its energy. It does have some inherent defects which show up after a while, and the light level emitted deteriorates gradually.

## Detectors

The detector (see Figure 16-13) is the device that converts the transmitted light back into electric energy. The detector is usually a PIN, or avalanche, photodiode.

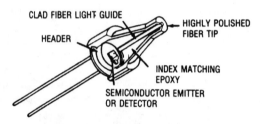

**Figure 16-13.** Fiber optics detector. (Copyright of Motorola, Inc. Used by Permission.)

The PN junction can be used to generate light in the LED. The PN junction can also be used to generate electricity when excited by light. When the PN junction is in the dark and reverse-biased, there is little current flow through it. If a light shines on the surface, the photoenergy of the light becomes absorbed. This causes hole-electron pairs to be created. If the carriers are created in or near the junction depletion region, they are swept across the junction by the electric field. This movement of charge carriers across the junction causes a current flow in the circuitry external to the diode. This current flow is proportional to the light power absorbed by the diode. This is how the light energy is converted back to electric energy.

# Uses for Fiber Optics

Wire systems of communications are being replaced by fiber optics. The fiber optics type is less expensive than the traditional systems. This is due primarily to the differences in terminals multiplexing equipment costs. The improvement in digital transmission alone makes it desirable to install the cables and systems. Undersea fiber optic cable systems are being constructed. These require some special engineering ability and maintenance procedures.

Optical fiber systems offer high bandwidth, freedom from external interference, immunity from interception by external means, and an inexpensive raw material, silicon.

# Summary

The laser is a device which uses radiant energy. It produces a narrow, highly intense beam of light that can be focused over long distances. Lasers are made in three types: solid state, gas, or semiconductor. Gas types are used for light shows by rock musicians. The solid state, or ruby, type is used by industry and the military for high-power intense light sources. The semiconductor type is used for producing light to be used in fiber optics systems.

There are many uses for lasers: industrial, military, surgical, and photographic to mention but a few.

Refraction and reflection of light inside a glass fiber becomes very important in a fiber optics system of communications. Refraction causes light to be bent. The speed reduction and the refraction are different for each wavelength in different materials.

Optical fibers are made of plastic or glass. The fiber is usually coated with plastic or glass. The core is the signal-handling portion of the fiber cable. Some light attenuation, or loss, in the fibers occurs, which limits their ability to handle long-distance communications without amplifiers. Fiber communications systems use only one fiber. All fibers are bundled for use in low-frequency, short-distance runs. Multiple-fiber cables can be used to enable multiple light sources to send information to a number of detectors. Multiplexing is often used, and more than one signal is transmitted on a single fiber.

The laser DH light source is used as is the high-radiance LED. The diode laser is preferred for moderate-band to wideband systems. The LED is a semiconductor PN junction device. It not only radiates light when properly stimulated, but it will also produce electric signals when light strikes it.

The PN junction LED is used as a detector for fiber optics systems of communications. Fiber optics hold much promise for future development.

# Review Questions

1. What does the word laser stand for?
2. How many types of lasers are there?
3. Which type of laser is used for fiber optics communication systems?
4. What is the frequency of visible light?
5. What is the difference between incoherent and coherent light?
6. Which type of light (coherent or incoherent) is used in lasers?
7. Of what is the DH semiconductor diode made?
8. What are some uses of the laser?
9. What is the difference between refraction and reflection?
10. What material is used to make optical fibers?
11. What is the main advantage of using the diode laser for moderate-band to wideband communications systems?
12. How much is a nanosecond?
13. How is light converted back to electric energy in fiber optics systems?
14. What is used as a detector in fiber optics?
15. List some uses for fiber optics.

# Chapter 17

# RADAR AND MICROWAVES

The word *radar* is a contraction of the expression *r*adio *d*etection *a*nd *r*anging. The italicized letters, as you can see, form the word *radar*. Radar is an application of radio principles to detect objects that cannot be observed visually and to determine the direction, range, and elevation.

Microwaves are those very short waves produced by extremely high and superhigh frequencies. Radar operates on these very short waves and thus is classified as microwave equipment. Microwaves are used for military, commercial, and police purposes. They are used for home cooking in the microwave oven. In this chapter we will take a look at these various applications of microwaves.

There are three distinct *bands* of frequencies where microwaves are used. The S band is where the microwave oven operates. It operates on 2450 megahertz (MHz). This means that the oscillator must complete a cycle 2450 million times per second. This was also the frequency used by police radar when it was first available in the 1950s. The military uses this S band for its Air Force land-based installations, and the Navy uses it for its land-based and some shipboard radar. The X band of frequencies are used by aircraft radar and by police speed radar equipment mounted outside the squad car on the window or inside on the dashboard. That X band radar for police use and for others in the experimental band is 10.525 gigahertz (GHz), or 10,525 MHz. They also use the K band (hand-held unit), which is in the gigahertz range. A gigahertz is 1000 megahertz (MHz). Since the gigahertz is 1000 MHz, it is 1000 million times per second. Police radar operates on 24.15 GHz. That amounts to 24,150 MHz (see Figure 17-1).

As you can see from these frequencies, they are not very high, or ultrahigh, but *super*-high in their designation. Remember the frequency designations?

- VHF (very high frequency) 30 to 300 MHz
- UHF (ultrahigh frequency) 300 to 3000 MHz
- SHF (superhigh frequency) 3000 to 30,000 MHz (or 3.0 to 30 GHz)
- EHF (extremely high frequency) 30,000 to 300,000 MHz (or 30 to 300 GHz)

## Principles of Operation for Radar

You have, no doubt, made a loud noise and heard an echo of that noise. Sometimes noises seem to come from different directions. Without looking, you know that there is a

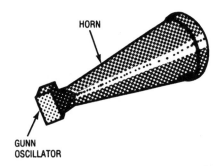

**Figure 17-1.** Microwave heart of police radar gun is the circularly polarized horn and the Gunn oscillator.

## TABLE 17-1

### IEEE RADAR BAND DESIGNATIONS

| Band | Frequency Range |
|------|-----------------|
| L | 1–2 GHz |
| S | 2–4 GHz |
| C | 4–8 GHz |
| X | 8–12 GHz |
| Ku | 12–18 GHz |
| K | 18–27 GHz |
| Ka | 27–40 GHz |
| V | 40–75 GHz |
| W | 75–110 GHz |
| mm | 110–300 GHz |
| $\mu$mm | 300–3000 GHz |

building, a hill, trees, or some other object nearby in the direction of the first echo, and other similar objects still further away in the direction of the second echo. Obviously the loud noise consisted of sound waves which started out at your location and, on striking the object, were reflected. As these reflected sound waves move by you, you hear them. According to the science of physics, sound travels at a constant rate. Therefore, it took a certain amount of time for the sound waves to get to the object and to return to you as an echo. And, still more time would have been required had the object been farther away. Furthermore, you are able to tell the direction of the object which caused the echo because of your ability to hear. With this faculty, your brain tells you that the object creating the echo is to the right, to the left, or in front of you.

Radar uses the same echo method to operate and tell you where an object is located. Radar units send out short, strong bursts of radio energy. Many of these bursts of energy strike objects in the vicinity and are reflected back to the site of the radar set (see Figure 17-2).

When the time required for the energy to go to the object and to return is carefully measured and translated into distance in miles or feet, it is possible to determine the distance of the object causing the echo. In early radar sets, two antennas were used to compare the strength of the reflected energy. From the comparison, it was possible to

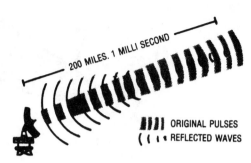

**Figure 17-2.** Determining distance by using radar.

determine the direction from which the echo came. For the same purpose, modern sets use various types of directional antenna. Thus, radar, like the method you used to determine the direction and distance of an object causing a sound echo, determines the distance and direction of an object creating a radio frequency echo.

The echo is called the *target* when used to designate the return in military installations. Modern radar techniques are highly sophisticated. Sets are unbelievably accurate and in many cases they are almost completely automatic. Yet, regardless of their stage of development, all depend upon this simple principle of creating and detecting a radio frequency echo.

The radar system consists of a transmitter which sends out the radio signals, a receiver which is located at the same site, and an indicator which gives a visual indication of echoes returned by a target.

When a radio signal which has a constant frequency is emitted by a radio transmitter, radio waves — in a manner similar to light and sound waves — travel out in all directions and are reflected by any object that they strike. On striking the object, components of the wave are reflected and likewise travel in all directions. Some of the reflected waves return to the site of the transmitter originating them, where they are picked up by the receiver, providing it is tuned to the correct frequency.

One of the problems which attends a reflected signal is having a powerful transmitter near the receiver and operating on the same frequency. In order for the receiver to detect the reflected signal and thus indicate the presence of the target, the transmitter signal must be prevented from affecting the receiver.

Another problem is measuring the distance. If a continuous constant frequency signal were transmitted, it would be impossible for the receiver to distinguish between various echoes at different distances, for they would all be alike. Thus, some means must be provided to eliminate the problems involved in using constant frequency signals.

Several systems satisfactorily solve these problems. They are the frequency modulation system, the frequency shift system, and the pulse modulation system. The last system is the most important and the most often used and will be discussed in detail here.

# Frequency Modulation System

In the frequency modulation system, two separate signals are fed to the receiver at the same time. For example, a signal transmitted at 440 MHz and a 420-MHz reflected signal reach the receiver simultaneously. When the two signals are mixed in the receiver, a beat note results. The frequency of the beat note varies directly with the distance to the object, increasing as the distance increases. A device that measures frequency can be calibrated to indicate range or distance to the object.

# Frequency Shift System

This system is based on the *Doppler effect* and is used in police radar. A familiar example is the change in pitch you hear in an automobile horn as the car approaches. Radio waves act in the same way. If the source of radio energy (in the case of radar, an object from which radio waves are reflected) is moving rapidly, the frequency of the radiated energy (the echo) changes. Radar using the Doppler effect to locate objects is based on the transmission of continuous or unmodulated radio waves toward the object. At the receiver the frequency of the waves reflected by the object is changed provided the object moves toward or away from the receiving point. Crosswise movement (which would describe a circle around the receiving point) would not change the frequency. The amount of change is proportional to the speed at which the object is moving toward or away from the receiving point. Since the receiver is located near the transmitter, it receives a signal from both the transmitter and the remote object. When the object is moving toward or away from it, the signal that is reflected is different from the transmitter frequency. The detector in the receiver responds to the difference in frequency. If the object is not moving or if it is moving crosswise to a radius drawn through it, the returning frequency is the same as the transmitter frequency, and the detector response is zero. It is therefore impossible by the frequency shift method to detect objects which are not approaching or moving away from the receiver.

The frequency shift principle is sometimes used in conjunction with a pulsed radar set to eliminate echoes from stationary objects. For example, at air bases in mountainous areas or in a city location with many buildings, the mountains or buildings cause radar echoes so strong (called *ground clutter*) that the weaker echoes from aircraft flying in the vicinity are obscured on the radar screen. By applying the frequency shift principle, we can differentiate the moving objects from the stationary objects. The echoes from the stationary objects can be eliminated, and the radar operator is able to see only the flying aircraft. In practice, the radar set itself uses the pulse system and the frequency shift detector device is a supplementary component attached to the radar set. This device is called a *moving target indicator*.

# Pulse System

This is the system used in almost all radar sets. In this system, the transmitter is turned on for short periods and off for long periods. During the period when the transmitter is turned on, it transmits a short burst of energy called a *pulse*. When a pulse strikes any object, part of the reflected energy is returned to the receiver, where it is displayed on the screen of a cathode ray tube. The cathode ray tube is a device capable of measuring periods of time as short as one-millionth of a second (a microsecond). Since the transmitter is turned off after each pulse, it does not interfere with the receiver as would be the case if a constant signal were used.

Complete location of an object in space by radar pulses depends upon two factors — the range or the distance of the target, and the direction, including both the azimuth and the elevation directions of the target.

# Range Finding

The pulse-modulated radar set measures distance in terms of time. When the radio energy is radiated into space, it continues to travel with a constant velocity. Its velocity is that of light, or about 186,000 miles per second (mi/s). In more useful terms, radio waves travel a mile in 5.38 microseconds (μs). They can go out a mile and return in 10.76 μs.

This constant velocity is used in radar to determine the distance or range of a target by measuring the time required for a pulse to travel to a target and return. Suppose, for example, a pulse of radio energy is transmitted toward a target some distance away and the radar echo returns in 538 μs. Since energy moves a mile and back in 10.76 μs, the distance of the target is

$$\frac{538}{10.76} = 50 \text{ mi}$$

In order to use the time-range relationship, a time-measuring device must be used. The cathode ray tube is useful for this purpose since it responds to changes as rapid as 1 μs apart. A time base is provided by using a linear sweep to produce a known rate of motion of an electron beam across the screen of the cathode ray tube.

Figure 17-3 shows the formation of the time base. In ①, a radar pulse is leaving an airplane. At the time the pulse is radiated, the spot on the screen of the cathode ray tube is deflected vertically for a brief instant, then it continues across the screen to the right. In ② and ③, the pulse is traveling toward the target, and the spot is moving across the screen. When the pulse strikes the target there is no deflection since energy is at the target itself. In ④ the reflected pulse is returning. In ⑤ the reflected energy has returned to the

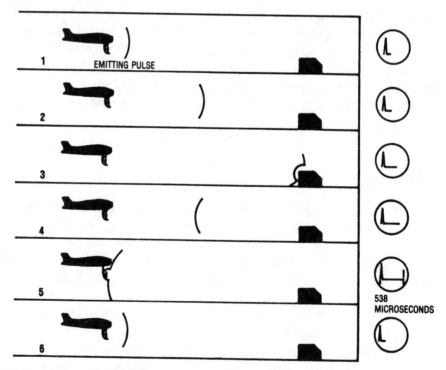

**Figure 17-3.** Formation of a time base.

receiver and there is a second vertical deflection of the spot on the right side of the cathode ray screen. The distance between the two upward reflections serves as the basis for determining the range of the target from the radar antenna. Assume, for example, that the set is designed so that the spot moves across the cathode ray tube in 700 $\mu$s. The spot was almost to the end before the echo arrived, its position indicating that it took 538 $\mu$s to reach that point. Since radar waves travel at 10.76 $\mu$s, the range of the target shown is 538 divided by 10.76, or 50 mi away. The last part of the illustration, ⑥, shows another pulse being transmitted and the start of the formation of a new time base.

## Finding Direction

Two dimensions must be considered when finding direction with radar echoes. One is azimuth, which is the relative horizontal direction of the target with respect to some direction reference expressed in degrees. For example, this direction may be expressed with reference to true north if the radar set is a ground installation or with reference to the heading of the airplane if the set is airborne. The other dimension is elevation, which, like azimuth, can be expressed in degrees. Elevation expresses the angular degrees that the target is below or above the radar set.

The determination of azimuth and elevation depend upon the directional characteristics of antennas and antenna arrays. Antennas for performing these functions are discussed later.

# Required Elements of a Pulse System

The basic elements in a typical pulse radar system are the timer, modulator, antenna, receiver, indicator, and transmitter (see Figure 17-4).

## Timer

Another name for the timer is *synchronizer*. It is the heart of all pulse radar systems. Its function is to ensure that all circuits connected with the radar system operate in a definite time relationship with one another and that the interval between pulses is of the proper length. The timer may be a separate unit by itself, or it may be included in the transmitter.

## Modulator

The modulator is usually a source of power for the transmitter. It is controlled by the pulse from the timer. It is sometimes called the *keyer*.

## Transmitter

The transmitter provides radio frequency (RF) energy at an extremely high power for a very short time. The frequency must be superhigh to get many hertz into the short pulse.

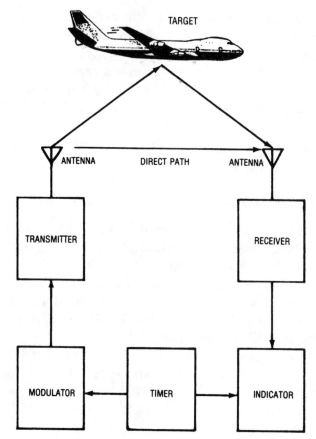

**Figure 17-4.** Pulse radar system block diagram.

# Antenna

The antenna is very directional in nature because it must obtain the angles of elevation and bearing of the target. To obtain this directivity at centimeter wavelengths, ordinary dipole antennas are used in conjunction with parabolic reflectors. Usually, in order to save space and weight, the same antenna is used for both transmitting and receiving. When this system is used, some kind of switching device is required for connecting it to the transmitter, where a pulse is being radiated, and to the receiver during the interval between pulses. Since the antenna only "sees" in one direction, it is usually rotated or moved about to cover the area around the radar set. This is called *searching*. The presence of targets in the area is established by this searching.

# Receiver

The receiver in radar equipment is primarily a superheterodyne receiver. It is usually quite sensitive. When pulsed operation is used, it must be capable of accepting signals in a bandwidth of 1 to 10 MHz.

# Indicators

The indicators present visually all the necessary information to locate the target on the indicator screen. The method of presenting the data depends on the purpose of the radar set.

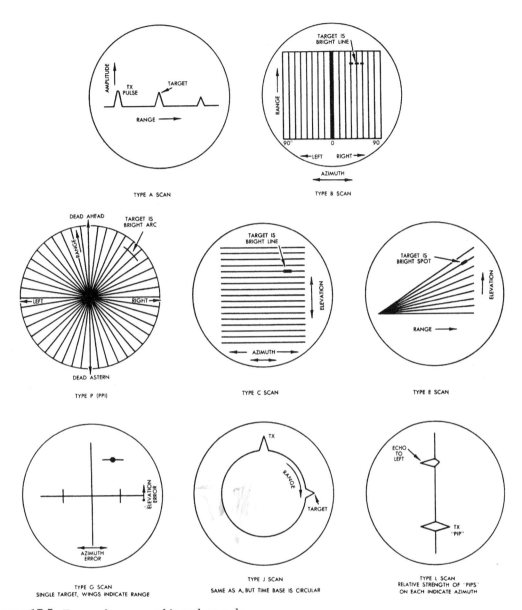

**Figure 17-5.** Types of scans used in radar work.

Since the spot "scans" the indicator screen to present the data, the method of presentation is often referred to as the *type of scan*. Figure 17-5 shows the types of scans available for use with radar sets.

# Power Requirements of a Pulse System

The maximum range which a radar set can attain obviously depends largely on the power output of the set. Enough power must be radiated so that at the maximum range the received echo signal will have a power level at least equal to the electronic noise level at the receiver.

Radar systems have been developed to the degree where they transmit the largest peak power ever radiated by any type of radio-transmitting equipment. This peak power is sometimes as high as 10 megawatts (MW), and seldom less than 20 kilowatts (kW).

In considering power requirements, you must distinguish between two types of power output — peak power, which is the power during a pulse, and average power, which is the average power over the pulse repetition period. While peak power is very large, average power may be small because of the great difference between pulse duration and pulse interval. The ratio of the pulse length to the pulse repetition period is known as the *duty cycle*.

# Waveguides and Cavity Resonators

The high frequencies of the microwaves make it impractical to use the ordinary coaxial cables normally used in radio frequency equipment to connect one piece of equipment to another, such as the output stage to the antenna. That is why a special type of device was designed for radar frequencies and ultrahigh television frequencies.

## Waveguides

Electromagnetic fields can transfer energy in a line which does not have a center conductor, provided the configuration of the fields is changed to compensate for the missing conductor. The area remaining, which is virtually a hollow pipe, is called a *waveguide*. A waveguide does not necessarily have to be circular in cross section. Practical waveguides, for example, are sometimes square, rectangular, or elliptical in cross section. Metallic walls are not necessary to guide electromagnetic fields in a waveguide, for the fields will be reflected whenever they encounter any kind of constant other than the substance in which they are traveling. For example, fields can be made to travel through a ceramic rod with little loss of energy. When they encounter the air at the surface of the rod, they are reflected back into the rod. Figure 17-6 shows some of the bends as they are produced for turning the signal in a waveguide without loss in signal strength. The waveguides bolt directly to these bends and carry the signal much the same a piece of coaxial cable carries the lower frequencies.

In order for energy to move from one end of a waveguide to the other without reflections, the size, shape, and dielectric material of the waveguide must be constant throughout its entire length. Any abrupt change in its size or shape results in reflections. Therefore, if no reflections are desired, any change in the direction or in the size of the waveguide must be gradual. When it is necessary that the change in direction or size be abrupt, then special devices, such as bends, twists, joints, or terminations, are used. Figure 17-6 shows the bends used for microwaves.

## Cavity Resonators

In most radio frequency circuits a resonant circuit consists of a coil and a capacitor connected in series or parallel. In order to vary the frequency of the circuit, it is necessary to

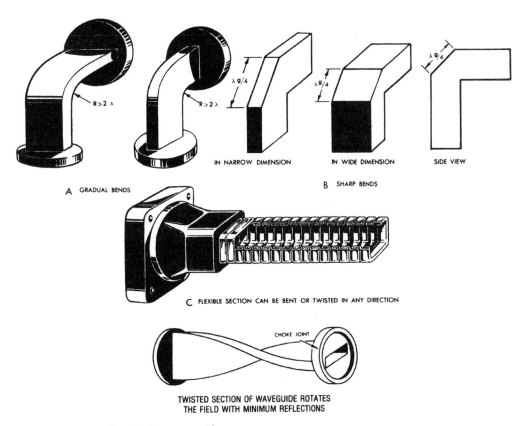

Figure 17-6. Types of bends for waveguides.

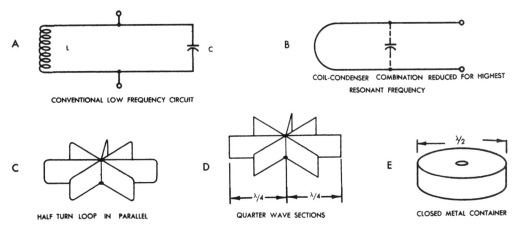

Figure 17-7. Development of a cavity for radar work.

change either the inductance or the capacitance. See Figure 17-7 for the development of a cavity from quarter-wavelength sections. In changing the coil and capacitor, you can reach a point where the inductance is a half-turn coil and where the capacity consists only of the stray capacity in the coil. At extremely high frequencies (where we are in radar operation) this resonant circuit would consist of a coil about an inch long and a quarter inch across. In this particular circuit, the current-handling capacity and breakdown voltage for the spacing would be low.

You can increase the current-carrying capacity of a resonant circuit by adding half-turn loops in parallel. This does not change the resonant frequency appreciably because it adds capacity in parallel and that lowers the frequency and the inductance in parallel which increases the frequency. As the effects of each cancel, the frequency remains about the same.

In Figure 17-7C several half-turn loops are added in parallel. In Figure 17-7D several parallel quarter-wave Lecher lines are seen. The Lecher lines are resonant when they are near a quarter-wavelength. When more and more loops are added in parallel, the assembly eventually becomes a closed, resonant box as seen in Figure 17-7E. This closed container is a quarter-wave in radius, or in other words, a half-wave in diameter. This box is called a *resonant cavity*.

A resonant cavity displays the same resonant characteristics as a tuned circuit composed of a coil and a capacitor. In it there are a larger number of current paths. This means that the resistance of the box to current flow is very low and that the $Q$ (merit) of the resonant circuit is very high. It is difficult to attain a $Q$ of several hundred in a coil of wire because $Q$ is found by dividing the $X_L$ by the internal resistance of the coil. It is fairly easy to construct a resonant cavity with a $Q$ of many thousand. Although a cavity is as efficient at low frequencies as at high frequencies, the large size required at low frequencies prohibits its use at those frequencies. For example, at 1 MHz, a resonant cavity would be a cylinder about 500 ft in diameter. When the frequency is in the vicinity of 10 GHz, the diameter of the cavity is only 0.6 in. This makes the cavity smaller than a conventional tuned circuit. Therefore, equipment which operates at a frequency of 3000 MHz or above usually employs resonant cavities as resonant circuits.

Cavities may have various physical shapes (see Figure 17-8). Any chamber enclosed in conducting walls resonates at several frequencies and produces a number of modes. The $Q$ of each cavity in Figure 17-8 is indicated. Of those shown, the cylinder cavity is useful in wavemeters or in frequency-measuring devices. The cylindrical ring is used in superhigh frequency oscillators as the frequency-determining element. The section of waveguide which is shown diagrammatically is used in some radar systems as a mixing chamber for combining signals from two sources. Resonant cavities are used in many ways in radar. The klystron and the magnetron both use resonant cavities to produce the needed frequencies for operation of a radar set.

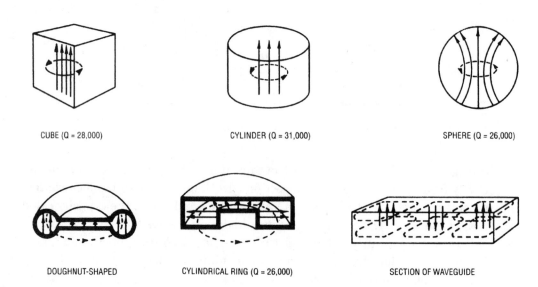

CUBE (Q = 28,000)　　　　CYLINDER (Q = 31,000)　　　　SPHERE (Q = 26,000)

DOUGHNUT-SHAPED　　　CYLINDRICAL RING (Q = 26,000)　　　SECTION OF WAVEGUIDE

**Figure 17-8.** Types of cavities.

# Radar Antennas

An *antenna* is a device that is used either for radiating electromagnetic energy into space or for collecting electromagnetic energy from space. In the radar transmitter the magnetron generates the high-frequency signal, but the antenna is needed to change this signal into electromagnetic fields which are suitable for propagation into space. The radar receiver will amplify any signal that appears at the input terminals, but an antenna is required to intercept the electromagnetic fields that are in space to change these fields into a voltage which the receiver can interpret.

## Antenna Reciprocity

Separate antennas are seldom required for both transmitting and receiving. Any antenna transfers energy from space to its input terminals with the same efficiency with which it transfers energy from the output terminals into space, assuming, of course, that the frequency is the same. This property of interchangeability of the same antenna for transmitting and receiving operations is known as *antenna reciprocity*. Antenna reciprocity is possible chiefly because antenna characteristics are essentially the same regardless of whether the antenna is sending or receiving electromagnetic energy. Because of antenna reciprocity, most radar sets installed in aircraft use the same antenna both for receiving and transmitting. The same is true for large installations of ground equipment. An automatic switch in the radio frequency line first connects the single antenna to the transmitter and then to the receiver, depending upon the sequence of operation. Because of reciprocity of radar antennas, this chapter treats antennas from the viewpoint of the transmitting antenna with the understanding that the same principles apply equally well when the antennas are used for receiving electromagnetic energy.

## Directional Properties

Usually, the most important characteristic of a radar antenna is its directional property, or simply its directivity. Directivity means that an antenna radiates more energy in one direction than the other. For that matter, all antennas are directional, some slightly, and others almost entirely. In radar operations, some antennas are required to send all energy in one direction in order that as much as possible of the electromagnetic energy generated by the transmitter will strike an object in a given direction. In other systems, it is desirable for the energy to be radiated equally well in all directions from the source. An example of an antenna system in which radar radiates energy in a given direction is the airborne navigation and bombing set. In this set, there is only a limited amount of power available at the transmitter. In order to achieve maximum benefit from this minimum power, all of it is sent in the same direction. Since the antenna in this set is also used for reception, it likewise receives electromagnetic energy only from one direction. Because of design features, it is possible to tell the direction of an object at which this directional antenna is sending energy or the direction of the object from which the antenna is receiving energy. Furthermore, the physical position of the antenna is indicative of the direction of the object. An example of a nondirectional radar antenna is the antenna installed in the radar beacon. This antenna must receive energy equally well in all directions in order that a radar-equipped airplane can ascertain its position regardless of its direction from the beacon antenna.

# Wave Propagation into Space

The electromagnetic field set up by an antenna travels away from the antenna in a straight line. There are many parts to this field and many directions in which energy travels. Figure 17-9 shows four of the large number of paths which energy can take. These and other paths which energy takes affect reception of radiated energy. If there is nothing between the emitting antenna and the receiving station ($R_1$), some energy will travel directly to it by way of path 3. Receiving station $R_2$ cannot receive energy because the earth is between the two points and because the energy cannot go through the earth any distance. Some of the energy will follow path 2 out into space. At a height some 60 mi above the earth is a heavily ionized layer of atmosphere. This layer, called the *ionosphere*, constitutes a change of media through which the energy must travel, and is sufficient to refract a wave. In the case of path 2, the refraction is sufficient to bend the energy wave out of the ionosphere back to the earth. Therefore, receiving station $R_3$, which is beyond the horizon and not located for receiving a direct ray, receives the reflected ray from the ionosphere.

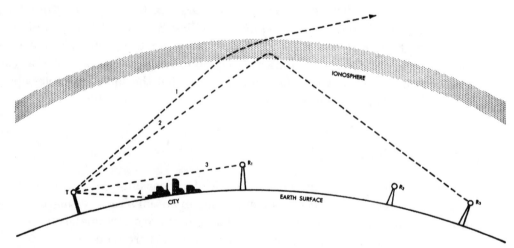

**Figure 17-9.** Effectiveness of radiated energy from an antenna.

It is the ionosphere that makes possible round-the-world communications. If the angle of incidence of an energy wave with the ionosphere is too great, the angle refraction may not be sufficient to bend the energy back toward the earth, and it may go on through the ionosphere and be lost, as seen in path 1. Paths 3 and 4 follow along the surface of the earth and are called *ground waves*. The waves bent back to the earth by the ionosphere are called *sky waves*.

Generally, medium- and high-frequency waves are reflected readily, whereas ultrahigh and superhigh frequencies are more likely to be lost through penetration of the ionosphere. Airborne radar operates in the superhigh-frequency (3000 to 30,000 MHz) part of the spectrum. Irregularity of the sky waves makes them unreliable for ground-based radar use.

Therefore, ground-based radar uses only the direct ground wave. Path 4, for example, where the ray goes directly to some object and returns by a direct path, is a direct ground wave.

## Parabolic Reflector

Radar uses the half-wavelength antenna. This *dipole* is used in front of a parabolic reflector most of the time. The parabolic reflector, Figure 17-10, takes advantage of the radio waves' ability to act the same as light waves. These radio waves can be focused into a narrow beam with a metal parabolic reflecting surface by placing the radiator at the focal point. It is important to shape the reflector properly so that the emitted wave can be formed into the desired 2° horizontal width and 15° vertical beam height.

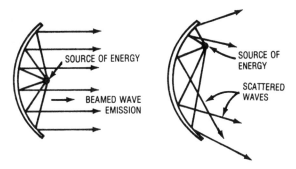

**Figure 17-10.** Parabolic reflectors for microwaves. Note the difference in A and B. A. Parallel beams from a properly located energy source. B. Scattered waves of energy from improperly located energy source.

# Uses of Radar

There are many uses of radar — most of them military applications. However, commercial airliners and ships at sea also use radar for finding where they are and how to get where they are going safely. We even make use of radar principles in the microwave oven to speed up the cooking of foods. Another rather important use of radar is in the enforcement of speed zones on the nation's highways.

Every time someone develops an electronic device to aid the enforcement of rules or laws, another device is also made which can be used to detect the presence of the enforcer. This is the case with police radar. There are receivers which can perceive the signal from the police radar *before* it can detect your car and register your speed. Such a receiver is a very sensitive device and represents some very advanced technology. Improvements made in this field every day keep the research and development engineers busy trying to keep up.

# Radar Detector

Figure 17-11 shows the schematic of a radar detector used to detect the presence of the X band and S band radar used by the police radar units. This particular unit is now obsolete

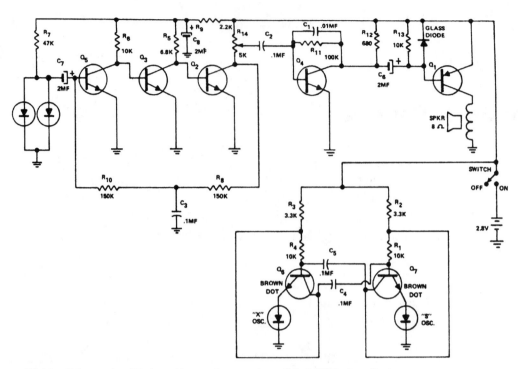

**Figure 17-11.** Schematic of Driver Alert radar receiver. (Mark IV Industries)

in terms of its front end, but serves to give you some idea as to how a continuous wave (CW) is detected and made to cause an audible alarm to sound. It also shows the use of the wave-guide antenna for detection and amplification plus the tuned slot for the S-band reception.

# Operation of the Circuit

The Driver Alert is a dual-band receiver for reception on the X and S bands. The X band covers 8200 to 12,400 MHz, and the S band covers 1700 to 2600 MHz. It is a self-contained unit with its own battery power supply (see Figure 17-12).

A resonant-slot antenna (visible in the exploded view, Figure 17-13) on the back panel of the case is tuned to the S band (or in particular 2455 MHz) which speed meters use. Whenever a microwave signal in the frequency range is received at the antenna, it is passed through a diode. Instantaneous forward conduction of the diode inhibits the resonant condition in the resonator. Therefore, when the diode is not conducting, it allows the resonant cavity antenna to build up to resonant conditions.

**Figure 17-12.** Driver Alert radar receiver. (Mark IV Industries)

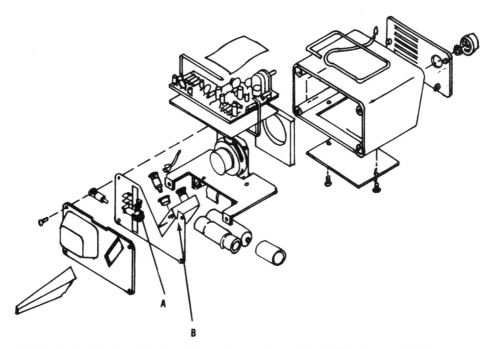

**Figure 17-13.** Exploded view of Driver Alert radar receiver. (Mark IV Industries) A. Resonant slot for S Band. B. Resonant cavity for X band.

Once the microwave signal is detected by the detector diodes, it is chopped into bursts of relatively short duration. These bursts, when detected, become an audio signal which is amplified by transistors $Q_1$ and $Q_4$. These transistors form an extremely high gain amplifier which amplifies up to 1000 times. $R_{14}$ is the volume control and controls the amount of signal fed to the audio amplifier stages.

The X band uses a tuned cavity antenna which receives the signal through a dielectric or microwave lens (the plastic rod which protrudes from the unit in the back). Since the antenna is mounted at a 45° angle, it is effective on both horizontally and vertically polarized signals.

Once the signal is received at 10,525 MHz, it is fed to the radar diodes for the X band and causes instantaneous forward conduction of the diode. This means that the resonant conditions of the tuned cavity are shorted by the diode, and the cavity ceases to function. Therefore, the diode has no signal to conduct and returns to a nonconduction state. As this happens, the resonant conditions of the tuned cavity return, and if a signal is present, the diode once again conducts, shorting the resonant cavity and repeating the previous operation. By conducting and not conducting, the diode causes short bursts to be fed to the flip-flop oscillator, causing it to oscillate at about 700 Hz, or in the audio range. This is amplified by the audio amplifier transistor to drive a speaker.

There is no adjustment for the sensitivity of the unit. If you turn on the *on-off* switch which is part of the volume control, you can hear a slight background noise. This sounds somewhat like radio static. Do not turn it up louder than necessary to hear. The volume control affects only the loudness of the signal produced and does not change the range sensitivity of the unit. Increasing the volume also increases the battery drain and shortens the life of the battery, just as it does in a transistor radio. Once the unit intercepts a radar beam it will automatically produce a warning sound similar to that which accompanies a TV test pattern. This gets loud very rapidly as you approach the source of the radar beam. It will stop as you pass out of the range of the transmitter.

# Microwave Ovens

Historically speaking, the microwave oven has not been around for long. In 1953 the first mass-produced microwave oven, offered by Raytheon Company for restaurant and hospital use, sold for $3000. This was an outgrowth of Raytheon's experience in making radar tubes for military radar during World War II. In 1958 vending machine operations were expanded by the use of push-button microwave ovens, bringing the price down to $2000. Then, in 1967 Amana, owned by Raytheon, brought out the countertop model for use with 115 volts (V) ac. Since 1974 the boom in sales has brought almost every household into the microwave oven age.

## Speed

Speed is the reason why people buy the microwave oven. They want to cook quickly. Combined with today's prepackaged foods, microwave cooking is dramatically quick and easy. A microwave can save up to 70 percent of cooking time in many instances. Microwaves instantly penetrate the food to a depth of about 1.5 in. for meats and 3 in. for other foods. Thus, the food starts to cook as soon as the oven is turned on.

## Cooking with Microwaves

Microwaves generate heat inside the food to cause it to cook faster. In conventional cooking, the heat is applied from outside. Microwaves cause millions of molecules of food to rub against one another. The rubbing of the molecules together generates heat by friction in the same way that heat is generated in two sticks of wood when rubbed together.

The frequency of operation of the microwave oven is 2450 MHz. This means that the molecules of the food are moved back and forth 2.45 billion times per second. Now it is possible to see why there is friction and heat generated.

## Generating the Frequency

Microwaves are generated the same in the microwave oven as in the radar sets previously discussed. A magnetron is used to generate the 2450-MHz operating frequency.

Figure 17-14 shows how the microwaves are moved from the magnetron and its antenna through the resonant cavity to the cooking cavity. A stirrer is used to bounce the waves and make them penetrate all parts of the cooking area.

Figure 17-15 shows how the magnetron is built. Note some of the areas where it can fail or cause trouble.

Note in Figure 17-16 in the schematic of a simple microwave oven how the magnetron high voltage is generated. A 2300-V secondary of the transformer produces the high voltage, but the magnetron needs dc. This is where the voltage doubler circuit is used. The voltage doubler is made up of the diode, high-voltage capacitor, and varistor. The magnetron needs about 3800 V. The charging of the capacitor and the rectification of the charging action produce the needed voltage from the original 2300 V. Note that the negative output of this circuit is attached to the filament of the magnetron and that the positive end is grounded as is the plate of the magnetron.

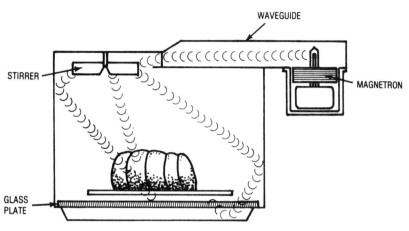

**Figure 17-14.** Microwave oven operation. (AMANA)

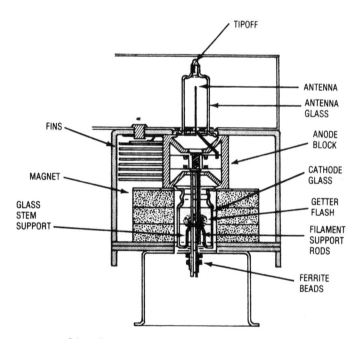

**Figure 17-15.** Magnetron used in microwave ovens.

By redrawing the circuit (see Figure 17-17), you can see how the high voltage is produced by the voltage doubler network. It would take two capacitors and two diodes to actually double the voltage, but in this case the 3800 V is all that is needed, and the one diode and capacitor is sufficient. The varistor is used to shunt the diode and to protect it in case the voltage becomes too high. The current is directed around the diode when the varistor breaks down or lowers its resistance due to extremely high voltage spikes. As mentioned before, this is the simplest of the microwave ovens schematically; you can easily see the essential parts. Other electronics packages have been added to boost sales and to appeal to those who have more money to spend. Some of the expensive additions include the timer and digital readouts of the time and cooking sequence. Basically, however, the cooking cavity and magnetron are the same.

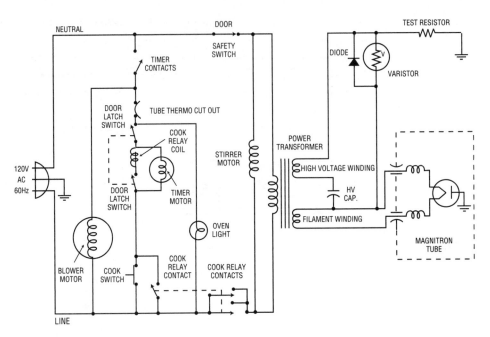

**Figure 17-16.** Schematic of a typical microwave oven.

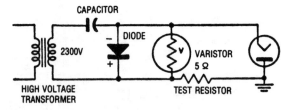

**Figure 17-17.** High-voltage circuit to produce magnetron plate potential.

## Summary

The word *radar* is a contraction of the expression *ra*dio *d*etection *a*nd *r*anging. Microwaves are those very short waves produced by extremely high and superhigh frequencies. Radar operates on these short waves and thus is classified under microwave equipment. Microwaves are used for military, commercial, and police purposes. They are also used for home cooking in the microwave oven.

Microwaves operate on three bands of frequencies. The S band is where the microwave oven operates; it operates on 2450 MHz. The X band and the K band are shared by many, but are used most often by the police to check the speed of drivers.

Radar uses the echo method to locate an object. Radar units send out short bursts of radio frequency energy and then wait for an echo to return. They then check the time and compute it into distance.

Problems inherent in radar are solved in a number of ways to the benefit of the user of the equipment. Frequency modulation, frequency shift, and the pulse systems are used to check the echo for distance and direction. The pulse system is used by high-power radar.

The frequency shift (also known as the Doppler effect) system is used in police radar and anywhere the target is moving toward or away from the source.

Radar displays its information on any number of indicators, but the cathode ray tube is the most accurate and quick to respond to changes. Antennas are usually of the half-wave variety. Resonant cavities and waveguides direct the energy toward the antenna for propagation into space.

Resonant cavities take the place of tuned circuits consisting of a coil and a capacitor in the lower-frequency circuits.

Radar antennas can be used for both transmitting and receiving. This calls for a T/R switch. The T/R switch is an electronic switch that changes the antenna from the *transmit* to the *receive* position.

Radar uses only ground waves since the energy it generates is not reflected with predictability by the ionosphere. A parabolic reflector is used to beam the energy from the antenna into comprehensive beams that are parallel to one another. It also serves as a directional device.

Very sensitive radar receivers can be made to detect the transmitted signal before it is utilized by the transmitting and receiving unit.

Microwave ovens use the same microwaves as radar. They cause the food molecules to rub against one another at a very high rate (2.45 billion times per second). This friction causes the food to heat up and become cooked. Microwave ovens are only one of the many uses for radar.

# Review Questions

1. Where did the term radar originate?
2. What are microwaves?
3. On what band is 2450 MHz located?
4. What is the X band?
5. What familiar device uses the K band?
6. What is the difference between superhigh and extremely high frequencies?
7. What is the Doppler effect?
8. What is the pulse system of radar?
9. How much is a microsecond?
10. What are the required elements of a pulse system?
11. Describe a radar indicator.
12. What is a magnetron?
13. What does a T/R switch do?
14. What is a waveguide?
15. What is a resonant cavity?
16. What is antenna reciprocity?
17. What is the most important characteristic of a radar antenna?
18. What is line-of-sight transmission?
19. What is a parabolic reflector?
20. How does a microwave oven cook food?

# Chapter 18

# COMPUTERS

T he world of computers is a fascinating one. Today, the personal computer has brought the advantages of the electronic computer to within the grasp of everyone. Schools are offering courses in the elementary grades, and local clubs offer all types of programs and information. Therefore, it would be rather difficult within the space limitations we have here to go into the programming and the languages available. What we will do is look at some of the basics of computing and its history, and at how electronics does the job so fast.

# History

People have been looking for easier ways to count ever since they had to keep track of things. The digital computer is not a new idea. It is merely an adaptation of something used for years in the Orient. The electron and its speed makes it possible for everyone to use the digital method of keeping track of things.

The *abacus* is a manually operated digital computer used in ancient civilizations and utilized to this day in the Orient. For those who think the computer is fast you only have to put a good abacus operator beside a computer operator to see which device is faster. The abacus wins almost every time.

The first adding machine was invented by Blaise Pascal, a Frenchman, in 1642. Then in 1662, an Englishman, Sir Samuel Morland, developed a more compact device which could multiply, add, and subtract. In 1682, Wilhelm Liebnitz, a German, perfected a machine which could perform all the basic operations (add, subtract, divide, multiply), as well as extract the square root. Liebnitz's principles are still in use today in our modern electronic digital computers.

# Electronics

Electronics arrived on the scene of computing in 1919 when W. H. Eccles and F. W. Jordon described in an article the electronic "trigger circuit" that could be used for automatic counting. But the Eccles-Jordon multivibrator was a little ahead of its time. Today, every digital computer uses these circuits, known as flip-flops, to store information, perform arithmetical operations, and control the timing sequences within the computer (see Figure 18-1). These flip-flops are part of a setup called a *bistable circuit*. Since the digital computer does binary arithmetic, the use of only two digits, 0 and 1, can easily represent electrically any device having an *off* and an *on* stage. This is how circuits like the bistable circuit work.

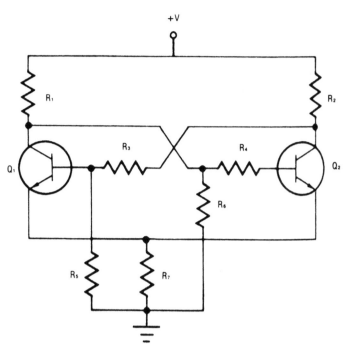

**Figure 18-1.** Flip-flop multivibrator circuit.

The bistable multivibrator operates in two states. (Figure 18-1 shows the schematic of this type of circuit.) Because of its actions, the bistable multivibrator is used as a master timing circuit in computers and in many home entertainment products.

Take a closer look at the schematic. $Q_1$ and $Q_2$ depend upon each other for the supply of base current. Assume that $Q_1$ is conducting when power is applied. This negative voltage at the collector of $Q_1$ is transferred to the base of $Q_2$ by way of $R_4$. This negative voltage reduces conduction through $Q_2$. The reduction causes the collector selector voltage of $Q_2$ to rise. This action continues until $Q_1$ is fully on and $Q_2$ is at cutoff. To change states, an external pulse is fed to the base of $Q_2$. This positive pulse developed at $Q_2$ starts $Q_2$ conducting. This action turns off $Q_1$. The multivibrator has changed states. To change states again, a positive pulse must be applied to the base of $Q_1$. (*State* means on or off.)

# Types of Computers

There are two types of computers: the analog and the digital.

## Analog Computers

Analog computers may represent data with a range of signal voltages. The range may be from –15 to +15 volts (V). An infinite number of actual signal levels may be used in analog computers. Analog computers are not as accurate as their digital counterparts. This is because a slight change in component values may make a large difference in output data. Analog signals also have a tendency to drift.

## Digital Computers

Digital computers represent data by using two signal levels. The levels are typically either 0 or +5 V.

Digital computers therefore do not have the drift problem that analog computers have. When the voltage drifts or changes from one value to another by one-tenth of a volt (0.1 V), it can give a different result in an analog computer. The digital is either on or off, and drift does not produce as great a problem.

Most problem-solving programs are now being written for digital computers. This is partly because a large number of programming aids has been developed for digital computers. Digital systems are far more popular than analog.

# Computer Circuits

Switches, relays, and diodes can be used to create logic circuits. The transistor can be used as a rapid, reliable switch. Figure 18-2 shows a basic AND circuit. The lamp is the output. Switches A *and* B must both be closed for the lamp to glow.

Figure 18-3 is a basic AND circuit with a transistor. When switches A and B are both closed, the transistor is energized. Conduction through the transistor causes current through $R$. This is in the same relative location as the lamp in Figure 18-2. The flow of current through $R$ produces a voltage drop. This voltage can then be passed along to the next processing stage in a computer system. In Figure 18-3 current can also be fed to another circuit by way of C. That circuit can also perform a function needed by a computer. Figures 18-2 and 18-3 both show AND circuits. The symbol for an AND circuit is shown in Figure 18-4.

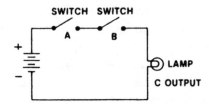

**Figure 18-2.** Basic AND circuit with lamp as output.

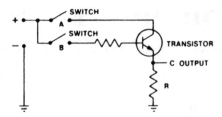

**Figure 18-3.** Basic AND circuit using a transistor.

**Figure 18-4.** Symbol for AND gate.

# Truth Table

Computers perform their functions on a binary basis. Binary values are 0 and 1. Binary functions of computers are based on switch conditions. Switches are either on or off. An *off* condition represents a 0. An *on* condition represents a 1.

A truth table shows how the inputs and outputs of a circuit are obtained. For instance, in Figure 18-2, if A and B are off, this means that C must be off. Since this is an AND circuit, both A and B must be closed for C to have an output.

Look at the truth table in Figure 18-5. The first line shows that A = 0, B = 0, and C = 0. On the second line, values indicate that A = 0, B = 1, and C = 0. Again, the lamp will not light. Remember that this is an AND circuit, and so both switches must be closed. The third condition shows A = 1 and B = 0; therefore, C = 0. Now look at the fourth line on the truth table. Both A and B have values of 1. Therefore, C = 1. Both switches are closed. Thus, current is delivered to the lamp.

| TRUTH TABLE | | |
|---|---|---|
| INPUTS | | OUTPUT |
| A | B | C |
| 0 | 0 | 0 |
| 0 | 1 | 0 |
| 1 | 0 | 0 |
| 1 | 1 | 1 |

**Figure 18-5.** Truth table for AND.

Since transistors and diodes are extremely small, it is possible to have these switching functions carried on in very small spaces. Complex problems can be solved by having a circuit either conduct or not conduct at the proper time.

A computer or calculator operates on pulses of electric energy. These pulses are called *bits*. The pulses can be converted to magnetic form and stored on memory devices: tapes or disks. In some memory systems, small bubbles are used to store bits of information. Magnetic tapes or disks can also be used to put in pulses needed to cause the circuit to operate.

## OR Circuit

Another computer building block is the OR circuit (see Figure 18-6). It has two switches in parallel and a lamp. If either A or B is closed, the lamp will glow. This feature identifies an OR circuit.

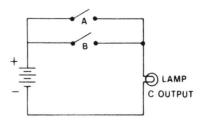

**Figure 18-6.** Basic OR gate with lamp as output.

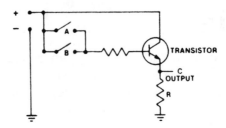

Figure 18-7. Basic OR circuit using a transistor.

Figure 18-7 is an OR circuit with a transistor. Closing either A or B will complete the transistor's base bias circuit. With either switch closed, the transistor receives and can conduct current. The transistor used in OR circuits is a sensitive device that can control the current in the load resistor $R$. With the base open, there will be little if any current through the transistor. When the base circuit is made positive, current (at a maximum or saturation level) will flow. The current flows through the transistor and the load resistor that is connected in series.

The symbol for an OR circuit is shown in Figure 18-8. Notice that it is shaped a little differently from the AND symbol.

A truth table for the OR circuits in Figures 18-6 and 18-7 is shown in Figure 18-9. If both switches, A and B, are open and have values of 0, then the output value, C, is also 0. If A = 0 and B = 1, then C = 1. If A = 1 and B = 1, then C = 1 as well. If A = 1 and B = 0, C = 1. Note that there are three conditions in this table in which C = 1. That is because it is an OR circuit. It can operate with either A or B closed. Remember, 1 means *on* or closed. Zero means open or *off*.

Figure 18-8. Symbol for OR gate.

| TRUTH TABLE | | |
|---|---|---|
| INPUTS | | OUTPUT |
| A | B | C |
| 0 | 0 | 0 |
| 0 | 1 | 1 |
| 1 | 0 | 1 |
| 1 | 1 | 1 |

Figure 18-9. Truth table for OR circuit.

# NOT Circuit

Another building block vital to the operation of computers is the NOT circuit (see Figure 18-10). A is the control switch; B is the output, a lamp. There are two power sources. One source controls the relay. The other controls the lamp by furnishing power when the relay contact is closed.

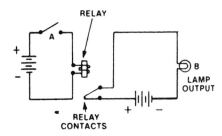

**Figure 18-10.** Basic NOT circuit with lamp as output.

The relay contacts **are** important. Notice that the relay contacts are closed when the switch is open. This **means** that A = 0 and B = 1. When switch A is closed, the relay energizes. This causes **the** relay contacts to open. Opening the contacts causes the lamp to go out. This is a **condition** represented by 0 in the logic circuits. Therefore, this is an *inverting circuit*. The **closing** of switch A causes lamp B to go out. This is the opposite of what usually happens **in** a circuit. Therefore, it is called an *inverting function*. (This circuit is also referred to **as an** *inverter*.)

Now look at the circuit in Figure 18-11. This is a NOT circuit with a transistor. Note that there are two power **supplies**. When switch A is open, no current flows in the circuit. When switch A is closed, **current** flows through the transistor to output B. From B, the output moves across a resistor *R*. In this type of circuit, the batteries have to be connected with the proper polarity. This connection of polarity determines whether or not the transistor conducts. Figure 18-12 shows the logic symbol for a NOT (inverter) circuit.

Figure 18-13 is a truth table. This shows only two conditions for this type of circuit. When A = 0, B = 1. That **is**, when A is off, B is on. The other condition is that when A = 1, B = 0. This means that **when** A is on. B is off. There are only two conditions the circuit has because it has only one **switch**. The switch can be either on or off.

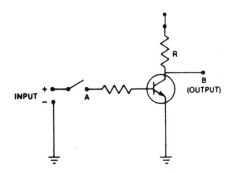

**Figure 18-11.** Basic NOT circuit using a transistor.

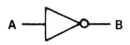

**Figure 18-12.** Symbol for **NOT** gate.

| TRUTH TABLE | |
|---|---|
| INPUT | OUTPUT |
| A | B |
| 0 | 1 |
| 1 | 0 |

**Figure 18-13.** Truth table for NOT circuit.

## NAND Circuits

A NAND circuit is an AND circuit with a negative structure. In effect, a NAND circuit does the same thing within a computer system as an AND circuit. The only difference is in the output. Outputs for a NAND circuit are exactly *opposite* from those of an AND circuit.

These relationships are shown in Figure 18-14. Included are the symbols for both the AND and NAND circuits. Figure 18-14 also presents truth tables for both types of circuits. Notice that the only difference in symbols lies in the circle added to the output side. This identifies a negative circuit structure.

In designing computer circuits, NAND structures are used because they are easier and less costly to make. Since NAND circuits can substitute for AND circuits, they are used in the majority of cases. Of course, with time, all this is subject to change as the technology improves the manufacturing techniques used for structures of this type.

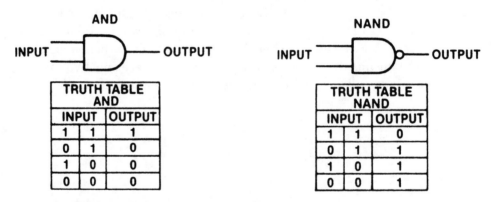

**Figure 18-14.** Symbols and truth tables for AND and NAND.

## NOR Circuits

A NOR circuit is a negative OR circuit (see Figure 18-15). This illustration shows that OR and NOR symbols are similar. The only difference is that the NOR circuit has a circle on the output. This indicates a negative circuit structure. Truth tables are also opposite for OR and NOR circuits.

The reasons for using NOR circuits are the same as those for using NAND circuits. That is, negative circuits are easier and less costly to manufacture. Therefore, where an OR function is needed, most computer manufacturers use negative circuits. Both OR and NOR circuits are available from semiconductor manufacturers. Using negative circuits saves money without sacrificing performance.

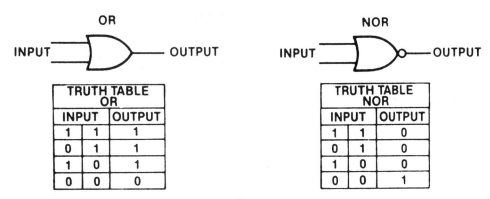

**Figure 18-15.** Symbols and truth tables for OR and NOR.

Five circuits have been reviewed in this section — AND, OR, NOT, NAND, and NOR. These represent the basic building blocks for computers and calculators. Various data can be stored on tapes, disks, or cards to cause pulses to operate the transistor. In all cases, the information is placed into the circuit as either a 1 or a 0. This binary coding makes it easier to direct a machine to make decisions. A computer is preprogrammed to react to commands from binary signals. The grouping of the circuits can cause a computer actually to make selections and comparisons.

# Binary Code

Uses for binary coding in computers have a number of names. The term *binary* is referred to as a bistable condition. This simply means that the computer circuits have two conditions. The bistable condition may be called on or off. It may also be described as either conducting or nonconducting. Still another set of descriptions is energized or deenergized. These terms are used to describe functions within a binary numbering or processing system.

Normally, we work with a base 10, or digital, numbering system. It takes 10 pennies to make a dime. It takes 10 dimes to make a dollar. The numbering system uses 10 digits, 0 through 9.

The binary system uses only two digits. In both the base 10 and the binary system, the position of the number is important. Consider the numbers 10 and 01. Both use a 0 and a 1. The *positions* of the numbers determine their value.

Scientific notation (powers of 10) is based on the positions of numbers. These offer a good example of the importance of the positions of numbers.

$$10^0 = 1 \qquad 10^1 = 10 \qquad 10^2 = 100 \qquad 10^3 = 1000$$

Now take the numeral 2. Raise it to various powers. This demonstrates the relationship between binary and decimal numbers.

$$2^0 = 1 \qquad 2^1 = 2 \qquad 2^2 = 4 \qquad 2^3 = 8$$

You should remember certain rules about this method of numbering. One rule is that any number raised to 0 power is 1. Also, any number raised to the first power is that number. Thus $2^0 = 1$, $3^0 = 1$, $10^0 = 1$; and $2^1 = 2$, $3^1 = 3$, $10^1 = 10$.

Through use of this principle, the binary numbering system can represent any quantity (see Table 18-1). Note how the relationships between decimal and binary values follow the same principle.

**TABLE 18-1**

**CONVERSION TABLE OF DECIMAL AND BINARY NUMBERS**

| Decimal Numbers | Decimal Value of the Binary Digit | | | | | |
|---|---|---|---|---|---|---|
| | $2^5$ (32) | $2^4$ (16) | $2^3$ (8) | $2^2$ (4) | $2^1$ (2) | $2^0$ (1) |
| 0 | 0 | 0 | 0 | 0 | 0 | 0 |
| 1 | 0 | 0 | 0 | 0 | 0 | 1 |
| 2 | 0 | 0 | 0 | 0 | 1 | 0 |
| 3 | 0 | 0 | 0 | 0 | 1 | 1 |
| 4 | 0 | 0 | 0 | 1 | 0 | 0 |
| 5 | 0 | 0 | 0 | 1 | 0 | 1 |
| 6 | 0 | 0 | 0 | 1 | 1 | 0 |
| 7 | 0 | 0 | 0 | 1 | 1 | 1 |
| 8 | 0 | 0 | 1 | 0 | 0 | 0 |
| 9 | 0 | 0 | 1 | 0 | 0 | 1 |
| 10 | 0 | 0 | 1 | 0 | 1 | 0 |
| 20 | 0 | 1 | 0 | 1 | 0 | 0 |
| 30 | 0 | 1 | 1 | 1 | 1 | 0 |
| 40 | 1 | 0 | 1 | 0 | 0 | 0 |
| 41 | 1 | 0 | 1 | 0 | 0 | 1 |
| 42 | 1 | 0 | 1 | 0 | 1 | 0 |
| 50 | 1 | 1 | 0 | 0 | 1 | 0 |
| 56 | 1 | 1 | 1 | 0 | 0 | 0 |

To convert binary notations to decimal values, just add the powers. For example, look at the decimal value 56. This is represented by 1 entry under each of the binary columns for 32, 16, and 8. Just add the values: $32 + 16 + 8 = 56$.

One of the advantages of binary coding is its simplicity. There would have to be 10 circuits to represent the decimal numbering system. In the binary system, only four circuits are needed to represent values of 0 through 9. With the binary system, computers can be smaller and less expensive than if the decimal system were used.

# Microprocessors, Personal Computers, and Special-Purpose Computers

Use of digital computing and data processing devices is growing rapidly. Part of the growth stems from the fact that costs have been reduced. These reductions have come with the introduction of printed circuit techniques. These methods have served both to improve efficiency and to reduce costs. Two important families of devices have made use of printed circuit techniques. These are microprocessors and personal computers (PCs). Microprocessors are individual printed circuit chips that perform specific processing functions. Each processor has a specialized job to do. An example of a microprocessor includes the so-called computer-on-a-chip. This is a miniature chip with circuits that perform arithmetic functions. Functions include addition, subtraction, multiplication, and division. Some processing chips are designed for communication. Other microprocessing functions include memory and logic. The microprocessor is a very valuable tool in the modern-day automobile which aids in a gas mileage and pollution control.

Personal computers are, generally, systems built around microprocessors and other printed circuits. Computer systems handle complete data or word processing jobs.

The personal computer integrates a number of machines for the control or production of complete jobs. For example, PCs are often used to control automatic electronic assembly equipment. In offices, PCs are used to process paperwork. PCs are also used in modern word processing systems. These systems generate letters and other business documents. (Figure 18-16 presents one of many types of computers available for home use.)

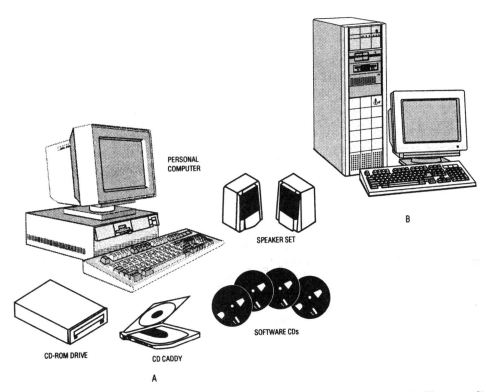

**Figure 18-16.** A. Personal computers with multimedia are available for home use. B. Floorstanding models are also available for "work at home."

# Microprocessors

A *microprocessor* is a logic device that can be programmed. Its function or logical operation can be altered by inputting instructions.

The term *microprocessor* has come to mean the central processing unit or CPU of a small computer system. The microprocessor cannot function by itself. However, once it is combined with a few support circuits, its characteristics are such that it can be called a computer.

The PC is a fully operational system. It is based on a microprocessor chip (see Figure 18-17). The chip has most of the ability of the processor located in the one integrated circuit device. The PC system resembles the larger computer inasmuch as it consists of:

# 1-Bit
# Microprocessor
# MC14500B

The MC14500B Industrial Control Unit (ICU) is a single-bit CMOS processor designed for use in systems requiring decisions based on successive single-bit information. An external ROM stores the control program. With a program counter (and output latches and input multiplexers, if required) the ICU in a system forms a stored program controller that replaces combinatorial logic. Applications include relay logic processing, serial data manipulation, and control. The ICU also may control an MPU or be controlled by an MPU.

- 16 Instructions
- DC to 1.0 MHz Operation at $V_{DD}$ = 5 V
- On-Chip Clock (Oscillator)
- Executes One Instruction per Clock Cycle
- 3 V to 18 V Operation
- Noise Immunity Typically 45% of $V_{DD}$
- Quiescent Current 5.0 μAdc Typical at $V_{DD}$ = 5 V
- Capable of Driving One Low-Power Schottky Load or Two Low-Power TTL Loads over Full Temperature Range

OUTLINE OF A TYPICAL ORGANIZATION FOR AN MC14500B-BASED SYSTEM

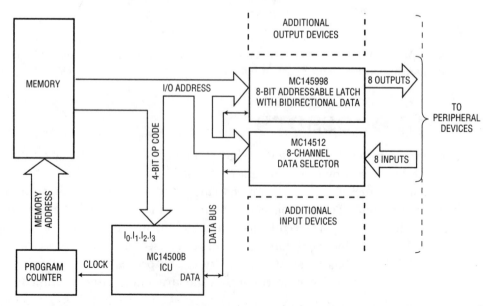

**Figure 18-17.** Microprocessor chip. Copyright of Motorola, Inc. Used by Permission.

- an ALU or arithmetic logic unit for performing arithmetic and/or logical operations. It can add, subtract, compare, and rotate left or right. It can also do logic operations such as AND, OR, NEGATE, EXCLUSIVE OR.
- a memory that is directly addressable. The memory may contain both data and instructional words.
- the ability to input and output data, usually in digital form. Data from the device can be exchanged between itself and input/output devices such as CRT displays, paper tape readers, floppy disk memories, magnetic tapes, teletypes, and various instruments.
- the ability to be programmed.

It is difficult to keep microprocessor and microcomputer separated inasmuch as the two terms are similar. Microprocessor is a chip. The *microcomputer* is known as the PC or personal computer today.

The microprocessor has made it possible to develop video games, intelligent computer terminals, process controllers, telephone switching, inventory control systems, better control of home appliances such as washing machines, microwave ovens, and dishwashers. The automobile uses microprocessors (a computer system) to improve fuel economy and to control emissions of pollutants.

Microprocessors are low cost, easy to use in designing a system, flexible, and easy to reprogram.

# How the Microprocessor Works

The microprocessor is made up of four different parts:

- memory
- arithmetic logic unit
- control unit
- input/output ports

## Memory

The memory has both instructions and data. These can be intermixed. Since each computer has a basic set of built-in capabilities, the sequence in which it uses the instruction set is placed in memory by a programmer. Once the programmer assembles the instructions properly, he/she can make the computer do various tasks, some very complex. As the control unit calls for each instruction located in the memory, one at a time, and interprets each instruction into control pulses, it is then sent out to cause the instruction to be executed. When the instruction is completed, the control unit will call for the next word or instruction.

## ALU

The arithmetic logic unit determines how the logic circuits implement fundamental arithmetic operations. It can do addition, subtraction, multiplication, and division. All the arithmetic and logic functions are performed by the ALU. Logic functions such as shifting and complementing and Boolean algebra operations are also accomplished by the ALU.

## Control Unit

The control unit serves a couple of important functions. It provides timing instructions and synchronization signals for all the other units. These signals can cause the other units to move data, manipulate numbers, and output and input information. All these operations depend on a program that is stored in the memory.

## Input/Output Ports

Communication with the computer is done through input/output (I/O) ports. The I/Os and interface circuits are needed to make the computer operate properly. The ports are tied into the main buses through which the computer sends its signals (see Figure 18-18).

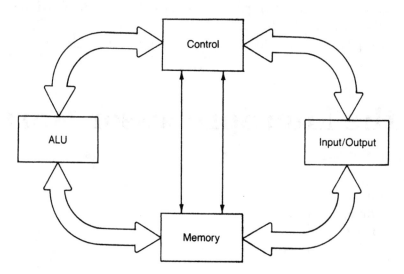

Figure 18-18. I/O ports — Main Bus.

# Communicating with the Microprocessor

There are four basic categories that microprocessors fall into when grouped according to their operations. These are data transfer, arithmetic functions, control functions, and logic manipulations.

## Talking to the Computer

The manufacturer of the chip or microprocessor determines the complexity of the instructions that will cause the device to do a particular job. The chip or processor is programmed to react to a certain language. This type of language is an abbreviated language called *assembly language*. Some assembly language instructions are branch, jump, jump-to-subroutine, and return-from-subroutine. Assembly language instructions have the form of initials or shortened words. These words represent microcomputer functions. Abbreviations are only for the convenience of the programmer. The program that the microcomputer eventually runs has to be in the form of binary numbers. When each instruction is converted to the binary code that the microprocessor recognizes, it is called a machine language program.

Assembly language instructions are called *mnemonics* [knee-mon-ics]. The assembly language mnemonics for the jump instruction is JMP. This typical microcomputer instruction is easier and faster to use. Assembly language makes things easier for the computer programmer. This is because the mnemonics are easier for a person to remember and write than the binary numbers that the computer has to use. However, the computer program must eventually be converted to the binary codes that the computer recognizes as instructions. A special program, called an assembler, can be run on the computer to convert the mnemonics to the binary codes they represent. This enables the programmer to write the program using words that have meaning to the programmer and produces machine codes that have meaning to the computer. Without going into any more detail as to how assembler programs work, let's take a look at some of the instructions that are used in mnemonic form with their equivalent meaning in everyday English.

| Mnemonic | English |
|----------|---------|
| JMP | Jump to new program location |
| JSR | Jump to subroutine |
| BRA | Branch using the offset |
| BEQ | Branch if accumulator is zero |
| BNE | Branch if accumulator is nonzero |
| BCC | Branch if carry bit is zero |
| BCS | Branch if carry bit is nonzero |
| BPL | Branch if minus bit is zero |
| BMI | Branch if minus bit is nonzero |
| RTS | Return from a subroutine |
| LDA | Load accumulator from memory |

# The Special-Purpose Computer at Work

When the appropriate control scheme program is available, the special-purpose computer has the ability to sample and control many inputs and outputs independently (see Figure 18-19). A good example of the computer at work is in the modern automobile. The automobile applications for computers involve control of both multiple systems independently and systems with multiple inputs and outputs. The automobile has what is referred to as multivariable systems. The engine has inputs for air/fuel ratio, throttle angle, and sparking times and several outputs such as torque, speed, and exhaust gases. All the outputs

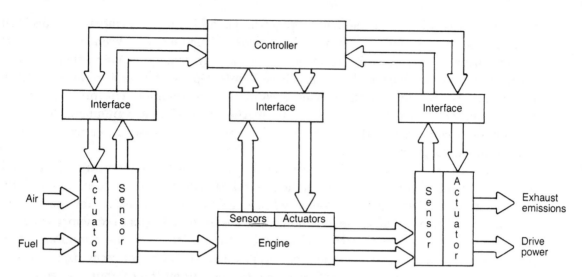

**Figure 18-19.** The automobile has a special-purpose computer with inputs and outputs.

must be controlled at the same time because some inputs affect more than one output. These types of controllers can be very complicated and are more difficult to create in analog circuitry. The increased computer complexity of these systems doesn't affect the complexity or cost of the computer system once the right size computer is chosen for the job. It only affects the job of programming the appropriate control scheme into the computer.

Many books have been written on computers, special-purpose computers, and microprocessors as well as digital electronics. They are available on every level and in great detail. The limits of one chapter in a book such as this make it possible only to show some of the basics and how they are utilized to make a working system. If you are interested in computers and how they operate electronically in terms of circuitry, then it is best to pursue the engineering approach to the subject. If you are interested in the repair and maintenance of the many computer systems, then it is best to study it from the technician's viewpoint that involves how the systems are designed and what they are designed to do rather than how to design the systems.

## Summary

The world of computers is a fascinating one. Today, the microprocessor has brought the advantages of the electronic computer to within the grasp of everyone.

The abacus is a manually operated digital computer that was used in ancient civilizations and is still utilized today in the Orient. The first adding machine was invented by a Frenchman in 1642. Then, in 1662, an Englishman developed a more compact device that could multiply, add, and subtract. In 1682 a German perfected a machine that could perform all the basic operations as well as extract the square root.

Electronics was introduced to the world of computing when flip-flops were invented in 1919 by Eccles and Jordan.

The bistable multivibrator has two states in which it operates: on or off. This can be used to time the computer and to cause its circuits to operate. The bistable multivibrator is another name for a flip-flop oscillator.

There are two types of computers: the analog and the digital. The digital computer is more accurate and is not subject to drifting when voltages vary slightly. It is the one more frequently used today.

Computer circuits consist of relays, switches, and diodes. The modern computers rely entirely on semiconductor devices. There are five basic circuits that form the gates or workhorse jobs for computers. They are the AND, OR, NOR, NAND, and NOT. Pulses of electricity serve as bits of information for the computer. In the binary system of counting, the position of the 1 or 0 makes a difference between a 2 and a 1 for instance. Scientific notation or powers of 10 form a good basis for understanding binary numbers.

PCs or personal computers are making a big impact on the home market. Use of digital computing and data processing devices is growing rapidly. Part of the growth stems from the fact that the devices are becoming less expensive with the passage of time.

Personal computers are, generally, systems built around microprocessors and other printed circuits. Computer systems handle complete data or word processing jobs. The special-purpose computer is being designed into many modern devices.

# Review Questions

1. What mechanical device is the basis for our modern computers?
2. When was the first adding machine invented?
3. Why is the flip-flop circuit so valuable for computers?
4. Who were Eccles and Jordan?
5. What is the difference between an analog and a digital computer?
6. Describe a NOR circuit.
7. Describe a NOT circuit.
8. Describe an OR circuit.
9. Describe an AND circuit.
10. What does the term binary refer to?
11. How do the powers of 10 figure into computing?
12. What is the difference between a microprocessor and a personal computer?

# Chapter 19

# APPLICATIONS OF ELECTRONICS

**W**here does the future lead in electronics? This is a good question. Perhaps the future is already present in today's newly acquired technology. We have seen the transistor replace the vacuum tube. Most of the equipment still operating today is of the transistorized version with very few vacuum tubes still around. However, there are some picture tubes left on television sets and cathode ray tubes being used as monitors for computers and word processors. These are due for replacement as soon as the liquid crystal display (LCD), like those used on calculators, are improved another degree or two. The color picture of television can already be displayed with liquid crystal technology. It is available as a hand-held set for under $100.

The transistor is being replaced as a discrete component by the integrated circuit (IC). Let us take a closer look at some of the applications of latest technology to see what they hold in store for us.

## The Phase-Locked Loop

Not too long ago nearly everyone applied for and got a license to operate a citizens band (CB) transceiver. The demand was so great that the channels which were available (23) were crowded, and it seemed as if everyone was trying to talk at the same time. The government responded through the Federal Communications Commission (FCC) to allow CB on 40 channels. The extra channels required different sets for those who wanted to use the last 20 channels. The cost would have been extremely high for a 40-channel set if it had not been for a newer method of producing the 40 frequencies with a high degree of accuracy (0.05 percent). That is where the phase-locked loop (PLL) came into being. It had been talked about back in the 1930s, but the high cost of the circuitry made its use prohibitive. However, the advent of the integrated circuit (IC) made a difference, and semiconductor technology finally paid off.

A synthesizer circuit allows all necessary frequencies for a CB transceiver to be generated by using a single-crystal controlled oscillator (see Figure 19-1). The three ICs make it possible to produce any frequency allocated to the CB band. You should be especially aware of certain devices since they are the primary components of the future and will be used more and more as newer products are developed. $D_1$ is a Ferranti varactor diode. This diode changes capacitance with a change in voltage. It is very useful in changing frequencies in

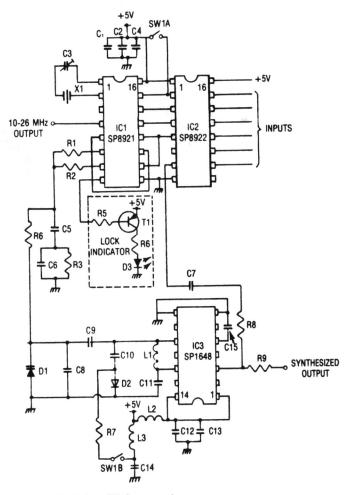

**Figure 19-1.** Circuit for synthesizing CB frequencies.

oscillators and in tuning circuits. $D_3$ is a light-emitting diode (LED). The LED has just begun to be used for pilot lights and in fiber optics systems. In the set shown in Figure 19-1 the only crystal used is $X_1$ at 10.240 megahertz (MHz). Remember, the CB band operates just below 30 MHz.

You can now change the frequency of the garage door opener by using a pencil or pen to change small thumbnail switches. As you change the combination of switches, you change the frequency you operate on and the modulation of the signal used to operate the opener. You do the same thing with the hand-held transmitter and the receiver to get them to operate on one frequency and not on another in the neighborhood. Take a look at Table 19-1 to see how the input code changes the frequency for a CB unit. The same digitized method is used for the door opener. All this uses frequency synthesizing to get the job done. There is a possibility of having 64 channels at 10-kilohertz (kHz) separation available from 26.895 to 27.525 MHz.

The phase-locked loop can be used to demodulate FM. It is also applicable to separating the Subsidiary Communication Authorization (SCA) multiplexed channel on commercial stereo FM.

TABLE 19-1

**INPUT CODE FOR CB SYNTHESIZER**

| Channel Number | Input Code FEDCBA | Output Frequency, MHz | Channel Number | Input Code FEDCBA | Output Frequency, MHz |
|---|---|---|---|---|---|
| 1 | 000111 | 26.965 | 21 | 100000 | 27.215 |
| 2 | 001000 | 26.975 | 22 | 100001 | 27.225 |
| 3 | 001001 | 26.985 | 23 | 100100 | 27.255 |
| 4 | 001011 | 27.005 | 24 | 100010 | 27.235 |
| 5 | 001100 | 27.015 | 25 | 100011 | 27.245 |
| 6 | 001101 | 27.025 | 26 | 100101 | 27.265 |
| 7 | 001110 | 27.035 | 27 | 100110 | 27.275 |
| 8 | 010000 | 27.055 | 28 | 100111 | 27.285 |
| 9 | 010001 | 27.065 | 29 | 101000 | 27.295 |
| 10 | 010010 | 27.075 | 30 | 101001 | 27.305 |
| 11 | 010011 | 27.085 | 31 | 101010 | 27.315 |
| 12 | 010101 | 27.105 | 32 | 101011 | 27.325 |
| 13 | 010110 | 27.115 | 33 | 101100 | 27.335 |
| 14 | 010111 | 27.125 | 34 | 101101 | 27.345 |
| 15 | 011000 | 27.135 | 35 | 101110 | 27.355 |
| 16 | 011010 | 27.155 | 36 | 101111 | 27.365 |
| 17 | 011011 | 27.165 | 37 | 110000 | 27.375 |
| 18 | 011100 | 27.175 | 38 | 110001 | 27.385 |
| 19 | 011101 | 27.185 | 39 | 110010 | 27.395 |
| 20 | 011111 | 27.205 | 40 | 110011 | 27.405 |

# Cable Television

The main objective of a cable television system is to deliver high-grade TV signals to a subscriber's home. Note the word subscriber, for it designates that the person pays for the privilege of watching a number of channels other than just those available locally from stations located nearby. The cable TV system is a system with a future. It will develop sporadically and then settle down to being the system used by most households to get their information. The next generation will turn to the cable channels for the weather, news, sports, and special-interest programming. They will also use the cable TV system in many other ways. High-speed DSL for those on the Web is the latest.

Figure 19-2 shows a complete television system. It also includes the cable link. The cable system deals with the radio frequency (RF) signal from the TV station. The head end is where the television channel is received and then processed for distribution throughout the cable network. The heart of the cable TV system is the coaxial cable transmission line. The other components of the system are used to get signals into or out of the cable or to overcome some of the basic limitations of the cable itself.

Since some losses are associated with the cable TV system, there is a need for amplifiers to rebuild the signal so that every receiver or subscriber gets a good-quality picture. Three categories of amplifiers are used in cable TV systems: (1) the *trunk*, or mainline, amplifiers spaced along the main trunk lines of the system to compensate for the cable

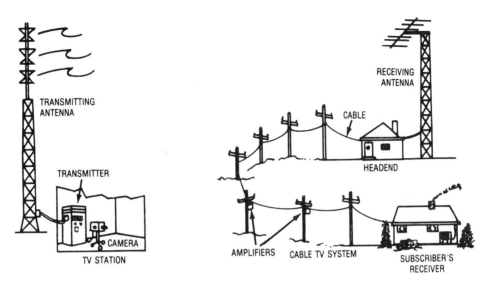

**Figure 19-2.** Cable TV system.

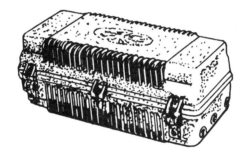

**Figure 19-3.** Cable TV amplifier.

attenuation, (2) the *bridger* amplifiers located either in the same case with the trunk amplifiers or in a separate place; they are used to take off the trunk line for distribution in feeder cables to the subscriber's neighborhood, and (3) the line extenders that are used along the feeder system where required. Some amplifiers can be used for any of the three purposes (see Figure 19-3).

Providing more channels will be the goal for cable TV in the near future. Cable TV producers originate more of their own programs now as the equipment for studios is less expensive. Satellite TV links are another future development. Smaller dish antennas for mounting on the side of your house are now available.

# Satellite Television

The space program has made it possible to put satellites into orbits that coincide with the earth's. The satellites orbit at an altitude of 22,300 mi. above the earth. They make one revolution every 24 h. They remain above the same spot on the surface and appear to be stationary in the sky (see Figure 19-4).

Figure 19-5 shows the satellite communications system. High-powered microwave transmitters are used to send the signal to the satellite. The signal is picked up by an

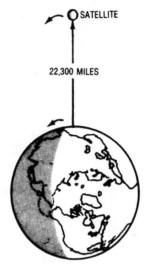

**Figure 19-4.** Satellite in synchronous orbit, rotates with the earth.

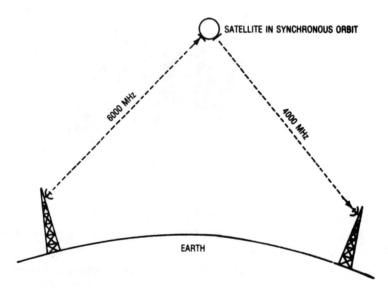

**Figure 19-5.** Satellite transmission system.

extremely sensitive receiver in the satellite and is rebroadcast toward the earth on another frequency. Most of the up-links are 6 GHz and most of the down-links are between 3.7 and 4.2 GHz. Equivalents are 6000 MHz for 6.0 GHz and 4000 MHz for 4.0 GHz. Many different signals are available from the satellites. Fourteen satellites are in orbit in Figure 19-6. These are called *geosynchronous* orbiting satellites. More are planned in the near future, and they will operate on the 12- to 14-GHz band. This system will replace the automobile road map. When you place a programmed disk in your car receiver, a display will show you exactly where you are on the map. This information is obtained by picking up the signals from the satellites in orbit above the earth and pinpointing your location in respect to the stationary satellites. These receivers were available on cars in the late 1990s. Oldsmobiles and Cadillacs now have this system available as an option. Toyota and others will add it soon.

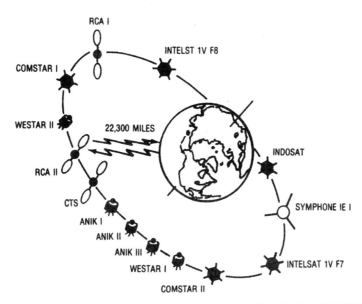

**Figure 19-6.** A. International agreement has placed satellites in the Clarke orbit (22,300 **mi**. above earth) in separation patterns of 4 geometric degrees (1833 mi. between satellites). As equipment design improves it will be possible to separate each satellite by 1 degree, or 458 mi. With the space shuttle it will be possible to put up satellites weighing about 5.5 tons.

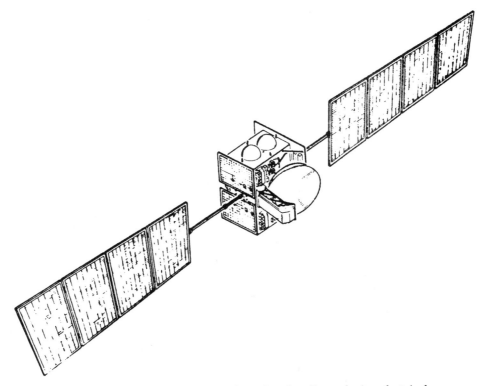

**Figure 19-6.** B. Satellite deployed in orbit with photoelectric cells producing electrical power.

Every domestic satellite over North America has several channels, called transponders, that process communications traffic. A satellite can have 24 or more transponders in operation, each capable of transmitting one TV signal or thousands of simultaneous telephone conversations. Other uses for the satellites are radio networks, teletext news, computer information services, and other data transmissions.

The future looks bright for satellite television. Satellites will have a great impact on the home and on everyone in the world. It will be possible to do your shopping by television, get college credit through educational programs without going to college physically, and enjoy audience participation programs. Access to people and places around the world will be greatly enhanced.

Higher frequencies and smaller antennas will permit inexpensive, truly portable satellite communications on both up-links and down-links. The 24-transponder satellite's capacity will seem small in comparison to tomorrow's communications space stations (see Figure 19-7 for a satellite earth station).

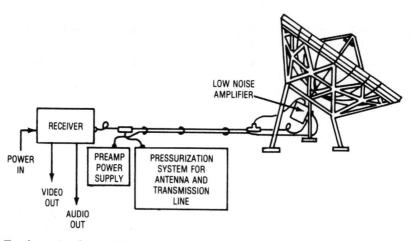

**Figure 19-7.** Earth station for satellite television.

# Global Positioning System

The ability of satellites to communicate with earth, one another, and with space vehicles makes them very important both militarily and commercially. A global positioning system (GPS) has been developed using satellites to make it possible for people on earth to be located to within 1 centimeter (cm), or 0.4 inch (in.).

This worldwide navigation system uses radio signals broadcast by satellites. A computerized radio receiver in an automobile, airplane, ship, or other vehicle uses the satellite signals to calculate its own location. People hiking in the wilderness and other people on foot may use a small, portable receiver. The U.S. Air Force operates the satellites, but the system has both military and civilian users.

The GPS has 24 satellites (more may be used in later applications), called Navstars, in six orbits with a height of about 12,500 mi. As many as eight satellites may be above the horizon when viewed from any point on earth.

The GPS receiver uses signals from at least four, and sometimes more, satellites. Each signal indicates the location of the satellite that sent the signal and the time it was broadcast. The receiver can determine its latitude and longitude by using only three satellites if its altitude is known.

Civilian GPS users can normally determine their location within 100 m (330 ft). A special technique called *differential GPS* can improve accuracy to within 10 m (33 ft). However, a technique called *carrier phase GPS* is accurate to within 1 cm (0.4 in.). The U.S. armed forces began developing GPS in the early 1970s so it could pinpoint the delivery of its atomic weapons.

The eventual marriage of artificial intelligence and space technology will begin a new era of robot satellites which can perform their own telecommunications, navigation, orientation, and repair. Interactive "intelligent" satellites operating on multiple bands will communicate with one another to relay signals directly around the world.

# Robotics

Control is the secret of robots. You must be able to control the movement of the remote device. The device may be located nearby or 22,300 mi. away in a spacecraft or satellite. That is where electrical and electronic control plays an important part in everyday robots.

# Robots

Robots have captured the imaginations of writers and movie producers for some time. However, only recently have they become useful in the production of quality products, and it is here that the greatest amount of time and effort is being spent in robot development. The ability to produce quality products is of utmost importance since the consumer benefits and the manufacturer stays in business. The "chip" that makes computers, cell phones, and television receivers possible is produced by robots.

### What Is a Robot?

There are a number of definitions for *robot*. But, for our purposes here, a robot is a reprogrammable, multifunctional manipulator designed to move material, parts, tools, or specialized devices through variable programmed motions for the performance of a variety of tasks. See Figure 19-8.

### History of Robots

A bit of history will place the robot into perspective and help explain its popularity. The robot is a relatively recent development.

Science fiction has featured robots for as long as they have been around, but the term *robot* came into common usage in 1921 when Karl Capek wrote *R.U.R.* Since Capek was Czechoslovakian, he used the word *robota* to describe the machine that performed like a human but did not have the senses of humans. The term *robota* means slave labor and was reduced to *robot* in English.

What is a true robot? A few ideas must be taken into consideration when you answer this question. Some of them are

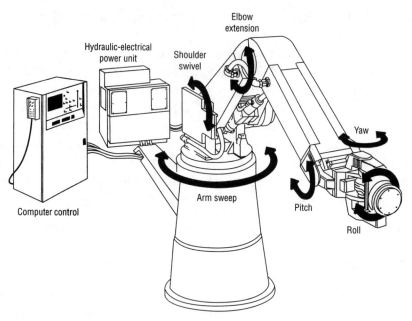

**Figure 19-8.** Complete industrial robot system.

- A robot is a device or system that is programmed by a human to perform human-like acts.
- A robot may sense various conditions and react in a preprogrammed manner.
- A robot may be able to react to various conditions in terms of the five human senses: sight, hearing, smell, taste, and touch. Sensors are available that allow all these senses to be inserted into a system.
- A robot is a system that can operate on its own without human supervision.
- A robot may make decisions by comparing information received from sensors and reacting in a preprogrammed way.

The invention of large-scale computers in the mid-1950s helped the robot to become a reality. Robots then became more popular with the advent of the personal computer or microcomputer in the 1970s. By incorporating a computer into the robotics system, it was possible to create a unit that could move, talk, lift things, see where it was going, and know what it was touching.

## Computer Programs

Special computer programs designed for specific jobs are used to control robots. Industrial robots, for instance, are designed to do a particular operation. This one operation may be done over and over again, but the robot, unlike humans, does not become fatigued or bored. A program is written to take into consideration the exact tasks to be performed. In some instances, it may take years to analyze the moves needed to perform a particular job. This information then must be fed into a computer in terms that it understands, and the computer then sends signals to the robot so that it will perform exactly as desired. In Figure 19-9 two robots are used to load and unload. By coordinating the motions of the two robots, the idle time of machine A is kept to under 1 second. The alternate path

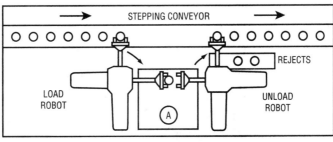

Figure 19-9. Robot with microprocessor designed to load and unload.

capability built into the command module allows a variety of sensors to detect reject parts and place them in an alternative location.

## Languages

Robots use a number of computer languages, which are designed for specific operations. Some of the languages used with robots are AL (Stanford's Artificial Intelligence Lab language), VAL, AML (developed by IBM), Pascal, and ADA.

## Microprocessors

A microprocessor is just that. *Micro* means small and a *processor* is a device that can process things. In this case, it is used to process information fed to it from an external source. A micro-controller is a dedicated small computer with a microprocessor as an integral part.

We use the term *microprocessor* to describe a special purpose chip or portion of a chip that gets its instructions from a keyboard, joystick, mouse, or any number of sensors (see Figure 19-10). The chip can do math, make logical decisions, and work with words and symbols.

The robot usually has a dedicated microprocessor designed for its special job requirements. It may be a simple chip, or it may be a large mainframe computer. Without the microprocessor, the robot is limited in its application. With the programmed microprocessor, it is possible to have the robot operate alone without connecting wires, other than for power purposes.

Figure 19-11 shows how a robot is used to pick up finished air conditioner units and pack them in shipping containers. This is what is called a pick-and-place robot. It can work 7 days a week, 24 hours a day without becoming bored or fatigued. Operating costs are low, and downtime is minimal.

# Industrial Applications

Robots are here to stay. A good example of a robot being put to use is in the painting of automobiles. They can be used for machine loading and unloading: placing parts where they are needed for machining or shipping. They are also handy for materials handling: packing parts or moving pallets. Fabricating processes use them for making invest- ment castings, grinding operations, and deburring; for water jet cutting; for wire harness

**Figure 19-10.** A. Microprocessor chip. B. Keyboard for programming. C. Joystick.

manufacturing; for applying glues, sealers, putty, and caulks; for drilling, fettling, and routing. They can paint cars, furniture, and other objects. Welding robots can easily be built to weld cars, furniture, and steel structures. They can also assemble electronics, automobiles, and small appliances. Inspection and testing can be done on any assembly line to make the results of quality control jobs of repetitive viewing and checking more reliable.

# How the Robot Works

There are some common components that may be examined for a better perspective on how a robot works.

The manipulator is classified by certain arm movements. There are, for instance, four coordinate systems used to describe the arm movement: polar coordinates, cylindrical coordinates, Cartesian coordinates, and articulate (jointed-arm, spherical) coordinates. Four of the basic parts of a robot are base, arm, wrist, and grippers.

## Base

The base of the robot is its anchor point. The base may be rigid. It is usually designed as a supporting unit for all the component parts of the robot. The base does not have to be

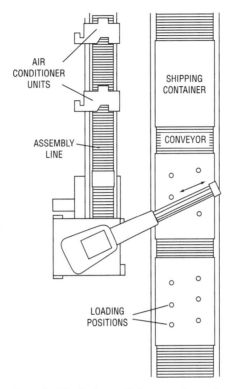

**Figure 19-11.** Robot used to pick up finished air conditioner units and pack them in shipping containers.

stationary since it may become part of the operational requirements of the robot. It may be capable of any combination of motions, including rotation, extension, twisting, and linear. Most robots have the base anchored to the floor, although because of limited floor space, they may be anchored to the ceiling or to suspended support systems overhead. A track or conveyor system may be used to move the robot along as needed. See Figure 19-12.

## Arm

Some type of arm is found on most industrial robots. It may be jointed and resemble a human arm or it may be a slide-in/slide-out type used to grasp something and bring it back closer to the robot. A jointed arm consists of a base rotation axis, a shoulder rotation axis, and an elbow rotation axis. This type of arm provides the largest working envelope per area of floor space of any design thus far. This is a six-axes type of arm, it requires some rather sophisticated computer control. Most arms now have some type of joint. From one to six jointed arms may be attached to a single base for special jobs. The expense of controlling this type of movement is rather large because of its complexity. See Figure 19-13.

## Wrist

Figure 19-14 shows how the wrist is attached to the jointed arm. The wrist is similar to a human wrist and can be designed with a wide range of motion, including extension,

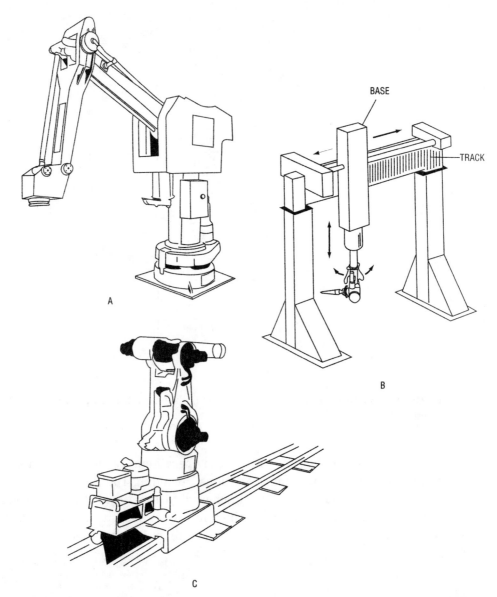

**Figure 19-12.** Three ways to mount a robot. A. Base mounted to the floor. B. Base mounted on a gantry for moving over the work area from above. C. Base mounted on a track for movement of the whole unit.

rotation, and twisting. This aids the robot in reaching places that are hard to reach by the human arm. It comes in handy especially when spray painting the interior of an automobile on the assembly line. It is also helpful in welding inside a pipe. This type of flexibility will improve the manufactured products we now enjoy and add others we were unable to fabricate earlier.

## Grippers
The grippers are at the end of the wrist. They are used to hold whatever the robot is to manipulate. See Figure 19-15. Pick-and-place robots have grippers to move objects from one place to another. Some robots have end-of-arm tooling instead of grippers. In such cases the robot is used primarily for one type of operation such as spray painting or

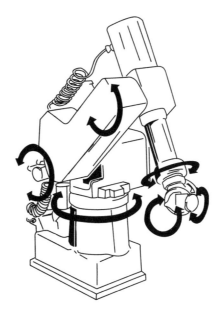

Figure 19-13. The six axes of a robot.

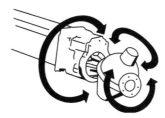

Figure 19-14. The wrist motions of a robot arm.

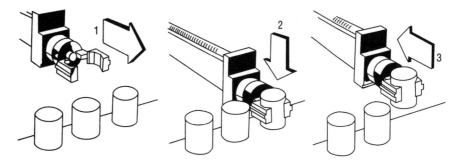

Figure 19-15. Robot arm (manipulator) with grippers for reaching out and picking up an object on a line **and bringing** it back to place it elsewhere.

welding. If **a** tool is attached, it is unnecessary to have a gripper on the end of the arm. A **pneumatic** impact wrench can be fitted at the end of the arm just as easily as the **grippers.** Various types of grippers are made to fit the job being done by the robot.

## Drive Systems

Drive **systems** for robots are classified as pneumatic, hydraulic, and electric. Each type has its **applications** due to physical limitations of the method used to do the work. Each type of drive has its advantages and disadvantages. Inasmuch as this book covers electronics we will deal primarily with the electric drive.

## Electric Drive

Electric drive systems use electric motors for their power source. The motors may be operated by direct current (dc) or alternating current (ac). Torque is developed by gearing the speed of the motor down to the required motion or movement of the arm. See Figure 19-16.

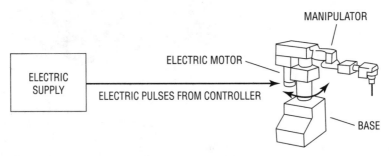

**Figure 19-16.** Electric-operated robot system.

Some advantages of electric drive systems are their ability to allow smooth startup of payloads and their smooth deceleration and stopping. Operation costs are minimal in terms of electrical power used and maintenance of the system motors. Electric robots have a higher repeatable accuracy than do hydraulic robots.

Electric drive robots cannot handle as heavy a payload as hydraulic drive systems. The payload for electric drive is between 6.6 and 176 pounds (3 and 80 kilograms). However, electric drive systems are very versatile in their operation.

Electric drive systems are used in arc welding, spot welding, machine loading and unloading, materials handling, and deburring as well as in assembly operations. Development of explosion-proof motors will make it possible to use them safely in spray painting atmospheres.

Electrical drive systems are used in high-technology robots.

# Robot-Computer Interface

A number of methods or systems are used to accomplish the mating of computer and robot. Figure 19-17 is a schematic diagram of the microcomputer system that is often used for this purpose. As shown, the three major parts of the computer are memory, the central processing unit (CPU), and input/output (I/O) ports.

*Memory.* The computer's memory is used to store data to be processed by the CPU or data resulting from processing.

*Central Processing Unit.* The CPU, sometimes called the microprocessing unit (MPU), contains the control circuitry, an arithmetic logic unit (ALU), registers, and an address program counter. This may be a larger computer in larger installations. The instruction code comes to the CPU from memory and gets decoded and executed. The

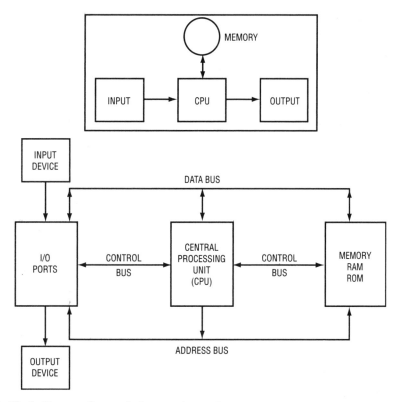

**Figure 19-17.**  Block diagram for a robot computer system.

CPU works in millionths of a second, enabling it to recall an instruction and process it without a slowdown in operation. Instructions are given in many computer languages. Each manufacturer specifies which language or languages the machine will use.

*Input/Output.* The connections that interface with the outside world of the microprocessor are called input/output ports. The input port allows data from a keyboard or other input device to be taken into the CPU. The output port is used to send data to an output device such as a motor. Bus lines carry the signals to and from the major parts of the microprocessor.

## Languages

Intelligent robot controllers have the ability to understand high-level languages. Almost every robot manufacturer has its own controller language, making the language unique for a particular controller from a particular manufacturer. Robots are usually preprogrammed for the benefit of the user.

VAL was developed for Unimation. HELP was developed for General Electric's assembly robots. AML, which is a manipulator language, was developed for IBM's assembly robot. MCL stands for manufacturing control language and was developed by McDonnell Douglas for an Air Force project; Cincinnati Milacron's T³ robot uses it. RPL stands for robot programming language and was developed by SRI. It is similar to languages such as Pascal and FORTRAN and is used for communicating with intelligent vision sensors. RAIL was developed by Automatix for robots and vision systems.

## Software

Computers are used to program robots, and the programming is done off-line. That means the robot is not directly involved when programming is taking place. The software can be recorded on floppy disks, magnetic tape, or other means and then checked on a robot and debugged or edited. Once the program is proven on a robot system, it can be duplicated and sold to others who want to use robots to perform similar tasks.

By using off-line programming, the programmer has greater flexibility to carry out complex operations, and the time spent in programming is reduced. In addition, the robot can remain in service while the programming is taking place, thereby increasing its productivity.

# Vision Systems

Vision systems can provide the robot controller with information about the location, orientation, and type of part to be handled. See Figures 19-18 and 19-19. Machine vision systems

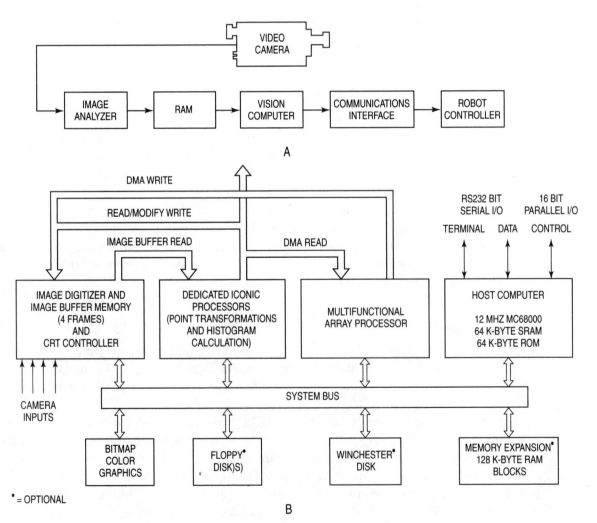

Figure 19-18.   Machine vision system for robots. A. Block diagram of vision system. B. Electronics for taking and presenting camera signals and processing information to a monitor and computer.

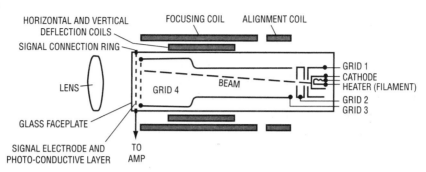

**Figure 19-19.** Video tube used in a robot vision system.

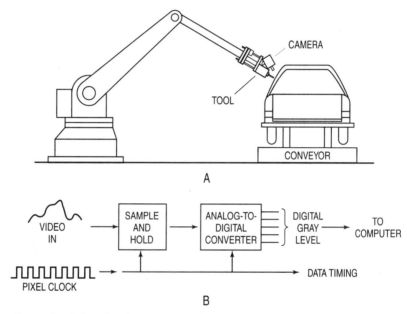

**Figure 19-20.** Example of the edge detection process. A. TV camera mounted on robot for visual inspection on auto assembly line. B. Pixel clock and data bit generation for a video signal fed to a computer.

are used for recognition and verification of parts, for inspection and sorting of parts, for noncontact measurements, and for providing part position and orientation information to the robot controller. Edge detection and clustering are the two processes used for identification by the computer of the parts viewed by the vision camera. See Figure 19-20. Much has to be done before vision is available for all robots at a reasonable price and with the reliability needed for daily production runs.

Robots which can assemble integrated circuits (ICs) have been built. These robots can attach the gold hairlike wires that go onto the IC devices before they are sealed. The robot unit also has a TV camera which causes the work area to be enlarged in size so that the operator can see it. Then the actual welding of the gold to the silicon chip takes place under controlled conditions. This one example where the robot can make its own brain.

The field of robotics has just begun. The future is unlimited. Anyone with a fundamental grasp of electronics will have a very good future in keeping these devices in operating condition until they can be designed to correct their *own* troubles.

# Videotape

Home recorders record off the air or can be connected to a camera, and local programming can be made to suit the family or client. More and more schools are using the videotape recorder for making lesson and topic presentations. The use of the videotape recorder is unlimited. Home models use the half-inch type. They can record up to 8 hours on one cassette (see Figure 19-21).

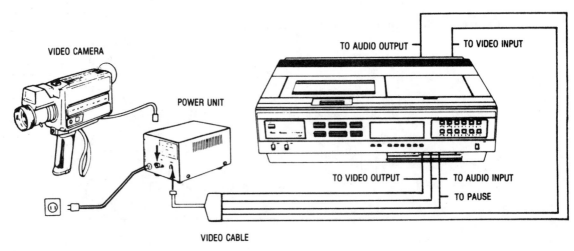

**Figure 19-21.** Recording from a video camera.

Figure 19-22 shows the hookup of the videotape recorder with the home TV set and separate UHF and VHF outside antennas. It can also be used with cable and pay TV. Copyright laws should be observed.

Magnetic tape is recorded by magnetizing the small particles of iron embedded on the tape with an adhesive. The alignment of the magnetic poles of the particles varies with the electric signal produced by picture, sound, and sync signals. The recording head moves so as to cover the entire tape as it moves across the head. The method of putting the information on the tape differentiates the Video Home System (VHS) from the Beta format. There are many methods available, but these two have been standardized to use the half-inch tape and to produce a good-quality picture and sound with fast-forward and reverse as well as pause or hold-a-picture capability. This device may cause a change in the future since commercials can be *speeded up*, and the program is then condensed for quick viewing. By eliminating the commercials, the viewer can thus see the program without interruption. This will, no doubt, change the way commercials are presented in the future as they try to cope with this speeded-up method of viewing the tape. This area holds some promising developments for the future and will be interesting to watch.

# The Compact Disc

The compact disc (CD) consists of a transparent plastic layer, an aluminum layer, and a protective plastic coating. See Figure 19-23. The most commonly used disc is a read-only

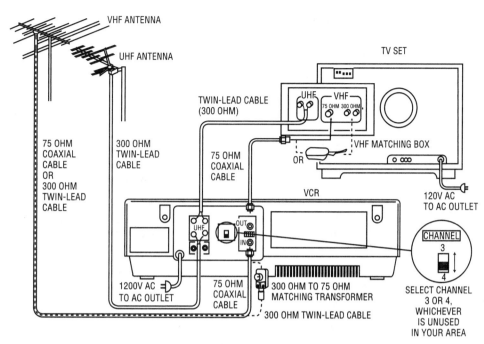

**Figure 19-22.** VHF/UHF reception area using separate antennas for recording.

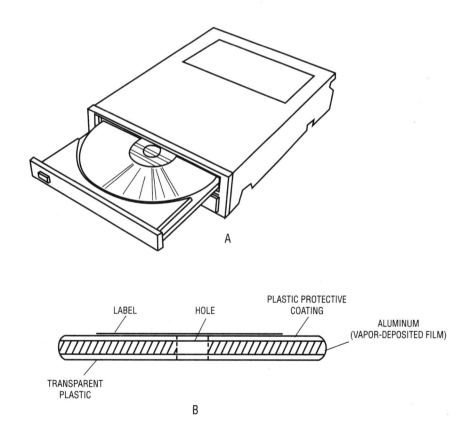

**Figure 19-23.** CD-ROM player. A. Disc drive fits into a PC slot, usually labeled D Drive. B. Cross-sectional view.

nonmagnetic data storage device in the form of a reflective disc that is 4.7 in. in diameter. Data stored can be audio, video, data, or a combination of these. Maximum storage is usually 650 megabytes (MB). A laser diode is utilized to read the CD. The CD has tiny reflective spots burned into its surface. The reflective spots take a pattern dictated by the information being stored.

The CD-RW *compact disc–rewriteable* is used to write, read, or erase and is made specially for the purpose of being reused. The CD-RW has a storage capacity of 650 MB and can be recorded and erased 10,000 times. Discs that are recorded with the MultiRead standard can also be used in CD-ROM drives and on compact disc audio drives. The compact disc–rewriteable was originally named the CD-E when it was developed in 1995; however, it wasn't offered to the public until 1997. By this time the name had been changed to CD-RW. Also see Chapter 10.

A multimedia education and entertainment system developed by Philips and Sony is the *compact disc–interactive* (CD-I). It consists of a proprietary compact disc player that connects to a televisions set. The CD-I system incorporates a dc real-time operating system, which synchronizes interactive video, audio, and text with real time.

The CD-R *compact disc–recordable* is a compact disc standard format, when used with a CD-R or CD-RW write drive, can record up to 650 MB of unerasable data at one time. The advantage of CD-R discs is that they record slightly faster, and cannot be erased or modified later. However, they can be copied to CD-RW discs. CD-Rs can also be played in regular CD-ROM drives and audio compact disc drives.

The CD-ROM XA is a *compact disc–ROM with extended architecture*. It is a version of the CD-ROM that was released in 1991. It enables software, audio, and video to be interleaved on the recorded tracks of the CD-ROM disc. To utilize this format, a regular CD-ROM player can be used, but an XA controller card is needed in the PC or computer.

# DVD

There are a lot of letters and letter–number combinations used in electronics today. Most of this is due to the need for abbreviation in a world of rapid communication. DVD is no exception. It is "shorthand" for *digital versatile disc*. It is a newer version of the common 650-MB compact disc. The DVD is capable of storing up to 17 gigabytes (GB) per side. DVD players can play newer DVD discs containing audio and video information and old audio-format CDs. It is a digital memory storage system that comes in the same footprint and hardware format as the CD-ROM. DVDs have a storage capacity of 4.7 to 17 GB of data on each side of the disc. The DVD disc drives are able to access the data at a rate of 1.3 megabytes per second (Mbps). There are a number of versions of the DVD for both audio and video. Whole movies can be put on a single disk. A version of the DVD disc drive that can read CD-R and CD-RW discs as well as audio CDs is the DVD2 or digital versatile disc, *second generation*. The DVD-RAM is a standard for DVD discs that can be written to and erased over 10,000 times. They have a storage capacity of 2.6 GB per side. At this time DVD-RAM formatted discs are not compatible with DVD-RW discs and vice versa. The DVD-ROM is a digital video disc with read only memory. The DVD-ROM is a standard that enables the recording of multimedia entertainment in improved formats over VHS video recordings and can also play audio CDs, CD-I discs, and CD-R discs. The standard compression format used to place video and data on them is MPEG-2.

MPEG-2 is the abbreviation for *motion picture experts group*. It is a widely used menu of standards for compressing video. MPEG-1 is a bit-stream standard for compressed video and audio optimized to fit into a maximum bandwidth of 1.5 Mbps but it has less resolution. MPEG-2 is intended for higher quality video-on-demand applications. It runs at data rates between 4 and 9 Mbps. Another standard is MPEG-4. It is a low-bit-rate compression algorithm intended for 64 Kbps connections. MP3 is for music.

The DVD-RW is a digital video disc that is ReWriteable. It is also a standard for DVD discs that can be written to and erased more than 10,000 times. They have a capacity of 3 GB per disc side and are not compatible with DVD-RAM formatted discs.

As you can see, once digital electronics was introduced, it opened the world to a new approach to storing information, pictures, and music. Digital electronics improved telecommunications in both switching and quality, and made possible the Internet. The Internet is a world of its own . . . and needs a few books to explain its many applications and operations.

# Space Manufacturing

There will be a number of opportunities for manufacturing in space. Manufacturing facilities in outer space will facilitate the development of purer medicines and better crystals for the production of semiconductors. This alone will provide an amazing array of new products for future use. The space shuttle has already accomplished many missions and has made many experiments possible. The space lab has the potential of doing many things not possible on earth due to the pull of gravity.

The space shuttle manipulator arm has the ability to recover payloads in orbit for repair and maintenance in space or for return to earth. The shuttle is able to return up to 14,545 kilograms (kg) (32,000 lb) of payload to earth.

As you can see, even space technology was made possible with the aid of electricity and electronics. The future for electricity and electronics is unlimited. Your only limitation is your imagination. Studying electronics opens up all these opportunities to you.

# Tape Scan Systems

One of the problems associated with obtaining good fidelity in picture and sound with a videotape machine is the speed at which the tape must be pulled across the recording and reproducing heads. The cylinder rotates at 1800 rpm and two heads are used to scan the tape as it moves rather rapidly over the heads (see Figure 19-24).

The video signal is 4 MHz and requires the best of equipment and signal processing to make sure it is not distorted or eliminated altogether. In the VHS system the tape head gap is only 0.3 of a millionth of a meter. To make sure the video is recorded on as little tape as possible, the information must be placed on the tape within as small a space as possible. Note in Figure 19-24 that the tape has the video signal recorded in a diagonal pattern. This puts more information in the same space that would otherwise be done if the tape was used in a horizontal fashion for signal recording. The tape comes in contact with two recording heads as it is pulled across the cylinder. The tape is held in place by two guide

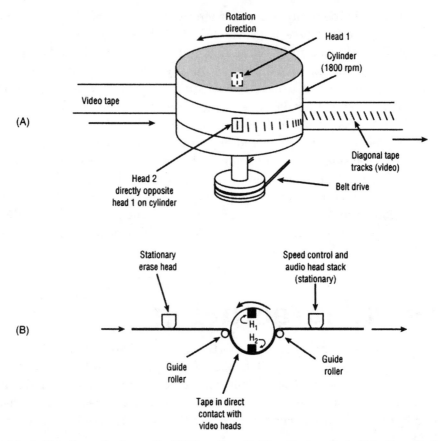

**Figure 19-24.** A. Rotating cylinder on the VHS system. B. How the tape is held to the cylinder in the VHS helical scan system.

rollers that must be very well aligned to produce the desired results. The cylinder rotates at 1800 rpm.

The two tape heads scan the tape as it moves across the cylinder at this speed. That means the tape moves at approximately 2 cm/s and gives the tape movement a relative speed of 580 cm/s, or 5.8 m or 19.11 ft. As you can see, it takes a lot of tape to record a one-hour program. Actually, the tape slant across the cylinder plus the increased relative speed of the tape form diagonal tracks on the tape during recording, and the system can handle the complete frequency span of the video signal.

# Home Video Systems

Beta and VHS home video systems have slight differences, and enough differences exists to make it impossible to use tapes recorded on one system to be played on the other. Both systems use 1800 rpm for the cylinder on the VHS and the drum on the Beta. Beta used 0.6 $\mu$m for its gap, and VHS uses only 0.3 $\mu$m for its gap. Beta is no longer available. All home systems are VHS. When two heads are used, each head records or plays back one TV field. This consists of 262.5 lines. When both heads then present their outputs, the whole 525 lines are shown as a complete video picture.

The erase head is disengaged during playback. Another head is used for audio recording and pickup. This head is combined with a speed control head that records control pulses on the outer track of the tape during recording. During playback the control head senses the prerecorded control pulses and thus regulates the speed. Usually the speed control and audio heads are stacked and, along with the erase head, are stationary.

Take a look at Figure 19-25 and examine the location of the 1-mm sound track on top of the tape. Note how the tape speed control track is located on the bottom of the tape. In 1979 stereo was introduced to television. That meant two tracks for audio was necessary along with a guardband. The 1-mm audio track was broken up into two 0.35-mm tracks for left and right channels and a 0.3-mm guardband.

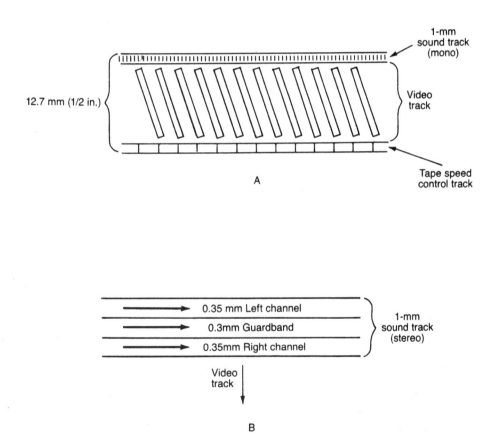

**Figure 19-25.** A. Video tape with all signals located. B. How stereo sound is produced on the older stereo system.

In 1984 the stereo was improved. It used two heads to superimpose the audio on the same track as the video. This took advantage of the increased relative speed that was produced by the rotating cylinder and the tape movement. The quality of the audio was improved. In order to make the older tapes compatible, the original upper audio stereo band was still retained. However, the fidelity of the audio signal was not improved, and the quality is somewhat lower.

# Summary

The transistor has replaced the vacuum tube, and the transistor has been replaced by the integrated circuit (IC). The future holds much promise for the field of electronics.

The phase-locked loop is used to generate all the frequencies needed for a 40-channel CB transceiver. It is also used to demodulate FM. Frequency synthesis is becoming the replacement for the *LC* circuit and resonance.

Cable television is a system with many possibilities as local programming by the franchise offers much broader coverage of local events. Banking and education will also benefit as cable TV is extended. Satellite television is booming at the present with its future ensured as people look forward to having 100 channels to choose from every day, 24 hours a day.

The eventual marriage of artificial intelligence and space technology will begin a new era of robot satellites which can perform their own telecommunications, navigation, orientation, and repair.

Robots are rapidly replacing the humdrum routine of assembly line work. They will require people who are up to date on their electronics and mechanics to keep them operational.

Videotape is unlimited in its uses and adaptations. Use in the home and school as well as in industry and business makes it an integral part of the growth of the electronics field.

Space manufacturing will allow us to make purer medicines and better crystals for use in semiconductor electronics. This field has just begun to tap some of its possibilities. The future looks bright for anyone in electronics.

Videotape is similar to audio tape. Videotaping scans the tape at the rate of 2 cm/s, giving a scan rate speed of 580 cm/s, or 19.11 ft/s. Stereo VHS is still lacking in audio quality.

The Compact Disc (CD) is capable of storing data in audio or video form. The CD is also available in rewriteable form where it can be used for recording as well as playback only. CD-RW is the terminology now used to designate the rewriteable CD. The CD-I system incorporates a dc real-time operating system which synchronizes interactive video, audio, and text with real time. The CD-R, compact disc–recordable, is a compact disc standard format that, when used with a CD-R or CD-RW write drive, can record up to 650 MB of unerasable data at one time. The CD-ROM XA is a compact disc ROM with extended architecture that enables software, audio, and video to be interleaved on the recorded tracks of the CD-ROM disc. An XA controller card is needed in the PC or computer to play back the contents.

DVD is the electronic shorthand for *digital versatile disc*. It is a newer version of the common 650-MB CD. The DVD is capable of storing up to 17 GB per side. DVD players can play newer DVD discs containing audio and video information and old audio-formatted CDs. DVD-RAM formatted discs are not compatible with DVD-RW discs. The DVD-ROM is a digital video disc with *read only memory*.

MPEG-2 is the abbreviation for *motion picture experts group*. It is a widely used menu of standards for compressing video, making it possible to record more picture and sound information on a smaller disc.

# Review Questions

1. What is a phase-locked loop?
2. What is the use made of the PLL?
3. What is the main objective of a cable TV system?
4. What is a trunk amplifier?
5. Where do you use a bridger amplifier?
6. How high up is a geosynchronous satellite?
7. How many geosynchronous satellites do we now have?
8. What is the specific use made for the geosynchronous satellites?
9. What does the GPS do for automobile owners?
10. How many satellites are used in the GPS?
11. What is a robot?
12. Who was Karl Capek?
13. How are robots controlled?
14. What computer languages do robots use?
15. What is a microprocessor?
16. What are two processes used for identification by the computer of parts viewed by its vision camera?
17. What are some uses for videotape recorders?
18. How can we obtain purer medicines and better crystals for our semiconductors?
19. What is the present limit on the space lab's return cargo?

# Chapter 20

# CELLULAR PHONES, PAGERS, FAXES, AND PRINTERS

Citizen band radio is available to the public but is limited in its ability to communicate for long distance, and the band has a tendency to become crowded with so many people owning transceivers. Operators do not need licenses — only the money to buy the equipment. Class D stations are the most popular, and they have 40 channels to work with. Each channel is 10 kHz wide, in the range 26.965 to 27.405 MHz. The transmitter output power for Class D stations is legally limited to 4 W for AM and 12 W PEP for single Sideband (SSB). Channel 9 is the official channel for emergency calls designed to bring aid to stranded motorists.

Amateur radio (ham radio) is designed for those who are interested in radio techniques and is not available for hire or any other purpose than experimentation by individuals licensed to operate on a band of frequencies. The FCC (Federal Communications Commission) licenses U.S. citizens, regardless of age. Each person must pass a test for a license. It is a test with technical questions and a test of your code receiving and sending skills of International, or Continental, Morse Code. The individual must have a certain level of technical competency and a basic knowledge of electronics. Code speeds are 5 words per minute (wpm), 13 wpm, and 20 wpm, depending on the class of license desired. All ham operators have assigned call letters to identify their stations. Various bands of frequencies are allotted for ham radio use. With the proper equipment it is possible for them to transmit and receive messages worldwide. However, commercial use of these bands of frequencies is not available. To obtain the latest in amateur radio license requirements see web page at: http://www.fcc.gov/cib/consumerfacts/amateur1.html.

With the advent of the integrated circuit (IC or chip) the ability to transmit from hand-held phones became a possibility. However, the distance covered by a small hand-held unit is somewhat limited. Therefore, it became necessary to develop a hand-held phone that could be carried in a purse or on the belt and used in much the same way as a fixed wired telephone. In order to accomplish this miniaturization and portability, cells were developed for the express use of what is now called cellular phones (see Figure 20-1).

**Figure 20-1.** Hand-held cellular telephone.

# Cellular Telephone System

A cellular telephone system is sometimes called *cellular radio*. It is a special form of mobile telephone that was recently developed. A cellular system consists of a network of repeaters. These stations are all connected to one or more central office switching systems. Individual subscribers are provided with, or buy their own, radio transceivers operating on VHF (very high frequency) or UHF (ultrahigh frequency) frequencies.

A network of *repeaters* is located so that most places are always in range of at least one repeater. An ideal would be for every geographic point in the country to be covered. When a subscriber drives a vehicle while operating the phone, operation is automatically switched from repeater to repeater without the operator being aware of the switching (see Figure 20-2). That is because control of repeaters is accomplished by computer. And the transfer signals from one cell to the next is so fast that it is not noticeable to the mobile operator using the system.

In the near future it is possible that all telephone communication will be by cellular radio. Worldwide communications of high quality and low cost, using entirely wireless means, might be achieved by the end of the twenty-first century.

There are many manufacturers of cellular telephone systems.

A cellular system can be designed to meet the demands of both large and small urban areas. It takes continuing expansion in areas of low population density and along travel routes. This flexibility provides the way to a nationwide mobile communications network. Cellular systems provide coverage by subdividing the service area into a network of coverage areas called *cells*. The cell contains a low-powered base station called a *cell site* and is assigned a set of frequencies. Cells that are adjacent to it are assigned different frequency groups to avoid interference. Cells that are far enough apart may use the same frequency groups at the same time. The FCC has authorized 666 channels in the 800-MHz band for cellular communicating. Each cell uses about 40 or 50 frequency pairs. Interference

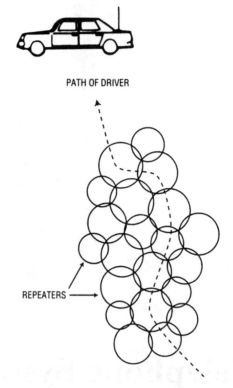

**Figure 20-2.** As the car drives along it may be switched automatically to a number of repeater stations.

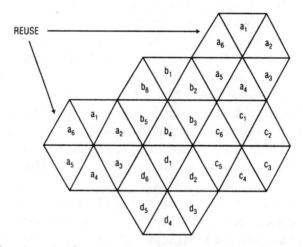

**Figure 20-3.** Physical location provides separation of same frequency cells. However, spacing allows for operation of both cells at the same time by different subscribers.

between cells does not occur because the frequency pairs are not used jointly in adjacent cells (see Figure 20-3). By limiting the cell site transmitter power levels to 100 W ERP (*E*ffective *R*adiated *P*ower), channel reuse is possible only five cells away. This allows reusing each radio channel for different conversations many times within a service area, which means efficient spectrum utilization.

A typical system uses cells ranging from 1 to 12 miles in radius. When the system is first started, the use of large cells is economical. This is because it minimizes the cost while providing the desired coverage.

# Cell Splitting

An advantage of a cellular system is the ability to expand as subscriber demand grows. New cells can be added to the outside of a system at any time. An existing cell can be subdivided by lowering the transmitter power and installing new cell sites. Minimum practical cell radius is around one mile. Smaller cells allow channels to be reused at smaller distances.

As service demand increases, frequency reuse and a process called *cell splitting* can be used to increase cell capacity. Cell splitting requires the addition of new cell sites between existing cell sites. They form a new configuration of smaller cells while using the same number of channels. Each of the new, smaller cells can serve about the same amount of traffic as the original, larger cell. The reuse capacity of the system is thereby increased. Frequencies are reused many more times with the small-cell configuration than with the large-cell configuration. More conversations per given area can be carried on at the same time.

# Processing a Call

*Call setup* and *call handoff* are two of the main functions that must be performed to process a call in a cellular system. For instance, when a mobile unit initiates a call, it sends a request for service and receives a voice channel assignment on one of a set of data channels. These are called *signaling control channels*. In the case of a land-originated call, the mobile is paged and replies to the data channels. If the volume of traffic threatens to exceed the paging capacity of the control channel, the cell site is capable of setting up a separate paging channel. This separate paging channel instructs the mobiles to monitor that channel for calls.

A *sector cell* is an area that has a unique frequency set. A vehicle leaving that area or "sector" requires a new channel assignment. This is called a *handoff*. Sector sharing is sometimes necessary. It is a method of assigning channels so that subscribers requesting service can be temporarily assigned a channel from an adjacent sector if all the channels in the desired sectors are busy. Signaling channel capacity can be increased by creating a new set of control channels and separating the data traffic so that only pages are sent on the original control channels. Channel assignments and mobile or portable originated calls are then processed on the new channels.

# The System

Cellular systems use state-of-the-art digital switching for both base site and subscriber equipment. Modular building blocks allow each system to customize to the characteristics of an area and the needs of the users. A combination of cells is used throughout the cellular geographic service area (CGSA). This provides the composite coverage needed for reliable service (see Figure 20-4).

The capacity of an individual cell is directly related to the number of busy-hour subscribers in the area covered by the cell. As the number of users changes, channels are added or removed from specific cells until the available spectrum allocation is fully utilized. The original cell layout of each specific system should take maximum advantage

**Figure 20-4.** Cellular geographic service area (CGSA). The Eastern Seaboard is one geographic area with Verizon servicing from Baltimore to Buffalo to Maine. Other companies have been allocated similar areas such as from San Francisco to Los Angeles and all along the western U.S. coast. The most heavily traveled interstate highways are covered by cells grouped into various regional company operations. Verizon now owns and operates all GTE systems.

of geographic separation, terrain, and other related factors to facilitate orderly growth with maximum frequency reuse.

Once basic frequency allocation is fully utilized, cells are then subdivided or split to increase the total number of channels in the system. Using directional (sector) antennas, it is possible for the cells reusing the same frequencies to be spaced closer together, thereby providing a significant increase in the system capacity. Limits on frequency reuse depend on the terrain and the geographic distribution of subscribers.

# DYNA T*A*C™

DYNA T*A*C is an acronym used for the Motorola system. The subscriber to the system's service must have equipment that can be used with the system. This includes

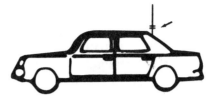

**Figure 20-5.** A vehicle-mounted antenna.

vehicle-mounted units and portable units. The equipment must be specially designed to operate within the 800-MHz cellular system. Whether mobile or portable, all subscriber equipment offers the same extensive user performance features. In addition, each product offers unique features, options, and accessories (see Figure 20-5).

## Cordless Telephones

A cordless telephone uses a radio link for connection to the base unit. This is usually FM. The FM link is used instead of a cord (see Figure 20-6). High frequency means the antennas can be small or short. Two small antennas, one at the main unit and the other at the receiver, allow use of the receiver as far away as 600 to 800 feet under ideal conditions. Some of the more advanced models now have 10 channels that are automatically switched and changed as needed. With almost every home equipped with a cordless phone it is possible that neighbors can interfere with one another if they all have phones that operate on

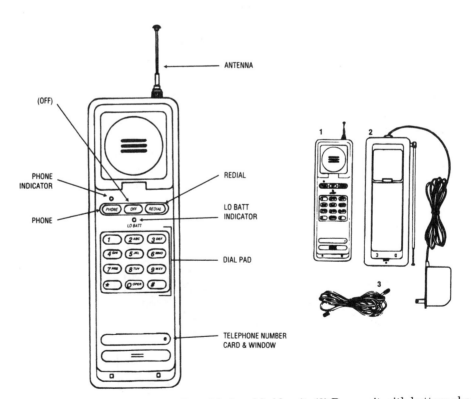

**Figure 20-6.** The cordless telephone. (1) Portable hand-held unit. (2) Base unit with battery charger and transmit/receive antenna. (3) Wire used to connect base unit to the wall phone outlet.

the same frequency. By being able to switch to a less crowded channel, it is possible to converse without interference.

Cordless telephones are available in various configurations. Some units are designed so that a call can be received in a remote place, but an outgoing call must be dialed from the main base. With other units, however, both incoming and outgoing calls can be controlled entirely from the receiver.

Cordless telephones are linked by low-power radio signals. This means they are subject to interference from other appliances as well as from stray electromagnetic impulses. Remember signal-to-noise ratio? However, the technology of the cordless phone is rapidly improving.

# Pagers

Pagers are about the size of a pack of cigarettes and fit into the shirt pocket. They weigh from 6 to 10 ounces and are *receive-only* devices. They may be activated by a series of audible beeps to alert the carrier to report to the base station or to call the base station by telephone. Frequency bands used by pagers are 30 to 50 MHz, 132–174, 406–512, and 929 to 931 MHz. High-gain antennas are placed within the plastic receiver case of most pager units.

Some of the newer pagers use microprocessors to control their capabilities. These devices use subaudible digital codes that permit several hundred different pagers to be selectively called. These newer pagers may emit audible beeps only, or they may beep and open the AF amplifier to allow the receiver to pass a voice message to the paged person.

Controls vary with different pagers. Depending on the pager type and manufacturer, pagers may have an on-off switch, a volume control, a manual squelch (*mute*) control, a beep/voice selection switch, or a push-to-hear switch.

A compact tone and voice FM radio pager is shown in Figure 20-7. This device has high receiver sensitivity and a variable volume control to produce loud, clear audio that allows messages to be heard, whether in a large building, an automobile, or a fringe area of operation. The electronic design of the unit reduces the current consumption so that, when the unit is powered by an AA alkaline battery, longer intervals of pager service are gained before battery replacement. An AA rechargeable nickel-cadmium battery can be used. This type of power source can easily be recharged in a single-unit charger.

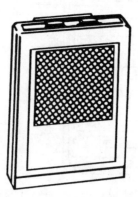

**Figure 20-7.** A shirt pocket–size pager.

The pager battery is monitored continuously. When the pager is turned on, it emits a double interrupted beep to indicate a good battery. If the battery becomes marginal, the pager sounds off with a distinct chirping alert, much like that given off by a smoke and fire detector when its battery is low. After this alerting signal, several more hours of reliable operation are available to allow sufficient time for battery replacement or recharge.

# Facsimiles (Faxes)

One of the latest machines available for the office and home use is the *facsimile*, or *fax*. The word facsimile means *(1) an exact reproduction or copy*, and *(2) the transmission and reproduction of graphic matter by electrical means as by radio or wire*. Since the price of fax machines has dropped in recent years, it is possible to find more and more of them in homes as well as in every office. The instant communication aspect of the fax makes it very useful for a number of commercial and personal applications.

One of the more important uses of the fax is to send weather information to ships at sea. The fax machine is used to transmit weather maps to ships by way of a polar orbiting satellite. Weather charts or maps can be transmitted by *HF* (high-frequency) shore-based stations and received from geostationary applied technology satellites (ATS) and can then be received aboard ships at sea. Only *HF SSB* (single sideband) and 135-MHz satellite fax emissions will be mentioned here, as there is a similarity to those machines used by offices to transmit legal documents and other graphic material. Fax copies are now considered legal and binding when they are properly signed.

## The Fax Picture

A facsimile picture is roughly comparable to a single frame of a TV picture, except that dark and light portions of each line are laid down across a sheet of special paper instead of across a CRT (cathode ray tube). The dark portions are developed electrochemically by applying a direct current (dc) through the paper as the line is being scanned. Instead of a one-thirtieth of a second required for a TV frame, it may take a minute or two to produce a fax picture or map. Some of the latest Fax machines produce "hardcopy" of their contents on plain paper copiers. They utilize the same technology that the photocopier, such as Xerox, uses with drop-in cartridges, which contain the light sensitive drum and toner needed to produce the finished copy in a long-lasting plain paper format.

In Figure 20-8 a basic flat-copy scanner is used to produce the video signals that are transmitted. In this system it is assumed that a standard 120-rpm line scanning rate and 96 lines-per-inch (LPI) are used. The receiving equipment is automatically started by radio with the transmission of a 5-second burst of a 300-Hz tone and stopped with a 5-second burst of a 450-Hz tone.

The picture information is produced through the use of a hollow, rotating, black-coated, 19-inch-long glass drum with a single-turn helical scratch on it from one end to the other. The thin scratch line permits light to penetrate into the drum. A lens system is used to produce an image of the map as the map moves along. It is projected up onto a flat metal surface with a very narrow slit across it, as shown in Figure 20-8. Light that passes through the slit falls on the rotating helix. As it turns once, at 120-rpm rate, the helix

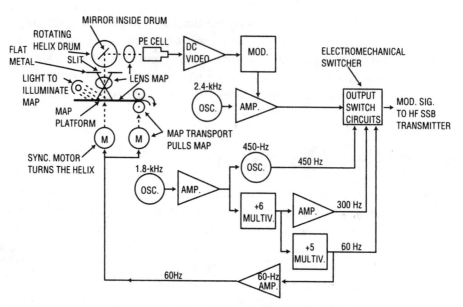

**Figure 20-8.** Facsimile transmitter used to send weather maps to ships at sea.

allows light-dark variations to pass through the rotating scratch on the surface. This generates the video signals for one horizontal line of the map. By using mirrors and lenses, the light impulses are passed through the helix opening and are reflected down the center of the drum to a photoelectric (PE) cell. Here they are amplified by the dc video amplifier. These video signals are used to modulate a 2.4-kHz carrier, with signals from black areas producing maximum carrier strength and signals from light areas developing less carrier strength. During transmission times the video-modulated 2.4-kHz carrier is fed as an AF signal to the AF input of a HF (2–16 MHz) SSB transmitter. Fax receivers pick up this signal.

An oscillator operating at 1.8 kHz (which is produced by a crystal or tuning fork) acts as a frequency standard. Its stabilized output is amplified and used to hold a *divide by 6 multivibrator* in synchronism to produce the ac needed for a 300-Hz start signal. It also locks in a 450-Hz phase shift oscillator to produce the ac for the stop signal. The 300-Hz ac synchronizes a *divide by 5 multivibrator* to produce an exact 60-Hz ac output. The 60-Hz ac is amplified and operates the synchronous helix motor as well as the map transport rollers that pull the map slowly under the video-pickup PE-cell system.

An electromechanical switcher acts as a timing device through cams and contacts to transmit first the 5-second 400-Hz start signal followed by 25 seconds of phasing signal (95 percent of eacn line is black, and 5 percent is white), then a 1-second burst of 60 Hz, and then the map signals start. At the end of map scanning, the stop signal is switched in for 5 seconds and the equipment is then shut down.

There are various transport lines-per-inch (LPI) speeds, helix rotation speeds (rpm), and carrier frequencies other than the standards mentioned. For example, if the starting signal is 853 Hz instead of 300 Hz, the helix motor will be switched to rotate at 240 rpm instead of 120.

The video-modulated 2.4-kHz carrier modulates a HF SSB transmitter. This results in a form of FSK (frequency shift keying). The black signals produce lower frequency sidebands than do the white signals. (This is reversed with some fax transmissions.) As a result, a receiver tuned to the transmitter carrier demodulates the video as tones that range from about 2300 Hz (white) down to about 1500 Hz (black). These frequencies sound to a listener like a warbling tone.

## Slow-Scan Amateur TV

A slow-scan amateur TV uses somewhat similar values: black = 1500 Hz, white = 2300 Hz, 128 lines/picture, 66.7 milliseconds per line including 5 milliseconds for horizontal sync, 30 milliseconds for vertical sync, totaling 8.53 seconds per picture).

## Reception and Printing (FAX)

Facsimile equipment aboard ships usually has its own HF receiver for SSB map transmissions from shore stations. There is also the possibility a 135.6-MHz down-link FM receiver is used to pick up transmissions that are relayed from geostationary satellites. These satellite transmissions are narrowband FM at 10 kHz and are relatively noise-free. The receiver feeds its signal to a facsimile recorder. Figure 20-9 shows a typical block diagram.

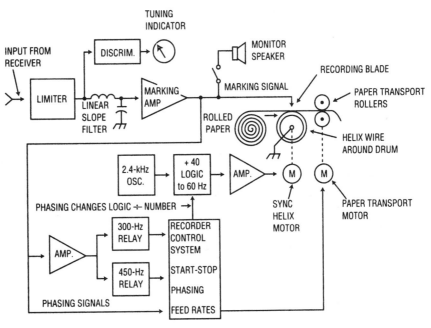

**Figure 20-9.** Facsimile unit that takes the received weather map signals and processes them to produce a weather map.

## Reproduction

Reproduction of maps is accomplished by using a moist, electrosensitive paper that is pulled slowly between a 120-rpm rotating drum and an essentially stationary, ruler-straight, steel electrode blade. The blade extends under and across the width of the moving paper. The paper is 19 inches wide. On the nonconductive revolving drum is mounted a single spiral or helical turn of wire. The whole rotating assembly is called a *helix*. The wire turns while functioning as one electrode, and the steel blade turns while functioning as the other. When paper is between the two electrodes and 60 volts is applied, the paper is electrochemically activated. This is done between the recording blade and the helical wire. That means the paper is marked at this point with a dark spot. Darkness of the marking is directly proportional to the voltage between electrodes.

Rotating the helix produces a left-to-right traveling point of contact between the helix wire and the blade. If a constant 60 volts is applied between the helix and the blade, a dark line is developed across the paper. By modulating the voltage it is possible for the horizontal line to consist of darker and lighter spots. As the right end of the helix wire passes beneath the upper electrode, the left end of the single-turn helix is back at the left side of the paper again and ready to start the next line. Inasmuch as the paper is constantly being pulled ahead slowly in step with the transmitted signal, one line is never laid down over the preceding one.

## Start and Stop Signals

After being detected, the 1500–2300-Hz signals from the receiver are passed through a limiter. The limiter takes off noise and interference that may have been picked up during transmission. Then the signals are passed through a linear-slope filter that passes 1500 Hz 10 times stronger than 2300 Hz, and are amplified. Keep in mind that the received signal serves three functions: (1) It has a marking voltage that marks the paper, (2) it is fed to a speaker for aural monitoring, and (3) it feeds control signals to the recorder control system. By passing it through an amplifier to boast its strength it is then used to actuate either a 300- or a 450-Hz resonant relay. These two frequencies are used to start and stop the recorder. The 25-second phasing signals follow the start signal and are increased by the divide-ratio of a logic circuit that normally divides the 2400-Hz crystal or tuning fork oscillator by 40 to produce 60 Hz ac. The helix motor operates on 60 Hz ac. The synchronous helix motor is in exact synchronism with the scanner helix at the transmitter. The phasing signals should reduce the 60-Hz frequency and slow the helix to move the left margin of the map to the left side of the recording paper. A manual control is also provided in case something does not work right. The manual control is actuated a few times to reduce the 60-Hz frequency and move the margin of the map toward the left edge of the paper.

## Synchronizing Signals

The 1-second, 60-Hz synchronizing signal that follows the phasing signals is not used in this system. However, it can be used to produce map alignment and start signals for other systems.

The 450-Hz stop signal is transmitted when the map is completed. The stop signal is transmitted to shut down the transmitter and, if on automatic mode, also the recorder circuitry.

The received map video signals can be considered either FM or FSK, with signals varying from 2300 to 1500 Hz. The signals, after limiting, can be fed to a 1900-Hz discriminator to a zero-center tuning indicator meter. The signals are tuned in correctly when the meter needle varies the least during map reception.

Excessive wear associated with the helix wire wiping against the blade can be reduced if the blade is made in the form of an endless steel loop moving along about 1/8 in. per minute. Reasonable duty cycle is three to six months with relatively constant use.

# Recorder Paper Feed

Recorder paper feed speed and the map feed rate of the scanner must be the same. Rolls of 19-in. wide electrosensitive paper are available in 170-ft lengths.

It is also possible to receive transmissions from parked satellites. They are relayed signals fed up to the satellites from ground stations. Cloud-cover pictures picked up from weather satellites are overlaid on charts of that portion of the earth's surface.

# Commercial and Home Fax Machines

Fax machines are replacing written communications that utilize the U.S. Postal System. Instant communications makes it possible to conduct business with much quicker response time. Machines made for this purpose are of a variety of designs. They may have a roll of thermal fax paper (image made by heating the paper) or utilize the computer printer technology (see Figure 20-10). Inkjet printers have been added to the fax machines to give better plain paper quality prints. The laser printer has also been adapted to fax use and produces an excellent quality finished product.

Figure 20-10. Plain paper home and/or office fax.

# Computer as a Fax Machine

A computer can be pressed into service as a fax machine by the insertion of a fax modem board. All the benefits of a computer can be made available to the advantage of the facsimile finished product. Figure 20-11 shows one of the latest upright plain paper faxes made to be attached to a computer. It can be used as a printer for the computer, a fax machine, and a copier.

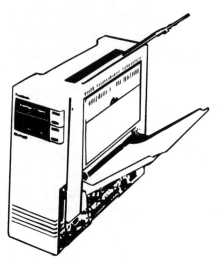

**Figure 20-11.** An upright printer that can be used as a fax or copier as well as a computer printer.

# Inkjet Printers

The ability to print with inkjets has been used since the early 1980s. Inkjet printers spray liquid ink instead of hammering pins on a ribbon to make an image. As nonimpact printers they are quiet and have few moving parts.

The inkjet is described as an ink spraying out onto paper. The print head has a series of little holes called nozzles. What occurs is that each of the nozzles actually heats up the ink, the ink bubbles, and, when the bubbles burst, ink sprays out (see Figure 20-12).

Inkjet printers are now widely used due to the fact that the ink has been tremendously improved. Ink used in early printers had a tendency to dry up in the nozzles, plugging them, and sometimes would run as it hit the paper.

Today, inks have been refined, and early model troubles no longer occur. Print quality of inkjet printers is, to an untrained eye, equal to laser printer output. Inkjet printers produce high-quality, 360 dots per inch (dpi) text, and are also excellent for printing graphs, charts, and other diagrams so often found in material being faxed. Because the dots produced by the inkjet spread slightly when they hit the paper, inkjet images look smoother than dot matrix images. For example, the tiny explosions occur thousands of times a second. It takes about 25 ink dots to dot the letter *i*. One convenient feature of inkjet printers is portability, a category where laser printers can probably never compete. Laser printers use a lot of mechanism to produce an image and getting it down to the size required to be a portable printer will be most difficult.

## Ink Cartridge

The consumable item for the inkjet is the ink cartridge. As for cost per page, one monochrome inkjet page costs just under a penny. This is a little more expensive than dot matrix but significantly cheaper than a laser-printed page. Best of all, prices for monochrome inkjet printers are rather low.

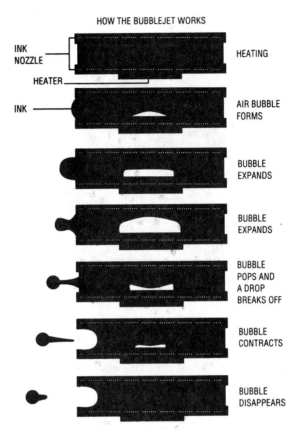

**Figure 20-12.** How the bubblejet printer works.

# Laser and LED Printers

Inkjet printers aren't the only nonimpact printers around. Laser and LED printers offer high-resolution printing for those with higher-end needs (see Figure 20-13).

Compared to the 360 dpi of dot matrix and inkjet printers, lasers only produce 300 dpi output. But if you compare dot matrix or inkjet output with laser output, laser output always wins.

The laser wins out because of the laser printing technology. Laser printing works with a *latent* image, or a kind of after-image. The actual image is first created on an electrically charged drum. Dry ink, or *toner*, sticks to the electrical charges. As the drum turns, it comes into contact with the paper, which is electrically charged too, and the dry ink then sticks to it.

If you lifted the lid of the printer and pulled out the paper before it hits the heat fusing rollers, you could brush off the toner with your hand. The toner has to go through the fusing rollers to make it stick to the page. Once paper goes through the heating unit, the heat melts the toner onto the paper and fixes it.

Because the toner actually runs together when it is fused, it makes the edges of the characters smoother. Instead of seeing a bunch of dots, you see solid black with smooth edges, smoother even than the inkjet edge.

The *laser* in *laser printer* comes from the fact that a laser beam is used to draw the

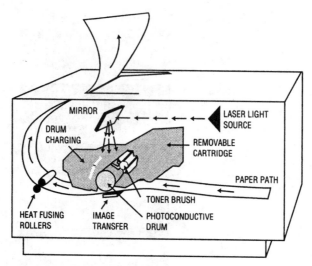

**Figure 20-13.** How the laser printer works.

electric charges on the drum. The laser is stationary, but a moving mirror reflects the beam onto the revolving drum. Wherever the laser beam shines, the drum is charged.

A similar type of printer uses light emitting diodes, or *LED*s, to charge the drum. In the LED printer, the LEDs turn off and on as the drum rolls past them.

Lasers need a motor to spin a mirror, so lasers have more moving parts than LEDs. Whether or not this makes laser printers less reliable than LED printers is a point of debate in the print industry.

Speed is measured in characters per second (cps) on the dot matrix and inkjet printers. Laser printer speeds are measured in terms of pages per minute (ppm). Dot matrix and inkjet printers print character by character, line by line. But laser printers create an image of a page and then print, page by page.

The slowest laser printers print at four pages per minute. Even at that speed, they are four to eight times faster than the 24-pin dot matrix printer in near-letter quality mode. Some laser printers can print at speeds in excess of 20 ppm. Printing graphics takes more time than printing text.

For laser printers the consumable is the cartridge. The cartridge includes toner and drum. Generally, cost for the cartridge averages about three cents per page. Average cost for the laser with all items included runs about 6 times as expensive as the old dot matrix type of printer.

# Summary

Cellular telephone system is sometimes called cellular radio. It is a special form of mobile telephone recently developed that consists of a network of repeaters. Network repeaters are located so that most places are always in range of at least one repeater. In the near future it is possible that all telephone communication will be by cellular radio. A cellular system can be designed to meet the demands of both large and small urban areas.

The FCC has authorized 666 channels in the 800-MHz band for cellular communicating. Each cell uses about 40 to 50 frequency pairs. Interference between cells does not occur because the frequency pairs are not used jointly in adjacent cells.

A typical system uses cells ranging from 1 to 12 miles in radius. When the system is first started, the use of large cells is economical.

Cell splitting is dividing an existing cell by subdividing and lowering the transmitter power and installing new cell sites. Minimum practical cell radius is around 1 mile.

Call setup and call handoff are two of the main functions that must be performed to process a call in a cellular system. For instance, when a mobile unit initiates a call, it sends a request for service and receives a voice channel assignment on one of a set of data channels. These are called signaling control channels.

A sector cell is an area that has a unique frequency set. A vehicle leaving that area or sector requires a new channel assignment. This is called a handoff. Sector sharing is sometimes necessary. The capacity of an individual cell is directly related to the number of busy-hour subscribers in the area covered by the cell.

A cordless telephone uses a radio link or connection to the base unit. This is usually FM. The cordless is effective for a distance of about 600 to 800 feet under ideal conditions. Ten-channel cordless phones are now available to reduce interference with neighbors or persons within the same household where more than one unit operates at the same time. Cordless telephones are linked by low-power radio frequency signals.

Pagers are about the size of a pack of cigarettes and fit into the shirt pocket. They weigh from 6 to 10 ounces and are receive-only devices. They may be activated by a series of audible beeps to alert the carrier to report to the base station or to call the base station by telephone. Frequency bands used by pagers are 30–50 MHz, 132–174 MHz, 406–512 MHz. UHF frequencies of 929–931 MHz are also used by pagers.

The word *facsimile* (fax) means an exact reproduction or copy, and the transmission and reproduction of graphic matter by electrical means as by radio or wire. A facsimile picture is roughly comparable to a single frame of a TV picture, except that dark and light portions of each line are laid down across a sheet of paper instead of across a CRT.

Facsimile machines are used in everyday communications, by police, and ships at sea for weather information.

Fax machines are replacing written communications that would normally use the postal system for delivery. Some fax machines use inkjet technology for printing the image. Others utilize the laser and its ability to give clearer, more detailed printouts. The computer can also be used as a fax machine if a scanner is attached and a modem board is installed in the computer. Plain paper fax is rapidly replacing the older thermal roll paper that was once associated with the old copier technology.

# Review Questions

1. How many channels are allocated to Class D citizen band radio?
2. What is the purpose of ham radio? What is another name for it?
3. What is another name for the cellular phone system?
4. What is a cell?
5. What is a repeater?
6. What is cell splitting?

7. Define: *call setup*, *handoff*, *sector cell*, and *signaling control channels*.
8. What is CGSA?
9. What is the usual range for a cordless telephone?
10. What is a pager?
11. What are the frequencies used for pagers?
12. Define facsimile.
13. Who uses facsimiles?
14. What does a fax machine do?
15. How are inkjet printers used in fax machines?
16. How are laser printers used in fax machines?
17. What is toner?

# Chapter 21

# ELECTRONICS IN THE FUTURE

The future of electronics seems to be tied to the development of superconductivity. The ability of a material to conduct more efficiently at low temperatures has caused some hectic research programs to come closer to making practical applications a reality. A Nobel Prize was awarded in 1987 for research in superconductivity.

Ceramic materials have become very interesting to the researchers. Yttrium-barium compounds have become the base material of superconductivity. Keep in mind that yttrium and barium are not good conductors; however, under certain conditions they behave differently. It is this behavior that is the subject of much study and development.

One of the first practical applications for the superconductor will probably be developed for the U.S. Department of Energy since it has assigned the Argonne National Laboratory in Illinois the major responsibility for developing a practical wire from the rapidly proliferating number of "high-temperature" superconducting ceramic materials recently discovered. The project's main objective is to produce a commercially usable wire that will become superconducting at liquid-nitrogen temperature (77 Kelvin or –321°C).

A wire for transmission lines is the goal. It has to be about 0.1428-inch thick and capable of carrying 100 amperes when chilled by liquid-nitrogen.

Other uses for superconductors are just beginning to surface. As more materials are tested and proven, more applications will be developed, and devices not possible at today's technology will emerge. Many are predicting that this superconductivity will do to the electronics field what the semiconductor did.

## Fiber Optics

Fiber optics have already been used to tell the driver of a Cadillac if the headlights, rear taillights, and other lights are operating properly. They have been made into long lines to carry telephone signals with great clarity. Newer developments have made it possible to move data along the fiber optics line for distances of 90 miles without having to install a repeater. This area of electronics is just opening with the advent of many optoelectronic devices.

## Digital Processing

Digital processing chips are available for doing a wide range of activities. In fact, the field is just beginning to heat up, and more products using digital processing are being developed to take advantage of all the circuit configurations available in chip form. NASA used it for enhancing the images of space probe pictures; now television sets are using the digital processing to get rid of "snow" and to improve the quality of pictures from weak signals. Image enhancement is but one of the uses for digital processing. Speech recognition for the operation of computers is but a short time away. Adaptive filtering using digital processing can be used to eliminate the echo effect sometimes found on telephone lines.

# Automobile Electronics

The automobile is using more and more electronics to do jobs once thought to be the sole responsibility of mechanical devices. Ignition is presently controlled by computers as are braking and steering. The near future holds promise for a collision avoidance system as well as for an electronically controlled suspension system to make the ride smoother. As the automobile becomes more sophisticated electronically, the demand for electronics technicians will increase.

# Communications

The ability to contact people on the phone no matter where they are will soon be economically possible. This will be due to the development of smaller transmitters and receivers that operate on frequencies that can be utilized more efficiently with single-sideband modulation on narrow band FM.

Data links are constantly being improved with more and more data being transmitted in shorter and shorter time periods. Better utilization of satellites will result in access to different types of markets for manufacturers.

# Home Appliances

The home appliance market has become digitized with the advent of the home computer and the timers on dishwashers, microwave ovens, and conventional ovens. Even the washing machine and dryer have digitized electronics for control.

Television receivers have flat pictures that can be placed on the wall and operated as a part of the room decor. Small receivers are already operational for use by those who want to take the television with them on their arm in a LCD picture with a 1-inch format. Color TV pictures using LCD are also available.

# Medical Electronics

The future of medical electronics is limited only by the imagination of those developing it. The *Star Trek* ideas of medical treatment are not far from implementation.

# CCD Television Cameras

There are three types of television cameras: the regular vacuum tube type (see page 294, Fig. 15-1), the MOS or metal oxide semiconductor type, and the CCD or charged coupled device type. Solid state MOS and CCD pickup devices are very similar in operation and performance; however, there are a few significant differences. Conversion of light to electrical energy occurs at each of the individual photodiodes, which produce a small electrical charge when light from the scene is focused on their exposed surface. A method of matrix scanning is used to repeatedly collect each of these charges and assemble them into a video signal. The scanning method used to collect these charges is one of the major differences between MOS and CCD pickup devices.

MOS devices use a scanning method that results in three or four signal output lines. These lines carry white, yellow, cyan, and green color signals; there is no green in the older three-line devices. One disadvantage of MOS devices is that the output signals are at a fairly low level (40–50 mV) and require low-noise preamps to bring the signals up to a usable level for standard signal color-processing circuits.

CCD devices use a scanning method that results in a single video output line. This signal contains all the necessary luminance and chrominance information required to generate NTSC (National Television Systems Committee) composite video. Also, the level of the output signal is high enough that no preamp is required. An advantage of CCD devices is that they have been more reliable than MOS devices. This type of camera or pickup is used in many installations where the Internet is utilized to carry the picture. See Figure 21-1 for a block diagram on how the CCD converts reflected scene lighting into electrical signals.

Solid state detectors such as those used in CCD devices utilize semiconductor photodiodes to detect light by causing photons to excite electrons from immobile, bound states of the semiconductor (the valance band) to a state where the electrons are mobile (the conduction band). The mobile electrons (those which can be moved easily) in the conduction band and the vacancies, or "holes," in the valence band can be moved through the solid with externally

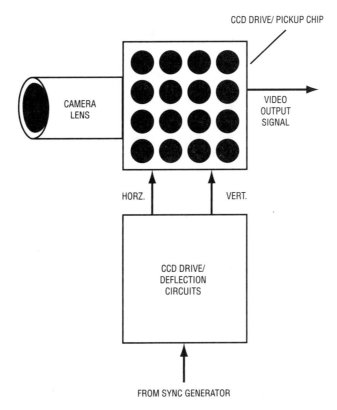

**Figure 21-1.** CCD camera.

applied electric fields, collected onto a metal electrode, and sensed as a photo-induced current. Microfabrication techniques developed for the integrated-circuit semiconductor industry are now used to construct large arrays of individual photodiodes closely spaced together. The CCD permits the charges that are collected by the individual diodes to be read out separately and displayed as an image for uses such as in compact and rugged television cameras.

# The Internet

The Internet is a rich on-line source of information, from bulletin boards and discussion groups to electronic mail and up-to-date news information. A number of providers connect subscribers to the worldwide web of computers owned by individuals everywhere. Each computer owner has an assigned e-mail address after the computer is connected to the network. The Internet can be accessed through an Internet service provider or a direct server connection. Then, the subscriber can use ftp and Telnet (browsing utilities that come with Windows-based programs) to browse the World Wide Web (WWW). Of course, subscribers need a local telephone company line connection and a modem to connect the computer to the line.

## Browser

An Internet browser or web browser is a computer program that allows users to download World Wide Web pages for viewing on their own computers. Two popular browser programs are Netscape Navigator and Microsoft Internet Explorer. The first browser program was called *Mosaic*. Today's browsers are able to provide graphics as well as text.

## Faster Access and Response

Sometimes it takes a few minutes to get connected to the Internet and downloading larger files to your computer may take an hour or so. When this happens, you then start to think in terms of making the connection instantly and getting the information you want just as quickly. That is where the digital subscriber loop, or digital subscriber line (DSL), becomes an important item in your search for speed. See Figure 21-2. Line is used by the public and loop is used by the technicians who work on the telecommunications equipment.

## Wide Area Network

The wide area network (WAN) is a network of computers or computing devices connected by telephone lines that extend beyond an area code's service area. See Figure 21-3. An example of a WAN application is a computer that accesses another computer in another state to acquire information. Popular ways for computers to connect over long distances are by using a dial-up modem, a frame relay circuit, an ATM circuit, an ISDN circuit, or a 56-kilobyte leased line. There are advantages and disadvantages to each of these services which are offered by long-distance telephone companies. The faster and more reliable the service, the more expensive it is. Frame relay is rapidly becoming the most economical WAN protocol for applications that frequently transfer data over long distances. See Figure 21-2. A protocol is a set of organized rules by which communications equipment transfers bits and bytes (data).

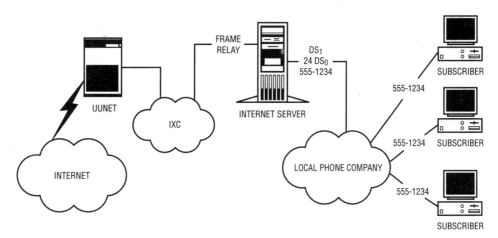

**Figure 21-2.** Internet service provider. IXC — Inter Exchange Carrier — a long-distance carrier like AT&T; DS1 — Digital Service, Level 1; DS0 — Digital Service, Level 0.

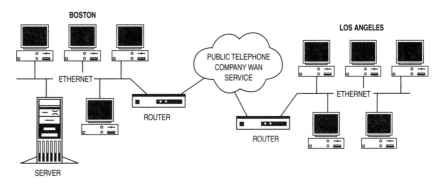

**Figure 21-3.** Wide area network. Ethernet — a family of local area network (LAN) protocols; router — a device used to forward packets from one device to another, based on address information; server — a computer dedicated to providing services to other computers.

The many communications protocols and layers of protocols that carry other protocols (called protocol stacks), include ISDN (Integrated Services Digital Network), Ethernet, token ring, POTS signaling, DSI, ATM, frame relay, and SONET. POTS means plain old telephone service line. SONET stands for synchronous optical network, a broadband transport system. It is implemented over fiber optic and is able to be configured in a ring, which allows it to reroute traffic with no interruption of service if a fiber is cut. ATM is asynchronous transfer mode and is a standard communications protocol. ATM is a frame–format communications protocol whereby data is transmitted and received 53 bytes or octets at a time. There are 48 customer bytes (payload) and 5 bytes for control and addressing.

# Summary

The future of electronics is bright. Electronics thrives on challenge. There will be many new demands to be met in the future. Change comes with the development of new devices. That means there will be a need to solve new problems created by change. Electronics will be there to rise to the challenge.

# GLOSSARY

*Absolute Units* — Units that have no prefixes. A good example is volts, amperes, ohms, and watts.

*Accuracy* — A term used in connection with a meter indicating how close to the true or actual value the meter reading actually is.

*Active Component* — A device whose characteristics can change during normal circuit operation.

*Alternating Current (ac)* — A current that is constantly changing in amplitude and direction.

*Ampere* — A unit of measurement for current. It is equal to one coulomb in motion. One coulomb per second is one ampere.

*Amplification* — A process whereby a small current controls a larger one in a transistor circuit, but where a small voltage controls a larger voltage in a vacuum tube circuit.

*Amplification Factor* — A ratio that exists between the base current and the collector current in a transistor.

*Amplifier* — A device made up of a number of circuits and designed to increase the size or amplitude of signals.

*Amplitude* — A term used to designate the height of an ac voltage or current.

*Analog Meter* — A meter that has the pointer deflection proportional to the current flow. The scale is placed on a clock-like dial and the pointer position indicates the value being read.

*Analog Multimeter* — A multimeter that displays its measured values by the displacement of a pointer over a calibrated scale.

*Anode* — The positive section of a diode. It can also be used to indicate the plate of a vacuum tube.

*Apparent Power* — The power that appears to be used in an alternating current circuit. It is measured in volt-amperes (VA). AP = volts × amps. The term is necessary to differentiate between the energy needed to power a circuit with inductance and/or capacitance and one of a purely resistive nature. True power or TP is used to represent power actually consumed by resistance. It is measured in watts.

*Attenuation* — Reducing circuit voltage by some proportion.

*Attenuator* — The device used to proportionally reduce voltage.

*Audio Frequency* — The range of frequencies that can be heard by the human ear.

*Audio Frequency Oscillator* — A circuit or in some instances a device used to produce sound in the audio range.

*Back EMF* — A force produced in an inductor that is opposed to that which produced it. It is also called counter emf.

*Bandwidth* — A number that describes the frequency range between the upper and lower half power points of a resonant circuit. A range of frequencies that has a resonant effect on *LC* circuits.

*Base* — One of the three terminals on a transistor. Its symbol is readily identified by a bar.

*Battery* — An electrochemical device that serves as a dc power source. A battery consists of two or more cells.

*Beat* — A term meaning to place two frequencies in a circuit so as to allow them to beat one against the other to produce totally different results — usually the sum and difference frequencies as well as the original two. Beat Frequency Oscillators are used to produce a frequency to beat against an incoming radio frequency so that the difference between the two is an audio frequency.

*Beta* — A term used to indicate amplification factor in transistors.

*BFO (Beat Frequency Oscillator)* — An oscillator used to generate a frequency that will beat against another to produce a difference or sum frequency. Used with CW reception to produce an audible note.

*Bias* — A fixed dc voltage that is applied to a circuit before a signal is injected. It establishes the operating characteristics of a particular transistor or vacuum tube.

*Binary* — A number system based on 2. In binary, only two numbers, 0 and 1, represent all possible mathematical values. Computers use the binary number system because it best represents what a computer understands — on and off.

*Bipolar-Junction Transistor* — A type of transistor that has the current between two terminals controlled by the current through a third terminal. It is abbreviated as BJT.

*Bit* — The smallest unit of data recognizable by the computer. Eight bits equal one byte.

*Bleeder Resistor* — A resistor connected across the output of a power supply filter circuit for the purpose of discharging the capacitors when the power is turned off. It also aids in the voltage regulation of the power supply.

*Block Diagram* — A drawing that shows the flow of information or the functions of a circuit in small blocks properly labeled.

*Breadboarding* — The assembling of experimental circuits so that they can be easily modified and tested before manufacturing.

*Byte* — Consisting of eight bits, a byte is the amount of storage space necessary to hold one alphanumeric character such as the letter A, the number 5, a comma, question mark, and so on.

*Capacitance* — That property of a capacitor that opposes any change in circuit voltage.

*Capacitive Filter* — A filter circuit that uses a capacitor connected in parallel with the load.

*Capacitive Reactance ($X_C$)* — The opposition a capacitor produces when the circuit has alternating current. It is measured in ohms.

*Capacitor* — A device that consists of two plates and a dielectric material.

*Cathode* — The n-type material section of a diode. In vacuum tubes it is the element that gives off electrons when heated sufficiently. It also has a negative charge. In semiconductor diodes it is the positive terminal.

*Cathode Ray Tube (CRT)* — A vacuum tube used in oscilloscopes for displaying Lissajous figures and in television sets as a picture tube. It is also used as a monitor in computer terminals.

*CD* — compact disc. A device used to store data or audio or video signals. Variations in the disc surface are "read" by a laser beam and converted to electrical impulses.

*CD-ROM (Compact Disc, Read-Only Memory)* — A form of data storage that uses laser optics rather than magnetic means for reading data. CD-ROMs read compact discs similar to the audio CDs available in music stores.

*Cellular Telephone* — A telephone system that utilizes a series of repeater stations to keep the subscriber in contact with the person on the other end of the line.

*Center Tap* — A connection to the center of the winding of a transformer.

*Ceramic Capacitor* — A capacitor made with a ceramic material used as its dielectric.

*Choke* — Another name for an inductor or coil of wire. It is given the name choke for its ability to hold back current flow in an ac circuit or its ability to choke or delay current flow.

*Circuit* — A path for electrons to move through components and wires.

*Circuit Diagram* — A drawing made in shorthand (using symbols) to show how components are connected.

*Coaxial Cable* — A type of cable used in television and radio. It handles high-frequency energy through a solid center conductor and an outer conductor of woven wire that acts as a shield when grounded.

*Coefficient of Coupling* — A term used to indicate the effectiveness of primary-to-secondary transfer of energy in a transformer.

*Coil* — A loop or series of loops of wire. Another name is a choke with the ability to hold back or choke the current flow in an ac circuit.

*Collector* — One of the output terminals of a transistor. It is represented by a straight line at an angle with the base terminal. Its symbol does not have an arrowhead.

*Collimator Lens* — A lens used in a CD player to cause the light beam to consist of parallel beams of light.

*Comparator Circuits* — Circuits that may be used in a digital multimeter that makes comparisons of voltages.

*Component* — A term that is used to refer to small electronic parts.

*Connector* — A device that permits electrical separation of one part of a system from another while providing convenience in disconnection and connection.

*Conventional Memory* — RAM (Random-Access-Memory) that DOS uses to run software programs. Conventional memory is limited to 640 kilobytes.

*Cordless Phone* — A hand-held phone that stays in touch with its base unit with the aid of a FM transmitter and receiver.

*Counter* — An electronic circuit that counts pulses.

*Current* — The flow of electrons from negative to positive.

*Cursor* — A short blinking line that appears underneath the space where the next character is to be typed or deleted. The cursor indicates that the computer is waiting for the user to input a command or information.

*Cutoff* — A condition of a transistor where minimum current flows.

*Cycle* — An older term used to represent frequency. Today, Hertz (Hz) is used for cycles per second. The term is still used to indicate the completion of a cycle or 360° of rotation of a generator.

*D'Arsonval Movement* — A type of meter movement that uses the moving coil along with a stationary permanent magnet to produce its reading.

*Deflection Plates* — Plates that are elements in a cathode ray tube, used to electrically control the beam position from left to right and up and down.

*Desoldering Tool* — Usually a vacuum-operated device that lifts molten solder from a terminal or a printed circuit board to aid in the replacement of a defective component.

*Dielectric* — The insulating material located between the two places of a capacitor.

*Dielectric Constant* — A number that indicates the relative quality of dielectric materials. Air is 1, and all other materials are compared with air to produce a constant.

*Dielectric Strength* — A number expressed in volts per mil. It indicates the breakdown voltage of dielectric materials.

*Digitizing* — The process of converting analog music or information into digits for use by computers or other digital circuited devices.

*Diode* — A two-terminal device. The di means two and the ode means electrode. Therefore, it has two electrodes. The diode is used to allow current to flow in only one direction.

*DIP* — An abbreviation that is used to indicate a type of package used for integrated circuits. It means Dual in-line package.

*Direction Current (dc)* — Current that flows in only one direction.

*Disc* — A flat surface used to store data, either electromagnetically or optically.

*Diskette* — A flat piece of flexible plastic covered with a magnetic coating that is used to store data (also called a floppy diskette). The existing standard for diskette size is 3.5 inches. Older drives also came in 5.25-inch sizes.

*DMM* — An abbreviation used to indicate a digital multimeter.

*Doping* — A process whereby a small amount of impurity is added to a semiconductor material (Germanium or Silicon) to change its properties and cause it to operate as a transistor or diode.

*DOS* — Disk Operating System. Software that translates the user's commands and allows application programs to interact with the computer's hardware. Supplies a file management system for efficient disk input and output.

*Dry Cell* — A primary cell that is made up of zinc and carbon with an electrolyte of sal ammoniac. Other types are also available.

*Eddy Current* — An induced current in the core of a transformer, inductor, or solenoid that is caused by the varying magnetic field produced by alternating current.

*Effective Voltage* — Another term used to describe an rms voltage. It is 0.7071 times the peak voltage.

*EHF* — Extremely High Frequency. The frequency range from 30,000 to 300,000 MHz or 30 GHz to 300 GHz.

*Electricity* — The flow of electrons along a conductor.

*Electronic Component* — A term that describes an individual part used in electrical or electronic circuits.

*Electrolytic Capacitor* — A capacitor that has polarity. It has more than 1 microfarad of capacitance and is marked with a + or −. It is made with a paste or liquid dielectric that produces a high capacitance per unit volume.

*Electromagnet* — A magnet that has a field that is produced by an electric current through a coil of wire.

*Electromagnetism* — A temporary magnetic field produced by electron flow through a coil of wire.

*Electromotive Force (emf)* — Another term used for voltage or the force that moves electrons.

*Electron* — A negatively charged particle of an atom.

*Electronics* — A term used to describe the physical actions of electrons in devices such as diodes, vacuum tubes, transistors, and integrated circuits.

*Electrostatic Field* — The energy field surrounding a charged body. The field is represented by lines of force.

*Element* — A number of definitions can be given. It is a basic substance such as iron or lead. It is also used to describe a part of an electronic device such as the cathode or anode element of a transistor or tube.

*Emitter* — A terminal of a transistor. It is identified by its characteristic arrow symbol.

*Expanded Memory* — Memory outside the DOS one-megabyte limit that is accessed in revolving blocks.

*Extended Memory* — Linear memory extending beyond DOS's one-megabyte limit. Extended memory is only available on 286 and above machines. It is off-limits to XTs.

*Facsimile (Fax)* — A copy or exact reproduction. A fax machine is a device that can receive over the air or by wire electrical impulses that can be converted into a printed page of text or pictures.

*Farad* — The unit of measurement for capacitance. The symbol is C. It is abbreviated by F. It is also broken down into microfarads (0.000 001 F) and picofarads (0.000 000 000 001 F).

*Feedback* — The process of feeding back to the input of a circuit part of its output. Feedback is needed to keep oscillators operating.

*FET* — An abbreviation for the Field Effect Transistor.

*FET Multimeter* — A multimeter that utilizes the high resistance of the FET transistor to its advantage to increase the sensitivity of the meter.

*Filter* — A circuit that is usually connected across the output of a power supply rectifier to reduce the ripple that results from full-wave or half-wave rectification of alternating current.

*Fixed Capacitor* — A term applied to capacitors that have a fixed or one value.

*Flux* — A term that can refer to the magnetic field of a coil or that can be used in reference to the material that is used in the core of solder to keep the heated area from becoming oxidized while being soldered.

*Flux Density* — The number of magnetic lines per unit area. The term used to measure flux density is the tesla (T).

*Format* — To prepare a disk or diskette so it can store information. Formatting organizes the tracks and sectors that store information. When you format a disk, you erase any information already stored on it.

*Forward Bias* — A term used to indicate a positive voltage to the p-type side of a diode or transistor and a negative voltage to the n-type side. Biasing controls the conduction or no conduction of a transistor or diode.

*Frequency* — A term indicating how many times something occurs in a given time period. In electronics it is the number of times something oscillates per second.

*Front-to-Back Ratio* — The ratio of reverse resistance to forward resistance in a diode.

*Full-Wave Bridge Rectifier* — A type of rectifier that uses 4 diodes in a diamond-shaped arrangement to produce full-wave rectification of an alternating current.

*Full-Wave Rectifier* — A type of rectifier that uses two diodes to convert ac to dc.

*Gain* — A proportional increase in voltage or current produced by an electronic circuit.

*Gate* — A terminal of a silicon controlled rectifier (SCR).

*Gauge* — (now used as gage) The size of a piece of wire in its cross-sectional area referred to as its gage number. A gage is also used to measure the diameter or number of the wire.

*Generator* — An electromechanical device used to produce electrical power.

*Gigahertz* — 1000 MHz or 1,000,000,000 Hz.

*Ground* — A term that can be used to designate the actual connection to earth, or that can be used to designate a common connection point in an electronic device.

*Grounding* — A term that means to establish a common connection to earth or ground.

*Half-Wave Rectification* — A single diode that converts only half of the ac voltage cycle to dc voltage. This produces a pulsating dc.

*Hard Disk* — A data storage device for personal computers that consists of rigid platters that are fixed inside a sealed casing. A hard disk can store more information and retrieve data faster than a diskette.

*Hardware* — Any part of a computer system that can be physically touched. Printers, keyboards, monitors (screens), and the computer itself are all hardware.

*Harmonic* — An even number multiple of a given frequency.

*Heat Sink* — A clamp or some other device used to dissipate undesired heat from the leads of heat-sensitive components during soldering. It can also refer to large fins placed on transistors and diodes to dissipate heat generated during operation.

*Henry* — A unit of measurement for inductance. H is the symbol for the henry. It may also be seen as millihenry (0.001 mH) and as microhenry (0.000 001 μH).

*Hertz (Hz)* — The unit of measure for frequency. It may also be seen as kilohertz (kHz) or megahertz (MHz).

*Heterodyne* — The beating together of two different frequencies to produce the sum and difference of the original two.

*Hybrid Circuit* — A circuit that is made with a combination of hard wiring and printed circuits or with integrated circuits.

*Hysteresis* — The magnetic loss in a transformer or inductor caused by residual magnetism of the core.

*Icon* — A visual symbol used to represent programs or documents in a graphical user interface that allows the user to point at a list of command options instead of typing a character-based command.

*IF* — Intermediate Frequencies. Those frequencies that result from the beating of two frequencies and are selected for use as a carrier frequency throughout a receiver in the heterodyning process.

*Impedance* — A unit of opposition that is also measured in ohms. It represents the total opposition to current flow in an ac circuit consisting of any combination of resistance, capacitance and inductance. The symbol for impedance is Z.

*Impedance Matching* — A term that usually refers to the use of transformers to match one amplifier stage to another or one device such as a microphone to another to allow for the maximum transfer of energy.

*Impedance Ratio* — The relationship between the impedance of the primary and secondary of a transformer.

*Induced Current* — The current caused by magnetic lines cutting a conductor.

*Inductance* — The ability of a coil, choke, or inductor to oppose any change in circuit current. It is represented by the symbol L and is measured in henrys.

*Inductive Filter* — A filter circuit that uses an inductor connected in series with the power supply load.

*Inductive Reactance ($X_L$)* — The opposition to current flow in an alternating current circuit caused by an inductor. It is measured in ohms.

*Inductor* — A device that is used to oppose any change in circuit current.

*Inkjet Printer* — A type of printer that uses a printhead that has a series of little holes called nozzles. The nozzles "spray" ink onto the paper.

*Input* — A term that can be the location where signals or information enters a stage or system. That information itself may also be referred to as the input.

*Instantaneous Voltage* — A term used to indicate the voltage at a specified point in the cycle.

*Insulator* — Any material that inhibits or slows the flow of electrons.

*Integrated Circuit (IC)* — A device that is made up of a collection of electronic components and their connections. They are manufactured by doping a slice of silicon.

*Kilo* — The prefix for one thousand (1000).

*Kilobyte* — A unit of measurement of computer memory equivalent to approximately one thousand (1024) bytes.

*Kirchhoff's Current Law* — A law that describes the relationships that determine current at a junction in the analysis of complex circuits. It states that the algebraic sum of currents at a junction equals zero. In other words, the current flowing to the junction flows away from the junction.

*Kirchhoff Voltage Law* — A law that states that the algebraic sum of voltage drops around a complete loop equals zero. Or, the sum of the voltage drops around a complete loop equals the applied voltage.

*Laser* — An acronym for *l*ight *a*mplification by *s*timu-lated *e*mission of *r*adiation. It is a device that uses radiant energy to produce a narrow, highly intense beam of light that can be focused over long distances. Other uses, especially in the medical field, are being developed every day.

*Laser Beam* — A beam of light produced for the purpose of "reading" information on a CD or for other uses.

*Laser Diode* — A diode used to generate a light source for a CD player.

*Laser Printer* — A computer printer that uses the elec-trophotographic method of printing (like a photocopier) with a laser beam as part of the light source. Laser printers produce high-resolution copy and are especially popular with business users.

*Law of Magnets* — A law that states that like poles repel and unlike poles attract.

*Left-Hand Rule* — A rule that states that if a current-carrying conductor is grasped with the left hand and the thumb pointing in the direction of current flow, then the fingers will indicate the direction of the magnetic lines around the conductor.

*Light Emitting Diode (LED)* — A semiconductor device that glows when it is forward biased. It is used to form the segments of a number in a digital display. It is also used as a pilot light.

*Lissajous Figure* — An oscilloscope waveform that shows the relationship between two frequencies.

*Magnet* — An object that possesses magnetism and attracts magnetic materials.

*Magnetic Field* — A force that surrounds a magnet. It produces magnetic lines.

*Magnetic Flux* — The field around a magnet produced by magnetic lines is called a flux field.

*Magnetic Lines* — Invisible lines that form loops around a magnet. They leave the north pole and enter the south pole.

*Magnetism* — The ability to attract nickel, iron, or cobalt.

*Maser* — A term for an experiment for the amplification of microwave frequencies.

*Measurement* — The ability to measure.

*Mega* — A prefix used to designate 1,000,000. (Also written as 1 M.)

*Megabyte* — A unit of measurement of computer memory equivalent to approximately one million (1,048,576) bytes.

*Memory* — An area where your computer stores data. Data can be permanently stored in ROM (read only memory) or stored temporarily in the computer's RAM (random-access-memory). A computer's RAM is emptied when the power is turned off.

*Menu* — A list of choices that appear in menu programs. You select an option by using a mouse or arrow key to highlight it. Some menus are pull-down menus; the top of the screen and options for that heading appear when choosing that category.

*Micro* — A prefix used to designate 0.000 001. (Also written as $10^{-6}$.)

*Microprocessor* — An integrated circuit containing all the central processing functions of a computer, also called a CPU.

*Milli* — A prefix used to designate 0.001. (Also written as $10^{-3}$.)

*Modem* — A piece of computer hardware that allows a computer to communicate with other computers (if they have modems attached) through telephone lines.

*Modulation* — The process whereby an audio frequency is impressed upon a radio frequency carrier.

*Molecule* — The smallest quantity of a compound made of elements.

*Multimeter* — A test instrument or meter used to measure voltage, current, and resistance.

*Multiplier* — A resistor used in a voltmeter to increase its ability to measure higher voltages. It is placed in series with the meter movement and is used to extend the meter range.

*Mutual Inductance* — The inductance that results when there is an interaction of adjacent inductors. It is measured in henrys.

*Nanometer* — Nano means $1 \times 10^{-9}$ or 0.000 000 001 meter.

*Needle-Nose Pliers* — A long, thin-jawed pair of pliers used for holding parts in areas where a wider jaw could not enter.

*Negative Ion* — An atom with a deficiency or excess of electrons.

*Newton* — a metric unit (SI) of measurement of force. It is the force necessary to accelerate a mass of one kilogram at the rate of one meter per second each second. One newton is equal to a force of 0.2248 pound.

*NPN Transistor* — A transistor made with a combination of n, p materials arranged in an n-p-n sequence. This type uses an n-material for the emitter, a p-type material for the base, and an n-type material for the collector.

*N-Type Material* — A piece of silicon or germanium that has been doped with donor atoms and has excess electrons.

*Neutron* — A neutral particle that has no charge and is located in an atom's nucleus.

*Nucleus* — The central part of an atom that contains neutrons and protons.

*Ohm's Law* — A law that says that the current is inversely proportional to the resistance in a circuit. It also says that the voltage and current are directly proportional.

*Open Circuit* — A term that means that the circuit has an interrupted flow of current. It may be caused unintentionally or intentionally with a switch.

*Optosensor* — A device used to sense light.

*Oscilloscope* — A device used to display and measure voltage and time.

*Output* — The location where signals or information leave a circuit or system and that information may itself be referred to as the output.

*Pager* — A compact receive-only device that alerts the carrier to call his or her home base and report in.

*Passive Component* — A device whose characteristics remain constant during normal circuit operation. In other words, it does not change as a result of power being applied to the circuit.

*Peak-to-Peak Voltage (p-p)* — A term that represents the highest positive and negative levels reached during a complete cycle of alternating current. It is represented as p-p on most meters. Rms voltage times 2.828 equals p-p voltage.

*Peak Voltage* — The highest positive or negative level reached by a voltage alternation.

*Permanent Magnet* — A magnet that retains its magnetism even after a magnetizing force has been removed.

*Permeability* — The ease with which magnetic lines pass through an object.

*Phase* — The relationship between zero-voltage crossings of two signals is expressed in degrees.

*Photodetector* — A device used to detect light.

*Piezoelectrical Effect* — A method of producing electricity by applying pressure on a quartz crystal or Rochelle salts.

*Pi Filter* — A filter circuit that uses an inductor in series with the output of a power supply. There are two capacitors across the line, one before the inductor and one after forming a pi configuration.

*PIN Diode* — A photodiode. A pin-type silicon photo diode type of photo-resistor. Photons of heat or light pass through the P Si layer and strike atoms in the intrinsic layer and produce free electrons and holes. Electrons move to the positive bias potential, and electrons fill the holes from the negative potential. The stronger the light, the more electron-hole pairs developed and the more current that can flow. This is not really a true diode but a fast-acting photo-resistive device.

*Pit* — A slight depression on an audio compact disc.

*PN Junction* — The place in a semiconductor diode or transistor where the p- and n-type materials are joined.

*PNP Transistor* — A transistor that uses p and n materials in a p-n-p order to form the device. P-type material is used for the emitter and collector with the base material being n-type.

*Polarity* — Two points of a voltage source that is normally identified as negative and positive. Polarity is also used to indicate the magnetic properties of a magnet.

*Positive Ion* — An atom with a deficiency of one electron is called a positive ion.

*Potentiometer* — A three-terminal variable resistor.

*Power* — Voltage times current equals power (P) and it is measured in watts (W).

*Power Factor* — A ratio of the power actually dissipated or used as compared to what appears to be used. It can be expressed in decimal form or in percentage form.

*Power Rating* — A rating of how much heat a resistor can dissipate.

*Precision Resistor* — A resistor with a tolerance of less than 1% is usually referred to as being a precision type.

*Preventive Maintenance* — A term used to indicate a scheduled process of inspections and corrections conducted to make sure that electrical or electronic devices operate properly.

*Primary* — The input side of a transformer.

*Printed Circuit* — A phenolic or fiberglass board with copper lines etched on it. The copper is bonded to the board material and then etched away to allow for the desired connections of component parts.

*Prism* — A device used to break up a beam of white light into its component colors.

*Prompt* — The DOS prompt usually takes the form of either C> or A>. The C refers to the hard drive located inside your computer and the A refers to the floppy diskette drive. Most computers sold today come with two floppy diskette drives; in this case, one is called the A: drive and the other is called the B: drive.

*Proton* — The positively charged particle in an atom's nucleus is called a proton.

*P-Type Material* — The semiconductor material that has a deficiency of electrons. It has been doped with acceptor atoms that produce holes which accept electrons easily.

*Q* — The merit of a coil or circuit. It is the ratio of inductive reactance to resistance. The quantity does not have a unit of measurement, just a number.

*Radio Frequency* — That frequency that exceeds the ability of the human ear to hear it. Radio frequency is generally accepted to be anything above 16 kHz.

*RC Time Constant* — The time it takes for a charging capacitor to reach 63.2% of its maximum value. $T = RC$. It takes five time constants to reach the practical maximum value of voltage in an *RC* circuit (99.3%).

*Reactance* — The means opposition to current flow in an ac circuit with either a capacitor or inductor or any combination of the two. Reactance is represented by the symbol X and measured in ohms. There is capacitive reactance or opposition to current flow in an ac circuit containing a capacitor, and inductive reactance or opposition to current flow in an ac circuit with an inductor, choke, or coil.

*Reciprocity Principle* — The process whereby the receiving and transmitting functions utilize the same antenna, as in many radar installations.

*Rectifier Circuit* — A circuit so designed to convert alternating current (ac) to direct current (dc).

*Reluctance* — The opposition to the flow of magnetic lines. It compares with resistance in a resistor.

*Residual Magnetism* — That magnetism left after a magnetizing force has been removed.

*Resistance* — Opposition to current flow.

*Resistivity* — The extent to which a material resists electron flow.

*Resistor* — A device designed to offer a specific amount of opposition to current flow.

*Resonance* — That condition of a circuit that occurs when the capacitive reactance is equal to the inductive reactance.

*Resonant Frequency* — The frequency at which $X_L = X_C$.

*Retentivity* — The ability of a material to retain magnetism.

*RF* — Radio Frequency. Any high frequency that is not within the audible range of the human being, usually the frequencies above 16,000 hertz.

*Rheostat* — A two-terminal device used as a variable voltage control.

*Right-Hand Rule* — The rule used to find the direction of motion in the rotor conductors of a motor.

*Ripple* — A series of pulses that ride on top of the direct current when ac is changed to dc.

*Ripple Voltage* — The peak-to-peak voltage variation produced by a rectifier circuit.

*Root Means Square (rms)* — The same thing as effective voltage or effective current.

*Rotor* — The moving part of an electric motor, an ac or dc generator, or the moving element in a variable capacitor.

*Saturation* — A term that indicates that a transistor reaches saturation when it is drawing maximum design current.

*Schematic* — A drawing that uses symbols to show how components in an electrical circuit are connected.

*Secondary* — The output side of a transformer.

*Sensitivity* — The specification of an analog voltmeter used to determine the resistance a meter will present to a circuit being tested. It is also used to refer to an oscilloscope specification that indicates its lowest voltage range. Sensitivity is used to indicate the weakest signal that can be detected by a radio receiver.

*Series Circuit* — An arrangement of a circuit whereby the total circuit current flows through every component.

*SHF* — Super High Frequency. The frequency range from 3000 to 30,000 MHz.

*Shielded Cable* — A cable with one of its conductors wrapped with a cloth-like outer covering made of finely woven wire that acts as a shield.

*Short Circuit* — An unintended alternate path of relatively low resistance that increases and bypasses the flow of current.

*Shunt* — Another word that can be used for parallel. It is often used to describe a resistor placed in parallel with an ammeter meter movement to extend its range.

*Sin* — An abbreviation for sine as used in trigonometric functions.

*Software* — A general term for all types of programs used to manage a computer's operations. Software is essentially a set of instructions the computer uses to perform a task. The solution of electrical problems is done almost always by engineers with the aid of software written especially for computer use. Complicated formulas and problems can be accurately solved every time with the computer programmed properly by software packages.

*Solenoid* — A coil of wire with a core. Usually wound with an iron core that is attracted to an armature to produce a switching action by closing switch contacts located on the end of the moving armature, the device is then referred to as a relay. Or the sucking effect of a coil (solenoid) can be utilized to cause a plunger to move and turn water or natural gas on and off.

*Spreadsheet* — A program that simulates an accountant's worksheet, which is made up of rows and columns. It's used to quickly calculate budgets and perform financial analysis.

*Static Electricity* — Electricity usually generated by friction that stands still or does not move until a path is provided for it.

*Stator* — The term used for a fixed element or plate in a variable capacitor.

*Steady State* — A condition that does not change with time.

*Sweep Circuit* — A circuit used in an oscilloscope or TV picture tube to produce a voltage that varies at a constant rate.

*Switch* — A device designed to quickly connect or disconnect an electrical circuit.

*Symptoms* — Behavioral characteristics of a circuit that indicate a malfunction.

*Terminal Strip* — A series of connection points that are insulated from each other and provide locations for components to be wired or joined.

*Tesla (T)* — A unit of measurement for the density of a magnetic field.

*Thevenin's Theorem* — A procedure that provides for the reduction of a complex circuit to simple terms. It provides for reducing to an equivalent voltage and an equivalent resistance in a series circuit.

*Tolerance* — The amount of + and – error between the actual and indicated values of a component.

*Transistor* — A three-terminal electronic device utilized to amplify and to act as a switch. It has base, collector, and emitter leads.

*Transformer* — An electrical device that changes high voltage to low voltage and low voltage to high voltage. It can either step up or step down voltages. It can also be used for matching impedances.

*Trimmer Capacitor* — A small variable capacitor often used across larger capacitors to make fine adjustments in tuned circuits.

*Troubleshooting* — The process of measurement and analysis used to determine a cause of electrical or electronic malfunction.

*True Power (TP)* — The power actually used in a circuit and dissipated in the form of heat energy. It is power dissipated by resistance.

*Turns Ratio* — The relationship between the number of turns in the primary of a transformer and the number of turns in the secondary is called the turns ratio.

*UHF* — Ultra High Frequency. The frequency range from 300 to 3000 MHz.

*Variable Capacitor* — A capacitor that can be adjusted to change its resistance is called variable.

*Varistor* — A type of diode used for voltage transient suppression, temperature compensating, and voltage compensating. The diode symbol is used in circuit schematics.

*Vector* — A description of a signal in terms of both magnitude and direction. It is any quantity having both magnitude and direction.

*Vector Sum* — The sum of two signals including both magnitude and direction.

*VHF* — Very High Frequency. The frequency range from 30 to 300 MHz.

*Volt* — The unit of measurement for voltage. It represents electrical pressure. Abbreviated as V.

*Voltage Divider* — A series arrangement of resistances connected to a dc or ac supply that is used to produce proportionally lower amounts of a source voltage by tapping across the various series resistor combinations.

*Voltage Drop* — The voltage that results from current flow through a resistor.

*Voltage Gain* — The ratio of output voltage to input voltage in a transistor circuit.

*VOM* — An abbreviation for an analog multimeter that can measure volts, ohms, and milliamperes.

*Watt (W)* — The unit of measurement for electrical power. It is the product of voltage and current.

*Weber (Wb)* — The unit of measurement for magnetic flux in the SI system.

*Wet Cell* — A cell made with a liquid electrolyte. It is usually referred to as a secondary cell and can be recharged. Wet cells use many combinations of materials for electrodes and electrolytes.

*Wicking* — A process used to remove solder by pressing a braid against the connection with a hot iron. The wick or braid absorbs the solder so it can be removed from the desired spot.

*Wire-Wound Resistor* — A fixed value resistor with turns of wire wrapped around a ceramic core and usually coated with an epoxy or a glaze.

*Zero-Adjusted Potentiometer* — An adjustment used in a VOM to adjust for the steady decline of a battery used in its ohmmeter circuit.

# ANSWERS TO REVIEW QUESTIONS

## Chapter 1

1. Electrical engineers are concerned with jobs that relate to the generation of electricity, its transmission, and its use.

2. Electrical engineers work for electrical and electronics manufacturers, in the electric light and power industries, in the aircraft and missile field, and in telephone broadcasting industries. They also consult and teach.

3. Increasing demands for qualified engineers can be expected in the computer field, fiber optics, and satellite communications.

4. In the telephone industry there are opportunities in many aspects of the field. There is room for those who want to do business on their own, and satellite communications offers opportunities.

5. In the TV repair field the demand for quality repair work is always high. The outlook is for continued growth in this field.

6. The broadcast industry is limited in its opportunities since it is a glamor industry. However, you can work your way up in the business, with the technical and engineering routes offering the best opportunities.

7. An engineer needs a minimum of a four-year bachelor's degree.

8. A technician needs to know a specific job or piece of equipment.

9. Information on electrical engineering can be obtained from the Institute of Electrical and Electronic Engineers in Washington, DC.

10. The electrical utilities job future looks good because of the constant turnover due to death or retirement of personnel. The job potential should continue to be excellent.

## Chapter 2

1. From the Greek *elektron*, which means *amber*.

2. An atom is the smallest part of an element.

3. An electron is a negative particle that orbits the nucleus of the atom.

4. Free electrons are found in some metals such as copper, silver, and gold.

5. An electron is 0.000 000 000 000 22 inches (in.) in diameter.

6. Five insulators are glass, dry wood, rubber, mica, and plastics.

7. Silicon and germanium are the most common semiconductor materials.

8. Voltage is electrical pressure needed to push electrons along a conductor. Current is the flow of electrons along a conductor.

9. A watt is the unit of measurement for electric power.

10. A kilowatt is 1000 W.

11. There are 746 W in 1 hp.

12. The symbol for ohms is the Greek letter omega ($\Omega$).

13. The symbol for a resistor is ‒\/\/\/‒ .

14. The carbon-composition resistors use the color code.

15. A kilo is 1000.

16. The term for 1 million is *mega*.

17. Ohm's law: The current in any circuit is equal to the voltage divided by the resistance.

18. Kirchhoff's law of voltages: The algebraic sum of the voltages around a loop is zero.

19. $R_T = \dfrac{R_1 \times R_2}{R_1 + R_2}$    or    $\dfrac{1}{R_T} = \dfrac{1}{R_1} + \dfrac{1}{R_2} + \dfrac{1}{R_3} + \dfrac{1}{R_4} + \cdots$

20. The fuse protects the circuit from overloads.

# Chapter 3

1. Inductance is the ability of a coil to oppose any change in circuit current.

2. A coil may have the name of coil, inductor, or choke.

3. The unit of measurement of inductance is the henry (H).

4. Mutual inductance occurs when two coils are placed near one another.

5. Five factors that influence mutual inductance are:

   - Physical size of the two coils.
   - Number of turns in each coil.
   - Distance between the coils.
   - Distance between the axes of the two coils.
   - Permeability of the cores.

6. Four factors that influence the inductance of a coil are:

   - Inductance is proportional to the square of the number of turns.
   - Inductance of a coil increases directly as the permeability of the core increases.
   - Inductance of a coil increases directly as the cross-sectional area increases.
   - Inductance of a coil decreases as its length increases.

7. Alternating current can be transported over long distances with low losses. The ac can be stepped up and down whereas dc cannot.

8. Public utilities are generally responsible for ac generation in the United States. There are some federal government projects.

9. Usually, ac is generated commercially at 13,800 V.

10. A waveform is the signature of an ac voltage.

11. Peak is 1.414 times rms voltage or current. Average is 0.637 times peak voltage or current. Rms is 0.7071 times the peak voltage or current.

12. Since there are no moving parts in a transformer, it is very efficient.

13. Eddy currents are small currents induced in the metal of a transformer by the changing magnetic field.

14. Hysteresis is the opposition given by a material when it is forced to change its magnetic orientation.

15. Copper losses are caused by the use of the wrong-size copper wire.

16. A capacitor is a device which has two plates and opposes any change in the voltage of a circuit.

17. Three factors that determine the capacity of a capacitor are:

   • Area of the plates.
   • Distance between the plates.
   • Dielectric material used to insulate the plates from one another.

18. The abbreviation WVDC means working volts dc of a capacitor or where can safely operate.

19. Electrolytics can be made nonpolarized by connecting them back to back. Connect them in series so that (+) is to (+) and/or (–) is to (–).

20. Capacitance is added when capacitors are connected in parallel.

21. A microfarad is one-millionth of a farad.

22. $X_C$ means capacitive reactance.

23. Capacitors and inductors can be used together to produce resonance, or they can be used as filters.

24. The formula is $X_{C_T} = X_{C_1} + X_{C_2} + X_{C_3} + \cdots$

25. The formula is $\dfrac{1}{X_{C_T}} = \dfrac{1}{X_{C_1}} + \dfrac{1}{X_{C_2}} + \dfrac{1}{X_{C_3}} + \cdots$

   also for 2 only capacitors

$$X_{C_T} = \frac{X_{C_1} \times X_{C_2}}{X_{C_1} + X_{C_2}}$$

# Chapter 4

1. The symbol for impedance is $Z$.

2. The vector sum can be found by measuring the vectors or by math. The vector sum is also a line representing the total of two or more vectors.

3. Impedance is measured in ohms.

4. Capacitors and inductors present opposite reactance values since one reacts with current and the other with voltage.

5. A phase angle is the relationship between voltage and current in an ac circuit with either an inductor or capacitor or both.

6. The impedance of a series $RCL$ circuit is found by

$$Z = \sqrt{R^2 + (X_L^2 - X_C^2)}$$

7. Current in an $RCL$ circuit is found by the formula $I_T = E_A/Z$.

8. Ohm's law for ac circuits aids in obtaining $X_L$ and $X_C$ or $E_L$ or $E_C$.

9. When the inductive reactance and capacitive reactance are equal, resonance exists.

10. The inductive reactance and capacitive reactance or the voltage across the inductor and capacitor must be equal to produce resonance.

11. The impedance of an *LC* circuit when the voltages are equal means it presents a short circuit for that frequency or zero opposition.

12. At resonance, a parallel *LC* circuit presents an infinite impedance.

13. $f_r = \dfrac{1}{2\pi\sqrt{LC}}$

14. The flywheel effect takes place when a capacitor and inductor are connected in parallel and the capacitor is charged. The circuit continues to have current flow through the components within the tank circuit.

15. A tank circuit is formed whenever a parallel connection of a coil and capacitor is made.

16. A damped oscillation is one that dies out, usually on its own. Each successive oscillation is lower in amplitude until it dies.

17. Impedance is the highest in a tank circuit when resonance occurs.

18. The phase angle of a series resonant circuit is the relationship between the voltage and the current.

19. In a parallel resonant circuit, frequencies that are higher than the resonant frequency find a low-impedance path to ground by way of the capacitor.

# Chapter 5

1. When ac was invented.

2. Copper oxide and selenium oxide.

3. To change ac to dc.

4. From a rectifier.

5. A selenium rectifier is larger than a germanium rectifier and less efficient.

6. Usually wherever they were needed in battery chargers and chemical plants.

7. Silicon and germanium.

8. Junction diode is made of two pieces of material: one P and one N. The point contact diode has a very, very sharp point touching one piece of the N or P material.

9. Reverse-bias is when the polarity opposite that called for is placed on elements.

10. Peak *inverse* voltage (piv) is the peak negative voltage that is applied to the plate during the portion of the hertz when the tube is not conducting.

11. A half wavelength puts out only half of the ac waveform as dc. A full wavelength puts out both.

12. A ripple voltage is what occurs after rectification and filtering, and rides on top of the dc.

13. Ripple can be smoothed by inductors and capacitors.

14. A bridge rectifier is a full-wave, diamond-shaped, four-diode rectifier.

15. A bridge rectifier has the ability to put out more from the same transformer.

16. The major advantage of the bridge is its lack of peak inverse voltage applied across the diodes.

17. A filter smooths out ripple.

18. In a capacitor input filter the capacitor is first in line as you view the filter coming from the rectifier.

19. A disadvantage of the capacitor input filter is that it acts like a short circuit across the rectifier while the capacitor is charging.

20. In an inductor input filter the inductor is first in line as you view it from the rectifier. It usually consists of an inductor in series with the output and a capacitor across the output.

21. The voltage output from an inductor input filter is lower.

22. The pi filter is two capacitors connected in parallel across the rectifier output separated by an inductor or a resistor.

23. The pi filter has the advantage of a steadier output voltage during varying loads. It is a smoother filter for the output is closer to dc than with the other types.

24. Percentage of ripple is 100 times the ratio of the rms value of the ripple voltage at the output of a rectifier filter to the average value of the output voltage.

25. The first one as seen from the rectifier.

26. It is cheaper and lighter in weight.

27. Where there is a high current drain from the rectifier.

28. It drops the voltage and the output is lower.

# Chapter 6

1. Silver sulfide varies inversely with the temperature; it started the work with crystals and electricity.

2. Because the electrons move through solid material instead of a vacuum as in the tube technology.

3. A diode is a device that allows current to flow in one direction only.

4. Junction diodes are made with two pieces of P and N material. They are joined together by heat.

5. Silicon and germanium.

6. An impurity is another element introduced into the pure silicon or germanium.

7. A donor is an impurity that gives up one electron.

8. An acceptor is an impurity that will accept one electron.

9. A carrier is N or P material that allows electrons to flow through it.

10. N material has an excess of electrons. P material has an electron missing and conducts by having holes.

11. Trivalent impurity is one that has one less electron than it needs to establish covalent bonds with four neighboring atoms.

12. A pentavalent (five) impurity is one that has one electron more than needed to establish covalent bonds with four neighboring atoms.

13. P material has hole flow.

14. N material has electron flow.

15. By using P and N material joined together.

16. Barrier potential is like the plate-cathode voltage of a vacuum tube diode. It is a difference or voltage across the junction of the diode or transistor that is in the neighborhood of a few tenths of a volt.

17. Forward bias is when the proper polarity is applied.

18. Reverse bias is when the opposite polarity is applied than is required to operate.

19. The point contact diode is made using one piece of P or N material and a very sharp contact from a piece of metal touching the material.

20. A zener diode is one that breaks down and then recovers.

21. A tunnel diode is one that appears to tunnel its way through the semiconductor material as it exhibits a negative resistance characteristic.

22. Avalanche breakdown is when the junction of the diode breaks down.

23. The diode is ruined when avalanche breakdown occurs.

24. A valence electron is one that is part of an atom and determines its characteristics.

25. Covalence bond means that the crystals are connected at four points and share electrons with four neighboring atoms.

26. Thyristor is the other name for SCR.

27. The gate controls the ability of the SCR to conduct from cathode to anode.

28. Transistor is a combination of *trans* from the word transfer and *istor* from resistor.

29. Point contact diodes are signal diodes and small current types. Junction diodes are usually used as power diodes and can handle more current.

30. The symbol for current gain in a transistor is $\alpha$ce.

31. Six types of transistors are: junction, alloy, grown-junction, microalloy, germanium mesa, and silicon planar field effect.

32. An FET transistor is a field effect transistor that operates as a result of one type of charge carrier.

33. FETs are used wherever they are needed to substitute almost directly for vacuum tubes.

34. An integrated circuit is one that has many resistors, capacitors, diodes, and transistors enclosed in one very small enclosure.

35. ICs are used in almost all types of electronics today. Computers rely upon them heavily.

# Chapter 7

1. As three-phase.

2. It is the shape of a triangle.

3. It is the shape of a Y.

4. Delta

5. Wye

6. Wye

7. Minimum of 3 with a wye using 4.

8. Interchange any 2 of its power line wires.

9. Waveform

10. 60 Hz

11. 50 Hz

12. ac or dc

13. Silicon Controlled Rectifier

14. It gets its name from the shading of one of its poles by a ring around it.

15. No starting mechanism needed.

16. The power line frequency.

17. It uses it to bring an inductive load back in phase and to correct the power factor to produce a smaller electric bill at the end of the month.

# Chapter 8

1. Distortion is the type of change the signal undergoes from the time it goes into the stage until it comes out.

2. Hum is another type of distortion.

3. Five classes of operation for an amplifier are class A, class B, class $AB_1$, class $AB_2$, and class C.

4. Two types of AB operation are $AB_1$ and $AB_2$.

5. The only difference between the two types of AB operation depends upon whether the grid current flows or does not flow during any part of the input signal.

6. Class C is used primarily in high-output amps like those in transmitters.

7. Two types of bias are forward and reverse, but in tubes it is fixed and self.

8. Grid-leak bias is when the bias voltage is zero and there is no signal voltage on the tube. Bias voltage is generated only when there is a signal applied between grid and cathode.

9. A resistor and a capacitor are needed for *RC* coupling.

10. The component needed for impedance coupling is a transformer.

11. The advantage of transformer coupling is impedance matching and proper signals for two tubes in a push-pull circuit.

12. Direct coupling is needed when dc is amplified.

13. Transistors have the same classes of operation as vacuum tubes: A, B, C, etc.

14. Transistor amplifiers use $RC$ coupling, transformer coupling, and direct coupling.

15. Common-base, common-emitter, and common-collector are three types of common transistor circuits.

16. The common-collector transistor is used for impedance matching.

17. Common-base amplifier has current gain of less than 1.

18. An op-amp is an operational amplifier. It is classified as either a linear amplifier or digital.

19. A differential amplifier is an ideal op-amp. It has both inverting and non-inverting inputs.

20. The op-amp in basic form is used in analog computers as an inverting, constant-gain multiplier. The op-amp is capable of operating as a noninverting constant-gain multiplier. It is also capable of being used as a unity follower with a gain of 1 or unity. It can be used for impedance matching.

# Chapter 9

1. Radio frequency amplifiers are used in both transmitters and receivers.

2. Radio frequency is above human hearing.

3. VHP is 30 to 300 MHz.

4. UHF is 300 to 3000 MHz.

5. Tuned circuit couplings have transformers in the input and output. The transformers have capacitors across them so they can be tuned.

6. Double-tuned is when the input and output are both tuned.

7. The RF amplifier has primarily tuned circuits.

8. Low Z input for RF amplifiers is obtained by tapping the transformers.

9. Power amplifiers have been used very well in transmitters.

10. RF amplifier stages usually operate class C for power applications.

11. The emitter is usually reverse-biased in a transistor power amplifier stage when it is operated class C.

12. Two types of feedback are regenerative and degenerative.

13. Positive feedback is regenerative, and negative feedback is degenerative.

14. A DIP is a dual-in-line-packaged integrated circuit.

15. MSI packages have up to 1000 components.

16. A linear amplifier is one that is analog in nature. There is a direct relationship between input and output signals.

17. An op amp is an operational amplifier which is a linear amp. It produces high gains in a frequency range from 0 to 1 MHz.

18. Op amps are popular because of high gain and small packages.

19. Closed loop means that the loop has feedback.

20. Differential amplifiers are used for circuits with two inputs that need to be compared.

# Chapter 10

1. Stereo means solid, or three-dimensional, space.

2. Two.

3. Compression is compressing, and rarefraction is when the compressed air is allowed to expand.

4. Human ear range is usually referred to as 16 Hz to 16 kHz.

5. Sensitivity of a microphone is in the power level delivered to its load.

6. Carbon microphone needs a power source; a dynamic microphone does not.

7. The dynamic microphone has a range of 30 Hz to 18 kHz.

8. The velocity microphone is fundamentally the same as that of the moving-coil model except that it has a strip of metal that is caused to vibrate in a magnetic field.

9. The crystal microphone operates by means of pressure of the sound on a crystal. The crystal produces electricity in proportion to the pressure applied.

10. The crystal is Rochelle salt or quartz.

11. A turntable handles only one record at a time; a record changer can handle up to 12 at one time.

12. Six types of styli are magnetic, moving-coil, ribbon, moving-magnet, crystal or ceramic, and frequency-modulated.

13. The magnetic recorder was invented by Valdemar Poulsen.

14. A cassette is smaller than a cartridge.

15. A CD is a compact disc designed to record and play back music of high quality. Recording takes place at professional production facilities and not by home equipment.

16. The CD is 122 mm or about 5 in. in diameter with a 15 mm center hole; 3-in. CDs are also commercially available.

17. The CD travels at a speed of 500 rpm near the center hole and about 200 rpms at the rim or outside edge.

18. Digitizing is the process whereby analog music is converted to a series of bits of information with synch and timing signals added.

19. A pit is a depression in the aluminized coating of the CD.

20. Encoding is the process whereby the digitized music is recorded with special codes to make sure it plays back at the same rate as it was recorded.

21. The CD pickup is mounted near the center of the disc and works outward using a laser beam to "read" the pits and coded messages on the CD.

22. A nanometer is one thousandth of a millionth of a meter or $1 \times 10^{-9}$.

23. A collimator lens produces a parallel beam of laser light.

24. An optosensor is a photodiode that is sensitive to light variations and used to change the light variations into electrical impulses to match the recorded music.

25. The laser diode produces 3 nW of power at 780 nm.

26. Handle your CD by the edges. Avoid fingerprint smears as well as heat, cold, and humidity.

27. A preamplifier takes weak signals and boosts them. A power amplifier needs a strong signal input before it can operate.

28. The push-pull amplifier has many advantages. It has even-harmonic cancellation and can put out about four times as much power as a single tube.

29. Two types of speakers are the electromagnetic and the permanent magnet.

30. The voice coil is a few turns of wire inserted into the magnetic field of the permanent magnet. It is fed ac from the amplifier.

31. The speaker enclosure is where the speaker is mounted.

32. Three types of speaker enclosures are the infinite baffle; the bass-reflex; and the flat, or plane surface, baffle.

33. A crossover network makes sure that the right frequencies go to the right-size speaker for maximum benefit.

34. A 12-in. speaker frequency range is usually up to 300 Hz.

35. Another name for the 12-in. speaker is the woofer.

36. The midrange speaker reproduces the midrange of sounds at 300 to 3000 Hz generally.

37. The horn speaker usually puts out frequencies over 3000 Hz.

# Chapter 11

1. An oscillator produces oscillations, or unstable frequencies.

2. Oscillators are used in transmitters and other devices.

3. An *LC* tank circuit produces oscillations as the capacitor charges and discharges through the inductor and the inductor's magnetic field collapses and puts the energy back into the circuit. This continues until feedback keeps it going or until the resistance of the circuit damps the oscillations.

4. A damped oscillation is one that dies out a little each time it oscillates, until it disappears.

5. You need feedback in an oscillator to keep it going.

6. Regenerative feedback is positive and adds to the circuit, whereas degenerative feedback is negative and subtracts from the bias voltage and decreases output of a tube or transistor.

7. Inverse feedback is degenerative, or negative, feedback. It is used in amplifiers.

8. Hartley oscillator has tapped inductance. Colpitts oscillator has tapped capacitance.

9. The crystal oscillator starts to oscillate as soon as an electric impulse gives it a tickle, and then feedback keeps it going.

10. The crystal oscillator is more stable in its output frequency.

11. The Clapp oscillator is a modified Colpitts device with a tunable capacitor in series with one of the tapped capacitors.

12. Wein is the name of the inventor, and the bridge circuit is a diamond-shaped configuration used in many circuits.

13. A multivibrator is essentially a nonsinusoidal two-stage oscillator in which one stage conducts while the other is cut off.

14. A blocking oscillator is one that conducts for a short period of time and is cut off (blocked) for a much longer period.

15. Microwaves are those frequencies that fit into the superhigh-frequency range.

16. A klystron is used to generate low- to medium-power microwaves.

17. A magnetron is a high-power-output microwave oscillator-amplifier.

18. A klystron is for low- to medium-power output with lower voltages used to operate than the magnetron.

19. A Gunn diode is used to produce low power levels of microwaves. It is used in hand-held police speed guns.

# Chapter 12

1. Modulation is the audio signal being impressed on an RF carrier.

2. AM changes the amplitude of the carrier wave, whereas FM changes the frequency with the audio signal variations.

3. Phase modulation is indirect FM signal modulation. The modulating frequency has an effect on the phase of the carrier wave.

4. FM is used in commercial broadcasting, TV sound channels, and narrow-band communications for commercial use by business and industry.

5. A carrier wave is that which is operating at RF levels and cannot be heard as it leaves the transmitter.

6. Frequency deviation is that part of the FM signal carrier that changes or deviates as a result of the modulation being impressed on it.

7. The modulation index is the ratio between the maximum frequency deviation and the maximum frequency of the modulating signal.

8. AM can use the diode detector or the crystal detector.

9. FM can use the ratio detector, the discriminator and the slope detector.

10. The AM detector uses a diode to rectify the AM signal with the modulation on it. The half-wave rectified signal is then put through a capacitor where the RF flows easily and the audio is pushed through the volume control or resistor which has a low impedance for it.

11. A discriminator separates the audio from the RF in an FM signal. It is used in an FM receiver.

12. A ratio detector separates the audio from the RF in an FM receiver.

13. A slope detector is an inexpensive way of demodulating FM.

14. The ratio detector does not need a limiter stage as the discriminator does.

15. The slope detector separates the audio from the RF in an FM signal.

# Chapter 13

1. A transmitter sends out or transmits radio frequency energy.

2. A transmission line takes RF from the transmitter to the antenna or from an antenna to the receiver.

3. The antenna for a transmitter produces the electromagnetic and electrostatic fields that carry the signal forth from the antenna. The antenna in a receiver receives or picks up the signal being transmitted by the transmitter.

4. AM is amplitude-modulated RF and FM is frequency-modulated RF.

5. An oscillator is needed to produce the basic frequency for transmission.

6. A buffer is placed after the oscillator to keep the oscillator from drifting under load.

7. A frequency multiplier does just that: it multiplies the frequency put into it.

8. A doubler stage increases the frequency by two times.

9. Frequency multipliers are needed to increase the frequency beyond the physical capabilities of a crystal.

10. Frequency deviation is that swing on both sides of the FM carrier wave caused by audio modulation.

11. FM covers 88 to 108 MHz.

12. A two-wire open line consists of two wires that are generally spaced from 2 to 6 in. apart. The twisted pair is just that; the two wires or conductors are twisted together.

13. A flexible coaxial line is constructed with a foam center supporting the conductor wire.

14. Waveguides are used for microwaves and radar.

15. A dipole antenna is one with two parts, usually half-wavelength.

16. A Hertz antenna is one that is half-wave in length.

17. A Marconi antenna is one that is a quarter-wavelength.

18. Wavelength is equal to the speed of light divided by the frequency or

$$\text{Lambda} = \frac{300 \text{ million}}{\text{frequency}}$$

19. The ionosphere is located above the earth a few miles and contains a layer of ionized gases.

20. The ground wave is transmitted and stays close to the earth.

21. The sky wave is the one that goes upward to strike the ionosphere.

22. Fading occurs because the ionosphere changes its location and because the signal is reflected at various angles back to earth and skips some areas as it moves. A number of things cause fading.

23. Frequency blackouts are closely related to fading; in some cases certain frequencies are completely eliminated and cannot be heard.

24. Frequency blackouts are caused by changing conditions in the ionosphere shortly before sunrise and shortly after sunset. Certain frequencies cannot be received at this time.

# Chapter 14

1. Transmission lines for receivers cut down on signal loss from the antenna.

2. SWR stands for standing wave ratio.

3. Standing waves are produced when the transmission line and the load are not matched to one another's impedance.

4. The reason why some AM receivers need an RF amplifier is that they are used to pick up very weak signals from faraway stations.

5. Most AM radios used for local stations do not need an RF amplifier.

6. The mixer takes the incoming frequency and mixes it with the oscillator signal, producing four frequencies.

7. The local oscillator is the one located in the receiver and is used to beat against the incoming frequency to the heterodyne.

8. An IF is an intermediate frequency.

9. A detector is usually one diode.

10. AM means amplitude-modulated. FM means frequency-modulated. An AM signal moves up and down. An FM signal moves back and forth when modulated.

11. FM needs a limiter when a discriminator is used to eliminate noise.

12. A discriminator is a detector or demodulator for FM.

13. A limiter is used only with discriminators to get rid of phase shift and noise peaks.

14. A ratio detector does not need a limiter.

15. A subcarrier is one that is impressed on the standard carrier frequency at a vacant spot in the modulation sidebands.

16. A pilot carrier is used in FM for stereo channels.

17. Multiplexing is the sending of two or more signals on one assigned frequency.

18. The AVC filters the audio output of the detector and feeds it back to IF stage to cut down on amplification when strong signals are received.

19. The AFC is the feedback that keeps the frequency of the local oscillator steady or stable.

20. Narrowband FM has a limited amount of modulation swing. It is used primarily for voice modulation and does not need large sidebands.

21. Foster-Seely refers to a type of discriminator used in FM.

22. FM stereo was authorized by the FCC in 1961.

23. Television uses AM for transmission of picture information.

24. Television uses FM for sound information.

25. It compares to the sidebands caused by modulation.

# Chapter 15

1. The cathode ray tube was invented by Vladimir Zworykin.

2. The first TV picture in the United States was transmitted from New York City to Whippany, New Jersey.

3. TV camera tubes are iconoscope, image orthicon, and vidicon.

4. A picture tube is also called a kinescope.

5. A vidicon is a type of television camera tube.

6. There are 525 lines on a TV screen.

7. Frame frequency of television in the United States is 30.

8. The field frequency is 60.

9. A frame is made up of two halves of the picture consisting of two complete scans of the tube from top to bottom, but since only every other line is picked up by the scanning, it takes two to give a whole picture. A field is one-half of a picture.

10. Scanning in a receiver is what the electron beam does to produce the picture on the kinescope. It moves back and forth and up and down scanning the whole face of the tube.

11. The subcarrier for color TV is 3.579 545 MHz.

12. Horizontal frequency for TV in the United States is 15,750 Hz.

13. The bandwidth for TV in the United States is 6 MHz.

14. Vestigial sideband modulation is when a vestige of one sideband is transmitted to trick the receiver into thinking it has both sidebands.

15. Red: 30%, blue: 11%, green: 59%.

16. Interleaving is using the spots left by normal modulation for putting in the color information on the carrier wave of a TV signal.

17. In the United States FM is used for sound on television.

18. Video is transmitted by AM.

19. Brilliance and luminance are the same.

20. The spectral colors are red, blue, and green plus orange, yellow, and violet.

21. Phase modulation aids in the color signal transmission and reception.

22. A phosphor is that part of the picture tube coating inside the front of the tube that glows red, blue, or green when hit by a beam of electrons from the cathode.

23. A shadow mask causes the three beams of electrons to merge before hitting the phosphors.

24. The tint or hue control adjusts the phase of the color signals.

25. High definition television.

26. Folded dipoles with directors and reflectors are usually used on rooftops for TV receivers.

27. The bow-tie antenna is just like that used for UHF and resembles a bow tie.

# Chapter 16

1. Laser stands for *l*ight *a*mplification by *s*timulated *e*mission of *r*adiation.

2. There are three types of lasers.

3. Fiber optics use a semiconductor laser.

4. The frequency of visible light is $430 \times 10^{12}$ to $730 \times 10^{12}$ Hz.

5. Incoherent light is scattered. Coherent light is a tightly contained beam.

6. Lasers use coherent light.

7. The DH semiconductor diode is made of gallium arsenide and aluminum gallium arsenide.

8. Some uses of lasers include industrial welding, distance measuring, surgical procedures, military applications, holograms, video disk playback, and communications.

9. Refraction means to be bent; reflection means to be turned back.

10. Fiber optics can be made of glass or plastic.

11. The laser used for wideband communications is the LED or light-emitting diode. It has a fast response time and a long lifetime.

12. A nanosecond is 0.000 000 001 s or 1000th of a millionth of a second.

13. Fiber optics converts light back to electric energy by using a PIN diode or avalanche photodiode.

14. The detector is either a PIN diode or an avalanche photodiode.

15. Fiber optics are used in medicine, in communications, and for decorations.

# Chapter 17

1. Radar means *ra*dio *d*etection *a*nd *r*anging.

2. Microwaves are in the superhigh frequency range or above 3000 MHz.

3. The 2450 MHz band is the S band.

4. The X band is microwaves in the 10,525 MHz range.

5. The K band is used by police radar to detect speeding vehicles.

6. Superhigh is 3000 to 30,000 MHz and extremely high is above 30,000 MHz.

7. Doppler effect is the difference in frequency when the frequency is reflected back to the receiver and what it was like on being transmitted.

8. In a pulse system, you send out a pulse, turn off the transmitter, wait for the return echo, and then send another pulse.

9. A microsecond is one-millionth (0.000 001) of a second.

10. A pulse system needs a magnetron, waveguides, and an antenna.

11. A radar indicator is usually a cathode ray tube.

12. A magnetron is a high-power-output amplifier-oscillator microwave tube.

13. A T/R switch is a transmit-receive switch that turns on and off with the pulses of energy radiated and that then switches over to the receiver for display of the pip, or echo.

14. A waveguide is the same as a transmission line for lower frequencies.

15. A resonant cavity is the same as an inductor and capacitor at lower frequencies.

16. Antenna reciprocity means that both transmitter and receiver can use the same antenna.

17. A radar antenna must be able to convert the microwaves into a voltage that the receiver can interpret.

18. Line of sight is just that: a straight line from transmitter tower to the receiver antenna. Microwaves do not bend or reflect when hitting the ionosphere.

19. A parabolic reflector is in the shape of a parabola and focuses the microwaves much as a headlight focuses light.

20. The microwave oven cooks by causing the molecules to rub against one another. Friction causes heat as the microwaves cause a rubbing in the neighborhood of 2450 million times per second.

# Chapter 18

1. The abacus is the mechanical device that is the basis of today's computers.

2. The first adding machine was invented in 1642.

3. The flip-flop can be used as a time base and as a pulse source in computers.

4. Eccles and Jordon came up with the flip-flop free-running multivibrator circuit.

5. Analog uses from –15 to +15 V while the digital uses an on-off 5-V signal for its pulses and circuit triggering.

6. A NOR circuit is a negative OR circuit.

7. A NOT circuit is a building block vital to the operation of a computer. The NOT circuit is used as an inverter.

8. An OR circuit needs two switches and will operate with either switch open or closed.

9. An AND circuit is one that has two switches that needs both closed to operate.

10. Binary refers to *two*. In computers it is 0 or 1, or on or off. It means bistable.

11. The powers of 10 are very interesting since they show the position of a number that determines its place or value.

12. A microprocessor is usually designed to do one function. A personal computer is a small computer and will do more than just one job. Personal computers are usually built around microprocessors.

# Chapter 19

1. A phase-locked loop is a frequency synthesizer, which can be used as an FM demodulator.

2. It can be used as an FM demodulator and to generate frequencies for transmitters.

3. The main objective of a cable TV system is to deliver a high-grade TV signal to the subscriber's home set.

4. A trunk amplifier is one that amplifies the signal along the cable to make up the attenuation due to distance and resistance and other factors in the cable.

5. A bridger amplifier is used usually in the line between the main cable and the subscriber's feeder cable.

6. 22,300 mi.

7. 14.

8. They are used for transmission of telephone messages, space flight information, television channels, and radio network news.

9. GPS aids in navigation. If you get lost it can pinpoint your location and tell you where you are and how to get to your destination.

10. 24.

11. A reprogrammable, multifunctional manipulator designed to move materials, parts, tools, or specialized devices through variable programmed motions for the performance of a variety of tasks.

12. Inventor of the term *robot* in his book *R.U.R.*

13. By humans and by specially designed computers.

14. AL, VAL, AML, Pascal, ADA.

15. A device that can process things. In most instances it is an information processor that processes information from an external source.

16. Edge detection and clustering.

17. The videotape recorder is used to tape TV programs off the air. It is also used by schools and program directors and filming crews.

18. Space manufacturing holds promise for gravityless manufacturing of purer medicines and better quality semiconductors.

19. 14,545 kg or 32,000 lb.

# Chapter 20

1. 40.

2. Ham radio is also known as amateur radio. It exists for experimental purposes and personal use by licensed operators.

3. Cellular radio.

4. A cell is a repeater unit.

5. A repeater is a base unit (cell) that contains a receiver and transmitter. It receives a signal on one frequency and usually re-broadcasts it on a different frequency and at a higher output power level.

6. Cell splitting requires the addition of new cell sites between existing cell sites. They form a new configuration of smaller cells while using the same number of channels. Each of the new smaller cells can serve about the same amount of traffic as the original larger cell.

7. *Call setup* occurs if the volume of traffic threatens to exceed the paging capacity of the control channel. The cell site is capable of setting up a separate paging channel. *Handoff* is the giving up of a weak signal to a cell closer to the signal source. One cells

hands off the subscriber to a cell closer to his location where his signal is stronger. *Sector cell* is an area that has a unique frequency set. A vehicle leaving that area or sector requires a new channel assignment. This is called a handoff. *Signal control channels* come into use when a mobile unit initiates a call. It sends a request for service and receives a voice channel assignment on one of a set of data channels. These data channels are called signal control channels.

8. Cellular Geographic Service Area (CGSA).

9. About 600 to 800 ft.

10. A pager is a small receiver that alerts the carrier to call the base station by telephone and report in.

11. Pagers use 30–50, 132–174, 406–512, and 929–932 MHz.

12. An exact reproduction or copy, and the transmission and reproduction of graphic matter by electrical means as by radio or wire.

13. Commerce, industry, private citizens, police, and ships at sea.

14. It produces an exact replica of any document or graphic sent by one machine to another of the same design.

15. Inkjet printers are used to give a finished product by using bubbles of ink to reproduce the image.

16. Laser printers similar to those used in computer printers are also designed for plain paper fax outputs.

17. Toner is the dry powder or ink used in copiers and laser printers.

# APPENDIX: ELECTRONICS JOBS*

# Jobs Directly Related to the Production of Semiconductors

## Bonder, Semiconductor Package (electronics)

Bonds connections between integrated circuit and package leads, using bonding equipment and gold or aluminum wire: Reviews schematic diagram of integrated circuit to determine bonding specifications. Turns dials to set bonding equipment temperature controls and to regulate wire-feeding mechanism according to specifications. Mounts spool of wire onto holder and inserts wire end through guides, using tweezers. Positions semiconductor package into magazine of automatic feed mechanism, and observes package, using microscope, or views package on equipment display screen, to ensure connections to be bonded are aligned with bonding wire. Adjusts alignment as necessary. Activates equipment that moves bonding wire to bonding unit and automatically bonds wire to specified connections on integrated circuit and package leads. Removes packages from bonding unit and places packages in work tray. May test tensile strength of bonded connection, using testing equipment.

## Die Attacher (electronics)

Attaches *dies* to empty integrated circuit packages to assemble complete semiconductor packages, using welding and gluing equipment: Places empty packages on heated chuck of equipment, using tweezers. Deposits bonding material, such as epoxy or gold, in center of empty packages, and positions and aligns dies on bonding materials in packages, using equipment or handtools. Removes packages with attached dies from chuck, using tweezers, and places packages in trays. May place loaded trays in oven to set gluing material. May view positioning of epoxy and die in empty packages, using microscope.

## Assembler, Semiconductor (electronics); Microelectronics Processor

Assembles microelectronic semiconductor devices, components, and subassemblies according to drawings and specifications, using microscope, bonding machines, and handtools,

*Many of these jobs have been available for less than 15 years. Because of limited space, some of the jobs mentioned in this appendix are not discussed in great detail. If more detailed information on one of these jobs is needed, it may be found in the *Occupational Handbook* printed by the U.S. Government Printing Office and available in the local high school guidance office or the public library.

performing any combination of following duties: Reads work orders and studies assembly drawings to determine operation to be performed. Observes processed semiconductor wafer under scribing machine microscope and aligns scribing tools with markings on wafer. Adjusts scribing machine controls, according to work order specifications, and presses switch to start scribing. Removes scribed wafer and breaks wafer into dice (chips), using probe. Places dice under microscope, visually examines dice for defects, according to learned procedures, and rejects defective dice. Positions mounting device on holder under bonding machine microscope, and adjusts bonding machine controls according to work order specifications. Positions die (chip) on mounting surface according to diagram. Presses switch on bonding machine to bond die to mounting surface. Places mounted die into holding fixture under microscope of lead bonding machine. Adjusts bonding machine controls according to work order specifications. Views die and moves controls to align and position bonding head for lead bonding according to diagram. Presses switch to bond lead and moves bonding head to points indicated in bonding diagram to attach and route leads as illustrated. Inserts and seals unprotected assembly into designated assembly container device, using welding machine and epoxy syringe, to protect microelectronic assembly and complete device, component, or subassembly package. Examines and tests assembly to various stages of production, using microscope, go-not-go test equipment, measuring instruments, pressure-vacuum tanks, and related devices, according to standard procedures, to detect nonstandard or defective assemblies. Rejects or routes nonstandard components for rework. Cleans parts and assemblies at various stages of production, using cleaning devices and equipment. Maintains records of production and defects. Bonds multiple dice to headers or other mounting devices. Important variations are kinds of equipment used, such as thermal compression wedge, wire ball, and wobble bonders, items assembled, or procedure performed.

## Saw Operator, Semiconductor Wafers (electronics); Scriber

Tends sawing machine that automatically scribes semiconductor wafers prior to separating wafers into individual *dies*: Flips switches and presses buttons to activate sawing machine. Places wafer mounted on laminated plastic onto movable chuck of sawing machine, using tweezers. Presses buttons and turns knobs and dials to adjust sawing specifications and to start sawing cycle. Monitors sawing cycle, using magnified viewer, to verify that wafers are scribed according to company specifications. Turns knobs and presses buttons to adjust sawing cycle if required. Removes scribed wafer from machine, using tweezers, and places wafer in container. Maintains production records. May mount wafers on plastic laminate, using mounting device. May visually inspect semiconductor wafers to identify misalignment of scribe lines. May maintain and clean sawing machine.

## Loader, Semiconductor Dies (electronics)

Loads *dies*, using vacuum wand, into carriers that hold and protect dies during fabrication of semiconductor packages: Pours dies from container onto filter paper. Picks up and places dies circuit-side up in indentations in carrier, using vacuum wand. Secures lid on carrier, using manual pressure. May clean dies prior to loading, using solutions and cleaning equipment. May sort semiconductor devices to remove defective devices marked in inspection department.

# Inspector, Semiconductor Wafer (electronics)

Performs any of following duties to inspect, measure, and test semiconductor wafers for conformance to specifications: Inspects wafers under high-intensity lamp to detect surface defects, such as scratches, chips, stains, burns, or haze. Measures thickness and resistivity of wafers, using electronic gauges or automatic sorting machine. Measures diameter and *flat* of wafers, using calipers. Inspects bow or flatness of wafers, using electronic gauges, or examines surface of wafers under high-intensity lamp. Tests for positive or negative conductivity of wafers, using electronic probe and gauge. Determines *crystal orientation* of wafers, using x-ray equipment. Encloses containers of inspected wafers in plastic bags for protection, using heat sealer. Records inspection data on production records or in computer, using computer terminal. May tend equipment that cleans surface of wafers [WAFER CLEANER (electronics)].

# Inspector, Crystal (electronics); Crystal Evaluator

Inspects, measures, and tests semiconductor crystal ingots to determine compliance to specifications, using a variety of measuring devices and test equipment: Transports ingots to work bench by cart or by hand. Measures ingot *flat*, diameter, and length, using calipers and ruler. Weighs ingots on scales and calculates weight loss due to grinding and sawing. Tests ingot resistivity, using electronic probes. Records weights, measurements, and other test results in logbook and labels and attaches labels to ingots. May operate saws to cut off ends of crystal or to cut sample wafers. May tend furnace that heat-treats ingots to alter resistivity to meet specifications. May determine *crystal orientation*, using x-ray machine. May operate machine to grind bevel on ends of crystals. May operate sandblasting machine to remove glaze from ingot surface. May tend equipment that etches ingots to remove surface material. May calculate proportion of impurities, using resistivity readings and specified formula.

# Inspector, Integrated Circuits (electronics)

Inspects integrated circuit (IC) assemblies, semiconductor wafers, and IC *dies* for conformance to company standards, using microscope: Reads work order to determine inspection criteria. Places group of items in trays on microscope stage, or positions items individually on stage for inspection, using vacuum pencil or tweezers. Turns knobs on microscope to adjust focus and magnification as required to view items for inspection. Views and inspects items according to company standards to detect defects, such as broken circuit lines, bridged circuits, misalignments, symbol errors, and missing solder. Discards defective items. May remove contaminants from items, using brush or airhose. May use magnifying glass to inspect electronic items.

# Tester, Semiconductor Wafers (electronic)

Tests electrical characteristics of circuits on semiconductor wafers, using test equipment: Reads processing documents to determine test specifications. Obtains specified probe cards from inventory and slides probe cards into designated slot on testing equipment. Activates testing equipment and places wafer on chuck of test equipment, using tweezers. Observes wafer through test equipment microscope and aligns wafer under probes of test equipment. Pushes buttons to activate test cycle. Records test readings from printouts or display screen in test log. Compares test readings with specifications' manual to identify wafers failing electrical tests. Maintains production records. May clean and maintain test equipment.

# Tester, Wafer Substrate (electronics)

Tests semiconductor wafer substrate, using testing equipment, such as probe tester, spectophotometer, and curve tracer, to evaluate electrical characteristics of wafer substrate: Places wafer, using tweezers, on test equipment that measures electrical characteristics of wafer substrate, such as resistivity, capacitance, and voltage. Starts equipment and observes equipment readout to determine wafer electrical measurements. Records measurements in production log. Compares measurements with specification sheets to determine if wafer substrate meets company standards. Sorts, boxes, and labels tested wafers. Delivers wafers and process sheets to production line workers. Maintains production records. May inspect wafer substrate surfaces, using ultraviolet lamp, to detect scratches and contamination.

# Group Leader, Semiconductor Testing (electronics); Production Aide

Assists SUPERVISOR, ELECTRONICS TESTING (electronics) in coordinating and monitoring activities of workers engaged in testing electronic devices (components), such as integrated circuits, transistors, and diodes, on semiconductor wafers, utilizing knowledge of equipment, procedures, and test specifications: Answers questions from workers pertaining to test procedures and test equipment operation. Monitors and expedites flow of materials through testing cycle. Enters test program into test equipment, using computer terminal, and examines test equipment printout to verify test equipment is functioning according to company and manual specifications. Notifies maintenance workers of test equipment malfunctions. Operates or tends equipment to substitute for absentee workers. Maintains production records. Trains workers in equipment operation and test procedures.

# Repairer, Probe Test Card, Semiconductor Wafers (electronics)

Replaces and realigns broken, worn, or misaligned probes on probe test cards, using hand-tools and equipment, to maintain probe cards for wafer probe (electrical test) equipment:

Reads specification sheets, manuals, and diagrams to determine probe card wiring and probe positions. Inspects probe cards, using microscope, to detect defects, such as loose wiring and loose, split, or worn probes. Removes, replaces, and re-solders defective wiring and probes, using tweezers and soldering iron. Observes probe card through microscope and aligns probes in specified position over test wafer, using tweezers. Aligns probes on even plane, using tweezers and test equipment. Maintains repair and replacement records. Maintains probe and inventory records, using computer terminal.

# Probe Test Equipment Technician, Semiconductor Wafers (electronics)

Sets up and maintains wafer probe and test equipment that test electrical properties of circuits on semiconductor wafers, following setup manual specifications: Reads process sheet and setup manual to determine test program required and setup specifications. Inserts test probe card in wafer probe equipment. Inserts floppy disk or magnetic tape into slot on test equipment and pushes buttons on test equipment or types commands on keyboard to enter test program. Places sample wafer from specified wafer lot on wafer probe equipment chuck, using vacuum wand or tweezers. Observes wafer through microscope attached to wafer probe equipment and turns knobs on equipment to align and position wafer circuit under probes on probe card, according to setup manual specifications. Activates test equipment to make trial run to ensure correct equipment setup and informs equipment operator that setup is completed. Removes and sharpens worn probes, using handtools and burnishing blade. Replaces defective probes or probe cards and solders joints on replacement probes, using soldering iron. Tests electrical continuity of cable connecting wafer probe and test equipment, using ohmmeter. Replaces defective connector cables. Removes, cleans, and refills reservoir of inker on wafer probe equipment that marks defective circuits with ink spot.

# Wafer Breaker, Semiconductors (electronics)

Breaks semiconductor wafers into individual *dies*, using equipment and handtools: Places semiconductor wafer on pad, using tweezers. Taps wafer, using pencil tip, awl, or tweezers to break wafer into quarters. Places semiconductor wafer quarter on plastic square. Presses semiconductor wafer, using roller or air pressure equipment, to break wafer quarter into individual dies. Places individual dies in containers, using vacuum wand or brush. May break wafers, using chemical solutions and breaking equipment.

# Sealer, Semiconductor Packages (electronics)

Positions and seals solder-coated lids or tops on semiconductor packages, using heated chuck or automatic furnace: Places lids on semiconductor packages, using tweezers. Places semiconductor packages on heated chuck in nitrogen box or places container of semiconductor packages on furnace conveyor to heat and seal semiconductor packages. Maintains production records.

# Programming Equipment Operator (electronics); Die Equipment Operator; PROM Burn-Off Operator

Tends equipment that automatically transfers programmed information onto integrated circuits (ICs) on *dies* or IC packages, such as PROMs (programmable read only memory): Reads work order that accompanies ICs for processing to determine programming specifications. Secures master PROM or die in equipment. Keys in codes, using computer keyboard, and presses buttons to activate equipment that reproduces master program in programmer. Observes digital display on equipment to ensure that master program is not defective and to determine when reproduction of program is completed in programmer. Positions blank PROMs and pushes lever to lock PROMs in sockets of programmer; or positions blank dies on platform of die probe equipment, views dies through microscope, and turns knobs on equipment to position specified areas of dies under probes prior to programming. Presses switches to activate programmer that reproduces program on blank PROMs or dies. Observes defect indicator lights to detect defective PROMs or dies and to determine completion of programming. Maintains production records. May load defective PROMs into equipment that exposes PROMs to ultraviolet light for timed intervals to erase programmed information in order to reuse PROMs.

## Tester, Semiconductor Packages (electronics)

Tends automatic equipment that test functions of semiconductor packages: Reads production documents to determine test specifications. Inserts specified test program (magnetic card) in test equipment to program equipment. Places tubular magazines containing semiconductor packages in feeder mechanism of equipment. Flips switch to activate test cycle. Monitors operation of test equipment to detect misfeeds. Sorts tested semiconductor packages into specified containers. Maintains production records. May tend centrifuge or temperature-cycle machine that places semiconductor packages under stress to cause circuit failure of weak sections during subsequent testing.

## Wafer Mounter (electronics)

Tends press that mounts semiconductor wafers in templates for polishing: Cleans templates, using cleaning solution and brush or towel, or tends machine that automatically cleans templates. Places cushion in template depressions and places wafers on cushion, or places wafers on waxed template. Positions templates in press and activates press that presses wafers into templates. May tend machine that automatically removes polished wafers from templates. May tend machine that laps templates to restore flatness. May tend press that attaches template to carriers before running wafers.

## Photo Mask Inspector (electronics); Mask Inspector

Inspects master or production photo mask plates used in fabrication of semiconductor devices for defects and to ensure conformance to specifications: Positions photo mask plate on stage of microscope or computer-aided inspection equipment. Focuses and aligns lens with predetermined coordinates on photo mask plate. Inspects photo mask plate for defects, such as misalignment of design, contamination, cracks, pinholes, and streaking, and compares photo mask pattern to original design to verify conformance to specifications. Enters commands to activate computer-aided inspection equipment that automatically scans surface of photo mask plate to locate and record defects. Records inspection information, and sorts photo mask plates according to inspection results. May measure critical dimensions of pattern on photo mask plate and compares dimensions to design requirements, using microscopes with attached calibrated viewing or image shearing apparatus. May photograph surface of defective photo mask plate, using instant-print camera. May operate laser-beam equipment to repair circuitry defects on photo mask plate [LASER-BEAM-TRIM OPERATOR (electronics)]. May tend equipment that removes contaminants and photoresist from surface of photo mask plate [PHOTO MASK CLEANER (electronics)].

# Laser-Beam-Trim Operator (electronics)

Operates computer controlled laser-trim system to trim excess material from electronic components: Reads production sheet to determine specific operation code and depresses keys on console control of computer to input code for particular operation. Inserts electronic component into vacuum chuck of laser-trim system, using tweezers, and depresses prescribed sequence of console buttons to actuate laser beam that automatically trims excess metal and glass from component. Observes light indicator on control panel to determine whether each component resets required specifications. Removes trimmed component from chuck and examines component under microscope for completeness of trim.

# Inspector, Semiconductor Wafer Processing (electronics)

Inspects semiconductor wafers, using microscope, to identify processing defects: Places wafers on stage of microscope, using tweezers or vacuum wand. Examines surface of wafer and test patterns on wafers, using microscope, and compares wafers to specification diagrams to identify processing defects, such as scratches, contamination, pattern misalignment, photoresist peel, and circuit bridges. Measures specified dimensions of test patterns on wafers, using electronic measuring devices attached to microscope at work station, to verify that wafers meet company specifications. Routes defective wafers to engineering department. Records observations onto processing sheets and log. Cleans inspecting area, using cleaning solution.

# Leak Tester, Semiconductor Packages (electronics)

Tends testing equipment that detects leaks in semiconductor packages: Loads tubes containing semiconductor packages into cannister. Places loaded cannister in equipment that replaces air in cannister with liquid freon or flushes cannister with radioactive isotopes. Pushes buttons and flips switches to activate equipment. Removes loaded cannister and places either cannister or tube of semiconductor packages in leak testing equipment, such as bubble pot or geiger counter tank, that detects gross and fine leaks in semiconductor packages. Inspects tube of packages in bubble pot, using magnifying glass, to identify packages with bubbles, indicating gross leaks. Observes control panel of geiger counter tank to identify radioactive semiconductor packages, indicating fine leaks. Removes tubes and cannisters from leak detection equipment and places defective semiconductor packages in containers, using tweezers. Places accepted semiconductor packages in tubes, using tweezers. Maintains production records.

## Die Tester (electronics)

Tests voltage of *dies* containing individual devices, such as diodes and transistors, using computerized testing system, to ensure that dies meet company and manufacturer specifications: Reads production sheets and computer code books to determine test instructions and codes. Keys instructions and codes into computer, using computer terminal. Pours dies into equipment bowl that automatically conveys each die to probe, tests voltage of each die, and sorts dies into specified bottles. Observes equipment operation to detect misfeeds, using built-in microscope. Maintains production records. Records voltage reading on bottle labels. Cleans and maintains testing equipment.

## Die Attaching Machine Tender (electronics)

Tends die attaching machine that bonds dies to semiconductor packages: Loads package carriers, magazines containing dies, and gold pieces, used to attach dies to packages, into designated slots of machine. Starts machine that automatically positions and melts gold on packages and positions dies in packages to form assembled semiconductor packages. Observes operation of machine to ensure that alignment of packages, gold, and dies meets company specifications, using video monitor. Turns knobs on machine to adjust alignment of parts assembled, regulate operation speed, and set temperature of molten gold. Removes magazines of packages for further processing.

# Jobs Directly Related to the Printed Circuit Boards

## Group Leader, Printed Circuit Board Assembly (electronics)

Assists supervisor in coordinating activities of workers engaged in assembly of printed circuit boards (PCBs), applying knowledge of PCB assembly techniques, specifications, and production scheduling: Confers with supervisor and reviews production schedules, specifications, and priorities to plan departmental work assignments. Requisitions, obtains, and distributes supplies and materials, such as PCBs, electronic components and parts, solder and flux, antistatic bags and wristbands, and schematic drawings and work orders. Assigns duties to assembly workers and oversees department activities. Revises work assignments as required by priorities and work availability. Explains and demonstrates PCB assembly line procedures and techniques to workers. Interprets schematic drawings, specifications, and work orders for workers. Assists workers in resolving technical problems and advises supervisor of complex production problems. Reads, prepares, collects, and maintains reports, such as individual and department production reports, employee time and attendance records, and component waste reports. May assemble sample PCBs for use as work aids by workers, using schematic drawings, handtools, and soldering equipment. May perform lead wires for electronic components to supply assembly line with preformed parts, using forming machines or handtools. May substitute for workers during worker absence or to relieve bottlenecks in work congested areas.

## Group Leader, Printed Circuit Board Quality Control (electronics)

Assists supervisor in coordinating activities of workers engaged in inspecting, testing, and repairing printed circuit boards (PCBs), applying knowledge of electronic theory, test procedures, repair techniques, and quality standards: Confers with supervisor and reviews production schedules to determine quantity and type of PCBs to be inspected, tested, or repaired. Assists in planning department work assignments. Collects and distributes PCBs and components for processing from receiving and assembly departments. Requisitions, obtains, and distributes supplies, materials, and equipment, such as electronic components and parts used during repair work, test and inspection specifications, and test equipment, fixtures, and handtools. Inspects test equipment to verify that equipment functions according to standards. Reports substandard equipment performance to supervisor or maintenance personnel. Compiles list of test and inspection personnel. Assigns duties to inspection, test, and repair workers, using knowledge of workers' experience and capabilities. Oversees department activities and revises work assignments as required by priorities and work availability. Explains inspection and test procedures and specifications. Demonstrates use of testing equipment and PCB repair techniques to workers. Interprets schematic drawings, procedure changes, and work orders for workers. Assists workers in resolving technical problems, utilizing knowledge of electronic theory, test procedures, and specifications. Reports unresolved problems to supervisor. Reads, prepares, collects, and maintains reports, such as individual and department production reports and test results, employee time and attendance records, and product waste reports. Maintains history files on vendors. May relieve workers during absence or substitute for workers to relieve bottlenecks in work-congested areas. May confer with technical personnel and other department supervisors to report and resolve problems concerning substandard assembly work and testing procedures.

## Supervisor, Printed Circuit Board Assembly (electronics)

Supervises and coordinates activities of workers engaged in assembling, testing, and inspecting printed circuit boards (PCBs): Directs training of workers in interpreting shop specifications, component recognition, equipment operation, and performance of job duties to ensure that assembly standards are met. Plans and coordinates assembly assignments to ensure that production goals are met. Examines boards during assembly process, applying knowledge of assembly and quality standards to ensure acceptability. Performs other duties as described under SUPERVISOR (any ind.) Master Title. May be designated according to stage of assembly as SUPERVISOR, FINAL (electronics); SUPERVISOR, POST-WAVE (electronics); SUPERVISOR, PRE-WAVE (electronics). May be designated according to function supervised as SUPERVISOR, ASSEMBLY (electronics); SUPERVISOR, QUALITY CONTROL (electronics); SUPERVISOR, TESTING (electronics).

# Wave-Soldering Machine Operator, Printed Circuit Boards (electronics); Flow-Solder Machine Operator

Controls and monitors operation of wave-soldering machine or system to solder electronic components onto printed circuit boards (PCBs): Reads production schedules and operations manuals and receives verbal instructions regarding sequential startup and operation of wave-soldering machines (bench models) or systems (modular floor models). Moves controls to activate machines or system modules, such as flux, preheater, wax stabilizer, chiller, lead wire cutter, wave-soldering units, conveyors, and PCB washing and drying machines. Adjusts controls of machines to regulate operating speeds and temperatures of machines and to regulate heights (stand or flow) of flux, wax, and solder waves, according to predetermined standards. Adds supplies to reservoirs of machines, such as flux, wax, solder, oil, and thinner. Verifies specific gravity of flux, using hydrometer. Adjusts dimensional and structural characteristics of conveyor pallets and rails used to hold and transport PCBs through wave-soldering process, using instruments and handtools, such as micrometers, dial indicators, pin gauges, wrenches, and screwdrivers, and knowledge of production schedules and standards. Places loaded master pallet (master fixture) and production pallets on conveyor line of wave-soldering machines or systems and observes pallets as they travel through process to verify that timing, temperatures, and dimensions are set according to standards. Observes heat-sensitive color marks or strips placed on PCBs to verify boards have been preheated to specified temperatures. Records machine settings before and after adjustments in daily logs. Observes meters, gauges, and indicator lights during operation of wave-soldering machines or systems to ensure that soldering of PCBs conforms to standards and to prevent damage to equipment. Removes samples of soldered boards to examine top and underside of boards to detect substandard soldering. Discusses quality of soldering with supervisor and makes adjustments to controls according to instructions received. Records data in daily operating logs, such as operating time, number of PCBs processed, number of pallets used, conditions of lead wire cutting blades, and hours of usage of flux and wax. Moves switches and turns valves to shut down machines and systems, following prescribed procedures. Drains and refills wax (used to stabilize components on boards prior to lead wire-cutting operation) and flux reservoirs, according to length of use and established replacement times. Skims dross from surface of molten solder pots and periodically drains and refills reservoirs. Removes machine parts, using handtools. Cleans parts and interiors of machines, using cleaning materials and supplies. Performs preventive maintenance tasks, such as examining and replacing lead wire cutter blades (on modular systems), tightening drive belts and pulleys, and lubricating motors, pumps, bearings, and pulleys, using handtools and lubricants, according to specifications and instructions received. Records cleaning and preventive maintenance activities in daily logs. May return substandard boards to workers for repairs. May feed and/or offbear board or pallets loaded with PCBs to or from conveyor system.

# Supervisor, Printed Circuit Board Testing (electronics); Supervisor, Electronic Testing; Supervisor, Functional Testing; Supervisor, In-Circuit Testing

Supervises and coordinates activities of workers engaged in repairing and testing performance of printed circuit boards (PCBs) during and after assembly, applying knowledge of electronic theory and circuitry, test procedures, and documentation: Directs training of workers in areas such as test procedures and equipment operation, troubleshooting and repairing, and interpretation of specifications and documentation. Coordinates testing and repair work to ensure completion of work assignment. Monitors repair activities to verify PCBs conform to documentation standards. Performs other duties as described under SUPERVISOR (any ind.) Master Title. May supervise activities of workers engaged in assembling PCBs.

# Inspector, Circuitry Negative (electronics); Film Inspector

Inspects circuitry film negatives (artwork) for conformance to specifications and touches up defects on negatives to prepare negatives for use in printed circuit board (PCB) fabrication: Compares size and location of hole images on circuitry negative with drilled holes in sample PCB panel to verify conformance of negative to specifications, using light table. Examines circuitry negatives to detect defects, such as under-exposure or over-exposure of negatives, circuitry width, and spacing, using light table and microscope. Fills nicks, scratches, and gaps in circuitry pattern on negatives with pen or plastic tape, removes excess ink from negatives with cotton swab, and cuts flaws and excess film from circuitry negatives to correct and prepare negatives for use in PCB fabrication, using utility knife. Measures circuitry negative for conformance to specified dimensions, using precision-measuring instruments.

# Functional Tester, Printed Circuit Boards (electronics)

Controls and monitors operation of computerized test equipment to test electronic functions of printed circuit board (PCB) assemblies, components, parts, and circuitry: Reviews schedule of board types to be tested, test procedures, and specification with supervisory personnel to determine specific instructions or modification in test procedures. Removes PCBs from carrier and positions boards in specified position on test unit. Connects cables from automatic computerized test equipment to PCBs to perform test. Observes CRT monitor during automatic testing to determine status of board functions relating to capacitance, resistance, tolerance, continuity and defects, such as shorts and leaks.

Compares instrument readings with specifications to determine board passage or failure. Removes and stacks PCBs that pass testing, for movement to next station. Observes CRT monitor to determine location of malfunction in defective circuit boards. Tears off paper tape printout and inserts printouts in designated slots on PCBs to accompany boards that pass and to facilitate repairs of defective boards. Removes and stacks PCBs that fail testing, for movement to repair station. May operate burn-in oven to heat PCBs to elevate temperatures preparatory to testing. May replace defective components, using desoldering tool, solder, and iron. May retest boards following troubleshooting by other workers to verify resolution of malfunctioning problems. May use multimeter or oscilloscope to determine exact location of defect on PCB [PRINTED-CIRCUIT TESTER (electronics)]. May complete production report to indicate number of units tested during shift. May periodically examine test equipment to verify that test equipment is functioning according to standards. May perform mathematical computations, using formula and figures obtained from screen, to determine that board measurements meet established criteria.

# Electronic Circuit Tester, Printed Circuit Boards (electronics)

Tests electronic circuits of printed circuit boards (PCBs), using computerized test equipment: Reviews test specifications to determine program disks and test fixtures required to test PCBs. Aligns specified test fixture in test equipment and secures PCBs on pins of test fixture. Inserts program disk in control panel and depresses keys of computer keyboard to start test equipment that automatically tests continuity of PCB circuits to detect short or open circuits. Observes control panel for signal lights or observes display screen for message indicating PCBs have passed or failed test. Tapes printout (paper tape) of test results on defective boards and stamps identifying number on boards passing tests. Maintains records of PCBs tested and number of boards passed or failed. May monitor computerized test equipment, using test boards, to verify that equipment functions according to standards. May repair defective circuitry of PCBs, using soldering iron.

# Film Touch-Up Inspector (electronics)

Inspects and repairs circuitry image on photoresist film (separate film or film laminated to fiberglass boards) used in manufacture of printed circuit boards: Inspects film under magnifying glass for holes, breaks, and bridges (connections) in photoresist circuit image. Removes excess photoresist, using knife. Touches up holes and breaks in photoresist circuitry image, using photoresist ink pen. Removes and stacks finished boards for transfer to next work station. Maintains production reports. May place lint-free paper between dry film sheets to avoid scratching circuit images on film.

# Printed Circuit Board Assembler, Hand (electronics)

Performs combination of following tasks in assembly of electronic components onto printed circuit boards according to specifications, using handtools: Reads worksheets and wiring diagrams, receives verbal instructions, follows sample board to determine assembly duties, and selects components, such as transistors, resistors, relays, capacitors, and integrated circuits. Twists, bends, trims, strips, or files wire leads of components or reams holes in boards to insert wire leads, using handtools. Inserts color-coded wires in designated holes and clinches wire ends, using pliers. Press-fits (mounts) component leads onto board. Places plastic insulating sleeves around specified wire leads of components and shrinks sleeves into place, using heat gun. Crimps wire leads on underside of board, using handtools or press. Applies sealer or masking compound to selected parts of board to protect parts from effects of wave solder process. Solders wire leads and points on underside of board, using soldering iron, to route and connect lead wires to board and between individual components. Installs heat sinks, sockets, face plates and accessories on boards, using handtools. May be designated according to unit installed as SOCKET ASSEMBLER (electronics) or stage of production as POST-WAVE ASSEMBLER (electronics); PRE-WAVE ASSEMBLER (electronics). May assist other workers in duties concerned with wave-soldering PCBs.

# Printed Circuit Board Component Tester, Chemical (electronics)

Tests electronic components for compliance to company standards, using chemicals and soldering equipment: Places droplets of acid on surface of petri dish to prepare for lead (metal) test, using eyedropper. Positions wire lead of component or pin on integrated circuit in acid, places acid-coated wire lead or pin on chemically prepared filter paper, and observes reaction on paper that indicates lead content. Soaks components that do not pass lead test in freon to remove acid from wire leads or pins and repeats test. Refers to vendor list to determine whether components are listed as acceptable. Dips wire leads of components in beaker of liquid flux, solder pot, and beaker of freon to conduct solderability test on wire leads, using metal tongs. Observes solder on wire leads for questionable conditions, such as bending of solder, discoloration, or no solder present, using magnification lamp. Sets aside components of questionable acceptance for approval by engineer. Maintains record of test results.

# Printed Circuit Board Component Tester, Preassembly (electronics)

Tests electronic function of components preparatory to printed circuit board (PCB) assembly for compliance to company standards, using test equipment: Selects and positions test fixture on test equipment panel according to type of component being tested. Keys data into computer keyboard or turns dials and presses buttons to set tolerances on equipment following company standards guide. Positions components in test fixture or mounts reels containing components on tape onto machine spindle and threads component tape through test machine guides. Activates equipment to start test and observes colored panel lights on equipment to determine whether components meet specifications. Replaces defective components. Verifies dimensions of components, using calipers and specification book. Obtains vendor history file for reference when excessive component defects result during testing. Maintains records of test results. May solder electronic components to test boards prior to testing. May be designated according to unit tested as CAPACITOR TESTER (electronics); HYBRID TESTER (electronics); INDUCTOR TESTER (electronics); RELAY TESTER (electronics); RESISTOR TESTER (electronics); TRANSFORMER TESTER (electronics).

# Printed Circuit Board Inspector, Preassembly (electronics)

Inspects printed circuit board (PCBs) preparatory to assembly for compliance to company guidelines, using test equipment: Examines PCBs on light table for breaks in copper foil circuits. Examines PCBs to detect circuit shorts, using magnification lamp. Measures dimensions of PCBs, using ruler. Views inside edges of solder holes to detect irregular surfaces, using eye loupe. Inserts leads to components in solder holes to verify spacing of holes and ease of component insertion. Verifies plating integrity of copper foil circuits and composition of solder through solder holes, using electronic test equipment. May record inspection results to maintain vendor history files.

# Production Repairer, Printed Circuit Board Assembly (electronics)

Performs any combination of following tasks to repair defects on printed circuit board (PCB) assemblies, using soldering and desoldering equipment: Examines PCBs under magnifying lamp and reads inspection tags to locate defects, such as insufficient or excess solder and missing, misaligned, or defective components and parts. Resolders connections, lead wires, joints, and terminals missed during soldering process. Removes excess solder from connections and clears solder from component holes, using desoldering equipment. Repositions misaligned components, such as resistors, capacitors, and integrated circuits (ICs), and removes and replaces damaged or missing components. Trims long leads, replaces eyelets and rivets, and repairs damaged substrate, using handtools and power tools. Modifies PCBs following shop instructions, using handtools and soldering equipment. May test soldered connections on PCBs to detect open circuits or shorts, using continuity tester.

# Reworker, Printed Circuit Board (electronics); Reworker

Repairs defective surfaces and circuitry on printed circuit boards (PCBs), using power and handtools, utilizing knowledge of electronic repair techniques: Reads work order to determine number of PCBs to be repaired, type of repairs required, and method and tools to be used in reworking. Positions PCBs under microscope to examine circuitry. Cuts and removes defective wires, using knife, and positions replacement wire on circuit, using magnetic tweezers. Repairs defective circuitry, using handtools and soldering or welding equipment. Repairs board surface faults, such as excess solder on board or gold plating on connectors, using handtools, and washes solder or liquid gold over specified areas to restore board surfaces. Cleans repaired boards with solvent, using brush and rags. Inspects boards to determine that repairs meet specifications. Removes and stacks repaired boards on racks for movement to next work station. Places PCBs in industrial oven to cure solder mask. May prepare production reports.

# Inspector, Printed Circuit Boards (electronics); Board Inspector; Circuit Board Inspector; Touch-Up Inspector, Printed Circuit Boards; Touch-Up Operator, Printed Circuit Boards

Inspects printed circuit boards (PCBs) for conformance to specifications and touches up defects: Inserts plug gauges into drilled holes in PCB panels to verify conformance to specified dimensions. Measures plated areas on PCB panels, using devices such as micrometers and dial indicators, to verify uniformity and thickness of plating. Brushes liquid photoresist on sections of circuitry pattern not completed and scrapes excess photoresist from panels, using artist's knife. Examines PCB circuitry, using *light table*, eye loupe, magnifier or binocular microscope, and specifications or artwork, to detect defects, such as shorts, breaks, excess or missing solder, scratches, cracks and incorrect layout, and scrapes excess copper and solder from board, using artist's knife. Measures PCBs for conformance to specified dimensions, using calipers, micrometers, dial indicators, rulers, and eye loupes. Rejects defective boards and records type and quantity of defects to keep record control. May inspect plating in holes and remove excess plating. May test adherence of photoresist to PCBs, using tape. May repair broken circuitry, using soldering iron or circuit bonding equipment. May test circuit continuity of boards, using bare board testers. May inspect inner layers of multilayer PCBs to verify that internal alignment and location of drilled component mounting holes meet specifications, using computer-controlled x-ray equipment, and be designated X-RAY TECHNICIAN, PRINTED CIRCUIT BOARDS (electronics).

# Preassembler, Printed Circuit Board (electronics)

Attaches preassembled components, such as terminal pins, connectors, and labels, to printed circuit boards (PCBs) and stamps identification codes on boards preparatory to further assembly: Examines PCBs for illegible issue numbers and removes and restamps numbers, using eraser and handstamp. Reads work specifications to determine type of labels, components, and locations on PCBs to attach items. Cuts component pins (wires) to specified length, using wire cutters, inserts pins in specified holes on PCBs, and secures pins to PCBs, using holding block and hammer or pneumatically powered stamping device. Affixes labels and tape to boards and stamps identification codes on boards.

# Touch-Up Screener, Printed Circuit Board Assembly (electronics)

Inspects printed circuit board (PCB) assemblies for defects, such as missing or damaged components, loose connections, or defective solder: Examines PCBs under magnification lamp and compares boards to sample board to detect defects. Labels defects requiring extensive repairs, such as missing or misaligned parts, damaged components, and loose connections, and routes boards to repairer. Performs minor repairs, such as cleaning boards with freon to remove solder flux; trimming long leads, using wire cutter; removing excess solder from solder points (connections), using suction bulb or solder wick and soldering iron; or resoldering connections on PCBs where solder is insufficient. Maintains record of defects and repairs to indicate recurring production problems. May reposition and solder misaligned components. May measure clearances between board and connectors, using gauges.

# Solder-Leveler, Printed Circuit Boards (electronics)

Tends processing equipment that prepares and applies solder to copper circuit areas of printed circuit boards or panels: Reads process specifications and activates equipment units, such as conveyors, sprayers, pumps, brushes, air knives (blower mechanism), and solder bath. Adjusts equipment controls to regulate processing factors, such as temperatures, pressures, and speeds, according to specifications. Feeds boards on motorized conveyor leading into series of processing units, such as sprayers, brushes, dryers, and flux rollers, that clean copper circuit areas and apply flux preparatory to solder application. Removes boards from precleaning equipment conveyor and clamps boards to feeding mechanism of solder-leveling equipment. Pushes button to activate equipment that automatically lowers board into molten solder bath, lifts boards out of bath, and moves boards past hot-air knives that smooth solder off non-copper areas. Monitors equipment operation and corrects minor malfunctions, such as board misfeeds, conveyor speeds, and temperature settings. Removes board from solder equipment and visually examines boards for completeness of soldering. Notifies supervisor of equipment malfunctions and board defects, such as discolored, missing, or unlevel solder. May reroute defective boards through processing equipment to correct defects. May maintain production records.

# Lamination Assembler, Printed Circuit Boards (electronics)

Assembles layers of laminating materials and printed circuit board (PCB) panels to prepare multilayer PCB panels for laminating process: Assembles and aligns inner and outer layers of PCB panels and laminating materials over guide pins in holding fixture to prepare multilayer PCB panels for laminating process, following standard procedures. Examines inner layers of panels as assembled for dents, chips, dust, or flaws. May tend laminating machine that bonds assembled panels to form multilayer PCB panels. May heat materials in oven to remove moisture prior to laminating process.

# Masker (electronics)

Applies liquefied soldering mask to designated terminal areas on underside of printed circuit boards (PCBs) to prevent molten solder from collecting on circuitry during wave-soldering process, using squeeze bottle of soldering mask and following pre-masked sample of PCBs and verbal instructions. Removes excess mask, using towel. Places masked PCBs in storage racks for movement to next work station.

# Wave-Solder Offbearer (electronics)

Removes printed circuit boards (PCBs) from discharge conveyor of automatic wave-soldering machine following completion of soldering process: Places empty carrier on workbench at discharge end of wave-soldering machine conveyor. Removes pallets containing wave-soldered PCBs from conveyor and places pallets in carrier. Examines boards to detect defects, such as missing solder, spots, or warped boards. Submits defective boards to supervisor for approval or correction of defects. Stacks acceptable boards on carrier or in tube for final assembly. May assist others in setting up and operating equipment for wave-soldering process. May assist in cleaning equipment, using scraper and airhose. May complete production report for supervisory use.

# Reflow Operator (electronics)

Tends reflow equipment that melts solder on printed circuit boards (PCBs) to redistribute and fuse solder on boards to improve appearance, hardness, and solderability for PCB component assembly: Immerses boards in cleaning solution or places boards in cleaning machine that cleans and removes contaminants from boards before reflow process. Starts reflow equipment and adjusts conveyor speed and temperature of infrared heating units to meet specifications. Feeds PCBs onto conveyor of equipment that applies liquid flux and infrared heat to boards to melt, redistribute, and fuse existing solder on boards and cleans, rinses, and dries boards following reflow process. Observes boards moving on conveyor to monitor flow and operation of reflow process. Removes boards upon completion of process and inspects boards for reflow process defects. May add chemicals to flux solution. May tend soldering equipment. May tend flow equipment that automatically cleans, rinses, and dries boards before and after reflow process.

# Electronic Equipment Setup Operator (electronics)

Sets up and operates equipment to fabricate single- and multilayer printed circuit boards (PCBs): Determines dimensions and tolerances of boards, sequence of operations, and tools and equipment required, according to diagrams and specifications, and measures and marks dimensions and reference points to lay out boards for machining. Sets up and operates equipment, such as shearing machine, laminator, and drills, to cut and form boards to specifications for use by other workers. Bevels edge connectors, using slotter, and verifies dimensional tolerances, using calipers. Cuts boards to size, using shearing machine. Sets up and operates drills to bore holes in fiberglass boards, and sets up and controls plating equipment to copperplate and apply photoresist to boards, using laminator. Photographs design image and exposes film to produce and develop acid resisting circuitry pattern, using photographic equipment. Etches and touches up circuitry on boards, using etcher. Bonds together two or more PCBs to form multilayer boards, using laminating press.

# Coordinate Measuring Equipment Operator (electronics); Mechanical Inspector

Operates coordinate measuring equipment to measure dimensions of printed circuit boards (PCBs): Reads blueprints to determine dimensions and tolerance specifications for selected PCBs. Positions PCBs on measuring table of equipment, views PCBs on monitor screen, and turns calibrated cranks on equipment to position control bar, with attached camera over specified sections of PCBs. Flips toggles to lock camera in position, activates digital readout equipment, and records specified measurement of PCBs on inspection sheets. Repeats process to cover all sections of PCBs including locations of specified drilled holes and samples of routed PCBs. Compares recorded measurements to blueprint dimensions and tolerances and notifies specified persons of discrepancies.

# Test Fixture Assembler (electronics)

Assembles test fixtures used to test electrical circuitry of printed circuit boards (PCBs), according to diagram instructions: Receives test fixture kits, work orders, and diagrams. Positions test fixture boards on worktable and inserts copper sleeves in designated holes of boards as indicated on diagrams. Positions pins flush with fixture board surface, using mallet. Places copper test pins in sleeves previously inserted and positions spacer blocks in prescribed pattern on top and bottom of boards. Inserts screws in spacer blocks and secures blocks to boards, using screwdriver. Positions wires against pins and blocks, following specified pattern, and secures wires around pins or blocks, using wire-wrapping gun. Positions fixture cover over spacer blocks on top of test fixture boards and positions metal back on assembled test fixture. Secures cover and back, using screws and screwdriver.

# Automatic Component Insertion Operator (electronics)

Tends computer-controlled machine that automatically inserts electronic components, such as resistors, capacitors, diodes, and integrated circuits, into holes of printed circuit boards (PCBs): Inserts specified program tape in machine or types commands on keyboard to enter program that directs machine to insert components in specified sequence and location on PCBs. Positions paper tape reels or plastic tubes containing components into feeding mechanism to load machine. Positions PCBs on machine bed holding fixtures, starts machine, and monitors machine operation. Removes assembled PCBs from machine bed and inspects PCBs for defects, such as missing components or incomplete insertion. Stops machine and notifies supervisor or maintenance mechanic of machine malfunctions. May make minor machine adjustments and clear machine jams. May manually insert missing components or route defective boards to repair department. May record production levels, machine downtime, and defects. May tend semiautomatic or manually controlled machines and position PCBs or components on board by hand. May be designated according to type of machine tended as VARIABLE CENTER DISTANCE OPERATOR (electronics); or according to component inserted as DUAL IN-LINE OPERATOR (electronics).

# Other Electronics Jobs

## Electronics Assembler (electronics)

Assembles electronic components, subassemblies, and systems by any one or combination of following methods: Reads work orders, follows production drawings and sample assemblies, and receives verbal instructions regarding duties to be performed. Positions and aligns parts in specified relationship to each other in *jig, fixture,* or other holding device, using tweezers, vacuum probe, or hand instruments. (1) Tends machines or uses handtools and power tools to crimp, stake, bolt, rivet, weld, solder, cement, press fit, or perform similar operation to secure parts in place. (2) Mounts assembled components, such as transformers, resistors, capacitors, integrated circuits, and sockets on chassis panel. (3) Connects component lead wires to printed circuit or routes and connects wires between individual component leads and other components, connectors, terminals, and contact points, using soldering, welding, thermocompression, or related bonding procedures. (4) Installs finished assemblies or subassemblies in cases and cabinets. (5) Assembles and attaches functional and cosmetic hardware, such as caps, clamps, and knobs, to assemblies, using handtools and power tools. (6) Performs intermediate assembly tasks, such as potting, encapsulating, cleaning, coating, epoxy bonding, curing, stamping, etching, impregnating, and color coding parts and assemblies. (7) Tends machines which press or shape component parts, such as contacts, shells, and insulators. (8) Tends automatic assembly equipment which joins components and parts. (9) Winds capacitors on manual or automatic winding equipment. (10) Adjusts or trims material from components, such as capacitors and potentiometers, to change electronic characteristic to specification measured by electronic test instrument. (11) Performs on-line go-not-go testing and inspection, using magnifying devices, fixed and movable measuring instruments, and electronic test equipment to ensure parts and assemblies meet production specifications and are free from defects. May perform assembly operations under a microscope or other magnifying device. Occupations related to assembly of printed circuit boards and fabrication of integrated circuit chips are defined elsewhere under separate definitions.

## Electronics Tester (electronics) II; Component Tester; Production Tester; Quality-Control Tester; Testing-Machine Operator

Tests electronic function of assemblies, components, and parts, using standard test equipment and procedures: Connects unit to test instrument, such as ohmmeter, voltmeter, ammeter, resistance bridge, or oscilloscope and turns switch. Reads instrument dial or scope that indicates resistance, capacitance, continuity, and wave pattern or defect, such as short or current leakage. Compares instrument reading with standard and rejects defective units. Records type and quantity of defect. May verify dimensions of parts, using standard gauges. May examine assembly, component, or part for defects, such as short leads, bent plate, or cracked seal. May tend equipment to subject unit to stress and strain prior to testing. May tend automatic test equipment. May operate x-ray photograph and closed circuit television viewing equipment to observe internal image of unit for assembly defect prior to testing. May adjust circuits in radio and television receivers for maximum signal response and be designated ALIGNER (electronics). May be designated according to unit tested as COIL TESTER (electronics); FILTER TESTER (electronics); TUBE TESTER (electronics).

# Electronics Inspector (electronics) II; Checker; Inspector, Component Parts; Inspector, Visual; Line Inspector

Inspects electronic assemblies, subassemblies, parts, and components by comparing them with samples or production illustrations: Examines unit visually for physical defects, such as broken wire, excess solder, holes in sealing material, unevenly wound coil, coating and plating blemishes, oil leaks, faulty resistance weld, scratches, and cracks. Compares hardware, such as eyelets, brackets, and lugs, on assemblies, subassemblies, and parts with parts list to verify installation. Examines hardware for specified contact with conductor area. Rejects faulty assembly, part, or component. Records type and quantity of defects to keep record control. May measure parts to verify accuracy of dimensions, using standard gauges. May sort defective components and parts for salvage or scrap. May be designated according to item inspected as CAPACITOR INSPECTOR (electronics); CATHODE-COATING INSPECTOR (electronics): CONNECTOR INSPECTOR (electronics); FILTER INSPECTOR (electronics); GRID INSPECTOR (electronics). Additional titles: RESISTOR INSPECTOR (electronics); TELEVISION-CHASSIS INSPECTOR (electronics); TUBE INSPECTOR (electronics).

# Solder Deposit Operator (electronics)

Cleans assembled semiconductor packages, coats package leads with flux, and applies solder to package leads, using solder dip machine, to complete semiconductor package assembly: Places packages in holding cases and immerses cases in series of acid baths to clean packages. Places packages in holding fixtures and immerses loaded fixtures in flux bath to coat leads of packages. Pushes button to activate solder pumps of solder-dip machine and places loaded fixtures on conveyor of machine that deposits solder on leads of packages to enhance electrical conductivity of leads. Removes fixtures from conveyor, inspects leads of packages for uniform soldering, and removes solder from top of packages, using needle-nose pliers. Places loaded fixtures in baskets and immerses baskets in series of acid and soap baths to clean packages. Removes packages from fixtures, places packages on trays, and positions loaded trays in oven to dry packages. Removes tray from oven and loads packages into lot box and cleans and maintains solder pot of machine. Maintains production records.

# Wire-Wrapping-Machine Operator (electronics)

Operates computer-controlled semiautomatic machine that wraps wires around electronic-pin connectors: Mounts connector panel on machine pallet that moves panel along programmed path, using wrench and screwdriver. Depresses specified button to start automatic programmed tape for pin sequence and observes panel lights that indicate size of wire prescribed in program. Selects and mounts specified wire on machine spindle. Threads wire through bit of wire-wrap gun, positions gun in support to align gun with pins on connector panel, and depresses trigger of wire-wrap gun to wrap wire on pins. Observes directional lights of machine to determine movement of pallet and gun support. Inspects wire-wrap of completed panels for tightness, neatness of fold, or broken wire.

# Wireworker (electrical equipment; electronics); Wire-Preparation Worker

Performs any combination of tasks involved in cutting, stripping, taping, forming, and soldering wires or wire leads of components used in electrical and electronic units, such as communication equipment, aircraft, ignition systems, electrical appliances, or other electrical or electronic control systems: Cuts wires to specified lengths, using wire cutter and ruler for measuring *jig*. Strips insulation from wire ends, using stripping tool. Twists wire ends and dips them into pot of solder to prevent fraying. Solders wires to specified connectors and terminals, using soldering iron, or crimps connectors and terminals to wire ends, using handtools. Wraps numbered or colored identification tape around wires. Rolls wires of identical number or color together and attaches tag to roll. Inserts wires into plastic insulation tubing. Bends, cuts, and crimps component leads to prepare component for mounting on printed circuit board or other assembly, using handtools. May insert wires in automatic numbering or color-coding machine to imprint part numbers or color codes. May insulate wire or component leads by dipping them into paraffin solution. May paint various protective coatings on wires with brush. May test wire and cable assemblies and repair defective assemblies. May use automated equipment or bench-mounted devices to cut, strip, bend, or crimp wire. May be designated by type of wire worked as COMPONENT LEAD FORMER (electronics).

# Wire-Stripping-and-Cutting-Machine Operator (any ind.)

Tends machine that automatically straightens, cuts to length, and strips insulation from ends of wire for use in electrical installation or assembly: Places reel of specified wire onto play-out spindle and threads end through feed rolls. Adjusts controls of machine to accommodate thickness of wire and to cut and strip wire to specifications. Starts machine and observes operation. Places boxes under ejection roll or manually removes and bundles or boxes finished pieces. Examines wire for conformance to specifications and removes defective pieces. Observes automatic counter or counts finished pieces and records data on tags or boxes. May install drive adjustment and tension or adjustment rolls in machine. May install ram and dies which bend finished pieces to specified angles. May replace dull or broken cutter or stripping blades. May tend machine that automatically bends wire to specified angle or prints identification onto insulation. May tend machines that crimp and attach connectors and terminals to wire ends. May be designated according to type of wire cut as LEADWIRE CUTTER (electronics).

# Sequencing-Machine Operator (electronics)

Sets up and operates sequencing machine to insert electronic components in programmed sequence onto paper tape for use in machine insertion of components onto printed circuit boards: Reads instructions to determine sequence, quantity, and types of components required. Mounts supply reels of components on spindles and threads component supply reel tape through machine. Installs and spaces holding devices on machine according to number of components to be sequenced, using wrench. Mounts paper tape reels onto spindles of machine to receive sequenced components. Keys data into computer terminal to enter program number that controls release of components in prescribed sequence onto paper tape. Presses button to activate machine and observes panelboard light indicators to verify sequencing procedures and to detect machine stoppages. Clears jams in holding devices and splices tapes, using handtools and adhesive tapes. Straightens bent wire leads or cuts damaged component from supply tape, using wire snips. Replaces defective components. Records production information. May operate sequencing machine equipped with verifier that automatically identifies faulty components.

# Electronics Assembler (inst. & app.)

Installs and wires electronic subassemblies, such as integrators, channel controls, and power sources for spectroscopic equipment, working from wire-prints, schematics, and verbal instructions: Routes wiring through assembly to terminals according to color code, and loops wires to panel. Strips insulation from ends of wire and cable, using wire stripper, and solders or wraps wires around terminals to provide electrical connection. Drills holes into frame and bolts or screws parts and assembled units to frame, using handtools.

# Electronics Worker (electronics)

Performs one or combination of duties to clean, trim, or prepare components or parts for assembly by other workers: Receives work directions from supervisor or reads work order for instructions regarding work to be performed. Cleans and deglosses parts, using various cleaning devices, solutions, and abrasives. Trims flashing from molded or cast parts, using cutting tool or file. Applies primers, plastics, adhesives, and other coatings to designated surfaces, using applicators, such as spray guns, brushes, or rollers. Fills shells, caps, cases, and other cavities with plastic encapsulating fluid or dips parts in fluid to protect, coat, and seal them [ENCAPSULATOR (elec. equip.; electronics)]. Prepares wires for assembly by measuring, cutting, stripping, twisting, tinning, and attaching contacts, lugs, and other terminal devices, using a variety of handtools and power tools and equipment [WIREWORKER (elec. equip.; electronics)].

Positions and fastens together parts, such as transformer laminates, glass tube laminates, electron tube mounts and cages, variable capacitor rotors and stators, paper loudspeaker cones, and shells and cases for various other components, using handtools and power tools. Charges rectifier plates, using current-generating device. Prints identifying information on component shells, using silk screen, transfer press, or electro-etch printing devices or equipment. Moves parts and finished components to designated areas of plant to supply assemblers or prepare for shipping or storage. Loads and unloads parts from ovens, baskets, pallets, and racks. Disassembles and reclaims parts, using heating equipment and handtools. Maintains records of production. May load mold with wound resistor forms and pour molten metal into mold to encase resistors. May be designated according to work performed as CAPACITOR ASSEMBLER, VARIABLE (electronics); CLEANER (electronics); CRIMPER (electronics); DIE CASTER (electronics); INSULATOR ASSEMBLER (electronics); METAL-BASE RECLAIMER (electronics). Additional titles: PAPERCONE MAKER (electronics) I; RECTIFIER-PLATE CHARGER (electronics); TAPER (electronics); TINNER (electronics); WIRE STRIPPER (electronics).

# Electronics Utility Worker (electronics)

Arranges layout of work station for other workers who fabricate, process, or assemble electronic equipment, and components, such as semiconductor devices, printed circuit boards, chassis assemblies, and wire harnesses and cables: Reads specifications, such as process guide, bill of material, wiring diagram, mechanical print, and schematic diagram, to determine equipment needed, such as piece parts, chemicals and gases, tools, test instruments, and *jigs and fixtures*, for work stations. Prepares and submits requisition. Positions equipment in specified arrangement at work stations. Notifies stockroom when stations need resupply of materials, such as chemicals, resistors, transistors, and color-coded wire, during processing or assembly operations. May assist workers to follow new work procedures and relieve operators from work stations. May set and adjust controls for processing, fabricating, and assembly line equipment, such as furnaces, process chambers, power supplies, timers, and multimeters. May test and repair assembled items by removing or adding piece parts or resoldering or rebonding defective connections, using handtools, production equipment, test machines, and test instruments, such as meters, resistance bridges, and automatic component testers.

# Test Fixture Designer (electronics)

Plots and draws schematics for test fixture heads used in testing printed circuit boards (PCBs) for electrical shorts and breaks, utilizing knowledge of printed circuit design, electronics, and customer specifications: Reviews blueprint to determine customer requirements and beginning and end points of PCB conductor paths. Confers with engineer to resolve questions concerning design of test fixture heads. Refers to blueprint and method sheets and chooses electrical points to be tested, making sure to include critical points along conductor path. Plots electrical points to be tested on layout sheet, using pencil. Chooses drill size to be used in drilling test head, according to test design and engineering specifications. Forwards layout of test points with requested drill size to numerical-control drilling department to be used as a guide to drill test head. Reviews blueprints to determine logical and most efficient method to wire test head to frame of test fixture. Draws schematics for TEST FIXTURE ASSEMBLER (electronics) to follow in wiring test head to test fixture frame. May review schematics with assembler to answer questions regarding correct wiring procedures. May monitor building of test fixture by assembler. May record data regarding completed job.

# Lead Hand, Inspecting and Testing (electronics)

Assists supervisor in coordinating activities of workers engaged in inspecting and testing electronic components for compliance to company standards: Receives boxes of components and records identifying information on specified forms. Collects random lot samples, according to lot volume and following company guidelines. Delivers samples to or instructs designated work to pull samples when job demands warrant attention. Distributes components to workers responsible for testing specific components. Instructs new workers in testing equipment operation, component testing sequence, and specification sheet interpretations. Modifies workers' assignments to meet priorities and to adjust work loads according to worker absenteeism. Completes requisition forms to obtain items requested and delivers items to workers. Observes calibration maintenance sheets to verify maintenance of test equipment and notifies designated personnel of unscheduled verification of equipment as warranted by questionable test results. Maintains records that indicate material discrepancies and records information on vendor history forms.

# Encapsulator (electrical equipment; electronics); Sealer; Tanker

Performs any one or combination of duties to encapsulate electrical and electronic parts, components, and assemblies with epoxy, plastic, or other material to protect them from damage and deterioration, using any of the following methods: (1) Positions module or transformer case under outlet of tank containing encapsulating fluid. Moves controls to force preset amount of fluid through opening in module or transformer case. (2) Places components on dipping rack or books, dips components in encapsulating fluid and places dipped components on drying frame or in oven to harden coating. (3) Fills capacitor shells or semiconductor caps with encapsulating fluid or applies encapsulating fluid to specified surfaces of component, using manual or electrically powered syringe or other applicator. (4) Tends compression molding machine to mold protective casing around component. May place component in mold and pour encapsulating fluid around it. May heat component in oven or curing device prior to and/or after encapsulating to remove moisture or cure encapsulating material. May inspect encapsulation for defects, such as chips and pinholes. May stamp encapsulated components with color code or trademark. May be designated according to method used as DIPPER (elec. equip.; electronics); FILLER (elec. equip.; electronics); POTTER (electronics). May weigh and mix materials, following formula, and be designated COMPOUND MIXER (elec. equip.; electronics). May apply bonding solution to bond flexible printed circuit boards to rigid boards, using potting equipment and be known as BONDER (electronics).

# Magnetic-Tape Winder (electronics); Cartridge Loader

Tends machines that wind magnetic tape into reels or cassette hubs for use in communication and control equipment, instruments, and computers: Positions tape supply reels or cassette hubs on letoff and windup spindles of tape-winding machines. Loops tape from supply reels through machine guides and into blank reels or hubs. Turns knob of footage counter devices that automatically cut tape and stop machines when specified length of tape has been wound into blank reels or hubs. Pushes switches to start individual machines and removes wound reels or hubs when machines stop. Scrapes detected surface defects from tape with knife. Splices tape ends together to form continuous loops in cassettes, using bench splicer. Inserts filled and blank hubs into cassettes and attaches covers. Packs reels and cassettes into containers and labels containers for shipment.

## Break-and-Load Operator (electronics)

Tends machine that breaks resistor plates along previously scored lines and loads resistors into magazines for further processing: Attaches guides of specified size to track on machine that feeds resistors into magazine, using Allen wrench. Snaps magazine onto machine. Breaks scrap edges from plates. Places plates into feed mechanism of machine and starts machine that automatically breaks plates into columns of resistors which slide down track. Taps column of resistors at bottom of track, using plastic rod, to break column into single resistors that fall into magazine. Clears machine jams, using plastic rod or tweezers. Breaks plates by hand to remove premarked, defective resistors. Fills out job order and production report. Cleans machine and work station, using brush. Carries magazines to storage room and picks up box of plates for next work order.

## Deflash and Wash Operator (electronics)

Tends equipment that automatically removes rubber *flash* from resistor ladders (rubber encased resistors) and cleans ladders: Loads resistor ladders into feed chute of machine and starts machine that automatically conveys resistor ladders through abrading chamber that removes flash from ladders. Visually inspects ladders to ensure that flash has been removed. Turns knob to adjust conveyor speed, if necessary, to increase or decrease abrading time. Clears jams on conveyor, using tweezers. Hits chamber of machine with mallet to loosen packed plastic abrading beads. Removes deflashed resistor ladders from conveyor, by hand, and places ladders in wire basket for subsequent cleaning and rinsing. Hangs basket of ladders on travel arm of wash equipment and starts equipment that ultrasonically cleans and rinses deflashed ladders in chemical baths.

## Electronics Tester (electronics) I; Quality-Control-Assembly-Test Technician; Technician, Test Systems; Tester, Systems; Test Technician; Troubleshooter

Performs a variety of electronic, mechanical, and electromechanical tests on electronic systems, subassemblies, and parts to ensure unit functions according to specifications or to determine cause of unit failure, using full range of electronic test instruments: Reads test schedule, work orders, test manuals, performance specifications, wiring diagrams, and schematics to ascertain testing procedure and equipment to be used. Performs functional tests of systems, subassemblies, and parts under specified environmental conditions, such as temperature change, vibration, pressure, and humidity, using devices such as temperature cabinets, shake-test machines, and centrifuges. Calibrates test instruments according to specifications. Connects unit to be tested to equipment, such as signal generator, frequency meter, or spectrum analyzer. Reads dials that indicate electronic characteristics, such as voltage, frequency, distortion, inductance, and capacitance. Compares results with specifications and records test data or plots test results on graph. Traces circuits of defective units, using knowledge of electronic theory and electronic test equipment to locate defects, such as shorts or faulty components. Replaces defective wiring and components, using handtools and soldering iron, or records defects on tag attached to unit and returns it to production department for repair. May write computer programs to control semiconductor device and electronic component test equipment prior to testing, utilizing knowledge of programming techniques, electronics, test equipment, and testing specifications. May examine switches, dials, and other hardware for conformance to specifications. May verify dimensions of pins, shafts, and other mechanical parts, using calipers, vernier gauges, and micrometers. May operate x-ray equipment to verify internal assembly and alignment of part according to specifications. May calibrate unit to obtain specified dial reading of characteristics such as frequency or inductance. May devise test equipment setup to evaluate performance and operation of nonstandard or customer returned units. May be designed according to unit tested as MEMORY-UNIT TEST TECHNICIAN (electronics); TELEVISION RECEIVER ANALYZER (electronics); TRANSMITTER TESTER (electronics); TUBE-TEST TECHNICIAN (electronics).

# Electronics Inspector (electronics) I; Quality Control Inspector; Systems Inspector

Inspects electronic assemblies, subassemblies, and parts for compliance with specifications following blueprints, drawings, and production and inspection manuals, using one or combination of following methods: Examines layout and installation of wiring, cables, subassemblies, hardware, and components to detect assembly errors. Compares assembly with parts list to detect missing hardware. Examines joints, using magnifying glass and mirror, and pulls wires and cables to locate soldering defects. Examines alignment of parts and measures parts for conformance with specified dimensions, using precision-measuring instruments, such as micrometers, vernier calipers and gauges. Twists dials, shafts, and gears to verify freedom of movement. Traces cables and harness assemblies, following cable print, to verify routing of wires to specified connections and conformance of cable lacing and insulation with manufacturing standards. Measures plated areas for uniformity and thickness, using micrometers or dial indicators. Verifies location of bolt and rivet holes, using templates, check fixtures, and measuring instruments. Examines parts to locate surface defects, such as chips, scratches, and pinholes. Records inspection data, such as serial number, type and percent of defects, and rework required. May stamp inspected equipment to indicate acceptance. May resolder broken connections. May perform functional and operational tests, using electronic test equipment, such as frequency meters, oscilloscope, and signal generator [ELECTRONICS TESTER (electronics) I]. May inspect and lay out optic axis of raw quartz crystals, using optical inspection equipment and be designated INSPECTOR, RAW QUARTZ (electronics). May inspect parts at random and be designated CHECK INSPECTOR (electronics). May inspect units on assembly line and be designated FINAL INSPECTOR (electronics) or IN-PROCESS INSPECTOR (electronics). May be designated according to kind of unit inspected as INSPECTOR, SUBASSEMBLIES (electronics) or INSPECTOR, TUBES (electronics).

# Saw Operator (electronics)

Operates battery of circular saws to cut plastic blocks containing ceramic-coated aluminum rods to length for processing into electronic resistors: Reads job order to determine length of rods required. Secures blocks in clamps of automatic-feed saw, preset to specified length. Closes safety cover and depresses controls to start saw and activate automatic feed. Loads additional saws and monitors sawing operations to detect machine malfunctions. Removes sawed blocks from saws at end of automatic cycle. Holds sawed block against stop of manual-feed saw, starts saw, and feeds block into blade to complete specified cuts. Holds cut blocks under running water to remove debris. Measures cut and washed blocks to verify conformance with job order, using calipers and micrometers. Periodically replaces cog in automatic-feed saw to change length of cut, following identification number on cog and using wrench. Replaces worn or damaged saw blades, using handtools. Cleans work area. Records specified data on production report. May assist maintenance personnel with saw maintenance and repair.

# Rod Tape Operator (electronics)

Tends rod tape machine that separates and tapes ceramic-coated aluminum rods and winds tape containing rods onto reels for subsequent processing: Places rods onto feed tray of taping machine that aligns rods in parallel configuration, applies tape to secure rods in place, and winds taped rods onto reel. Monitors operation of machine to detect malfunctions. Replaces empty tape rolls and removes full reels of taped rods from machine. May perform duties as described under SCREENER OPERATOR (any ind.) to separate defective slugs (rods cut to specified lengths, approximately one inch or less).

# Supervisor, Electronics (electronics)

Supervises and coordinates activities of workers engaged in assembling, testing, and inspecting electronics systems, subassemblies, components, and parts: Analyzes manufacturing schedule to plan work assignments for group on basis of priority and worker skills. Reviews work orders, product specifications, technical instructions, and facility requirements, and requisitions tools, equipment, and supplies for section. Assigns duties to workers and evaluates worker performance. Demonstrates work methods, such as wire routing, assembly, inspection sequence, and test equipment operation. Recommends and initiates personnel action, such as promotion, transfer, or discharge. Examines product according to knowledge of manufacturing and quality standards to ensure acceptability. Reviews and compiles operations records, such as yield, scrap, and maintenance costs. May be required to participate in work of subordinates by performing such tasks as machine and equipment setup, adjustment, and minor repair. Performs other duties as described under SUPERVISOR (any ind.) Master Title. May be designated according to function supervised as SUPERVISOR, ELECTRONICS ASSEMBLY (electronics); SUPERVISOR, ELECTRONICS INSPECTION (electronics); SUPERVISOR, ELECTRONICS TESTING (electronics).

# Supervisor, Hearing-Aid Assembly (per. protect. & med. dev.)

Supervises and coordinates activities of workers engaged in assembling and repairing hearing aids: Reads assembly orders and repair tickets to determine work priorities, and assigns work to those workers responsible for assembly or repair. Records production, such as type of hearing aids to be assembled, number of units, and date assigned. Monitors work in progress to determine whether production is maintained according to schedule. Confers with workers' representatives to resolve grievances. Demonstrates job tasks to train new workers and assigns experienced workers to assist in training. Compiles weekly and monthly production reports of hearing aids assembled and repaired. Requisitions supplies and equipment as required. Performs other duties as described under SUPERVISOR (any ind.) Master Title.

# INDEX

# MOVE TO THE HEAD OF YOUR CLASS
# THE EASY WAY!

Barron's presents THE EASY WAY SERIES—specially prepared by top educators, it maximizes effective learning while minimizing the time and effort it takes to raise your grades, brush up on the basics, and build your confidence. Comprehensive and full of clear review examples, **THE EASY WAY SERIES** is your best bet for better grades, quickly!